Application of Ambient Pressure X-ray Photoelectron Spectroscopy to Catalysis

Application of Ambient Pressure X-ray Photoelectron Spectroscopy to Catalysis

Franklin Tao
Department of Chemical and Petroleum Engineering
University of Kansas
USA

WILEY

Registered Office(s)
John Wiley & Sons, Inc., 111 River Street, Hoboken, NJ 07030, USA
John Wiley & Sons Ltd, The Atrium, Southern Gate, Chichester, West Sussex, PO19 8SQ, UK

For details of our global editorial offices, customer services, and more information about Wiley products visit us at www.wiley.com.

Wiley also publishes its books in a variety of electronic formats and by print-on-demand. Some content that appears in standard print versions of this book may not be available in other formats.

Library of Congress Cataloging-in-Publication Data
9781119845447

Cover Design: Wiley
Cover Image: Courtesy of Mark Biesinger; Courtesy of Franklin Tao

Set in 9.5/12.5pt STIXTwoText by Straive, Pondicherry, India

Contents

Preface

The technique of X-ray photoelectron spectroscopy (XPS) originates from the major contributions of two Nobel Prize laureates: the theoretical explanation of the photoelectric effect by Albert Einstein and the experimental and instrumental development work of Kai Siegbahn. The pioneering work of these two physicists led to the development of XPS analysis of various types of materials. Work originating with Kai Siegbahn and his coworkers and benefiting from the instrumentation improvement and developments of scientists at the Lawrence Berkeley National Laboratory, the Fritz Haber Institute, and other research groups, has resulted in the availability of ambient pressure X-ray photoelectron spectroscopy (AP-XPS). This instrumental approach provides a significant surface analysis technique which can be used in numerous topics of real world including studies of heterogeneous catalysis, advanced materials and devices, energy science, and environmental science.

The application of AP-XPS has been driven to a large extent by the interest in exploring the environment-induced or environment-maintained dynamic surface that differs from the static traditional surface often explored in high vacuum surface science. Studies of an environment-induced or -maintained surface require a technique that has the capability of characterizing these surfaces in the ambient environment. Here an environment is broadly defined as a gaseous environment at a certain pressure or even a liquid environment where a sample such as a real catalyst might be located.

Catalysis is the basis for chemical production in various important applications. Although fundamental studies of catalysis do not directly produce commercial value, they provide open, scientific knowledge helpful for discovering and developing catalysts with potential to be used in the production of value-added chemicals. A significant aspect of fundamental studies of heterogeneous catalysis is the accurate characterization of the catalyst, particularly the authentic surface of the heterogeneous catalyst in its working environment. In general, information about the surface includes both surface chemistry and surface structure. Here, surface chemistry refers to the constituting elements, the quantitative composition, the electronic state and chemical environment of the atoms of the catalyst surface, the bonding between catalyst atoms and adsorbates, and the interaction between catalyst surface atoms and gaseous molecules of the environment. AP-XPS is an appropriate, powerful technique for uncovering the chemistry of an environment-induced or -maintained surface of a catalyst under the reaction conditions of catalysis. AP-XPS has been widely used in the field of heterogeneous catalysis for over a decade.

This is the first research monograph focused primarily on the method and application of AP-XPS. Before the unique features of AP-XPS are introduced, the principle and instrumentation of XPS and the interpretation of XPS data are described briefly in Chapters 1–4. The features of AP-XPS are presented in Chapters 5–9. These chapters provide a basis for understanding how AP-XPS observes and tracks the surface chemistry of a catalyst under real catalytic conditions. The second half of this book (Chapters 10–21) presents the application of AP-XPS to fundamental studies of over ten different catalytic reactions. In these chapters, the focus is on how AP-XPS can provide key information for understanding catalytic mechanisms. The takeaways in each chapter are in italics. There have been a great number of examples of AP-XPS studies of catalysis published in the literature, but it is not possible to discuss all of them in this book. I do appreciate them. Numerous excellent works from the communities of catalysis and AP-XPS have not been included. Many excellent studies could not be discussed in detail or even cited as examples here. The particular works discussed as examples provide an overview of the sorts of information and mechanistic understanding that have been obtained using AP-XPS in the study of catalysis. The literature is filled with many more examples that could have been included.

Franklin Tao
University of Kansas
2022

1

From Surface of Model Catalyst in UHV to Surface of Nanoparticle Catalyst During Catalysis

Surface has been one of the most important states other than the routine three states or called three phases including solid, liquid, and gas. Surface of a catalyst can be considered as a special solid state since most heterogeneous catalysts are at a solid state. The reason that it can be called a special solid state is its structural imperfection. The imperfection is laid on two aspects including its topmost atomic layer and subsurface. An imagination for producing such a special solid state in terms of a surface is to cut a crystal from the middle. This cutting creates two surfaces. The coordination environment of atoms of the topmost layer (Figure 1.1a) is distinctly different from those of an atomic layer in bulk.

The difference in coordination environment including coordination number and interatomic distance between the topmost atomic layer/subsurface and atomic layers in bulk is the origin of the surface activity or reactivity. The coordination environment of an atom of the topmost atomic layer can be termed undercoordination. The broadly defined undercoordination is the origin of catalytic activity of a heterogeneous catalyst since a catalytic reaction involves bonding to surface of a catalyst and is performed on the surface. The origin of catalytic activity at undercoordination is true even for catalysis on a bent internal surface in a microporous material. Notably, the internal surface of a micropore creates confinement effect that is an additional origin for catalytic activity in a great number of high temperature catalytic reactions of hydrocarbons and low-temperature chemical transformation of methane.[2–4] Origin of catalytic activity of a microscopic aluminosilicate catalyst with bent internal surface is an important topic. Here we focus on catalysis on the open surface of a solid catalyst.

Pt(111) is the simplest model surface (Figure 1.1a). Each Pt atom on the surface coordinates with six Pt atoms in the same atomic layer and three in the next layer. It has three open coordination spots and thus exhibits a tendency to adsorb an atom or a molecule. For instance, C atom of a CO molecule forms a Pt—C bond; this binding is enhanced by a backdonation of *d* electrons of the Pt atom to the antibonding orbital of CO molecule, $2\pi^*$. The formation of a surface bond between a reactant molecule and the topmost atom of a catalyst weakens the intramolecular bond of a CO molecule. The weakening of this C≡O bond can either facilitate dissociation of CO or trigger an attack from an atom, molecule or intermediate such as a binding of a H atom to either C or O atom of the activated CO molecule. Obviously, the surface openness in terms of undercoordination of the surface atoms initiates the surface reactivity or catalytic activity.

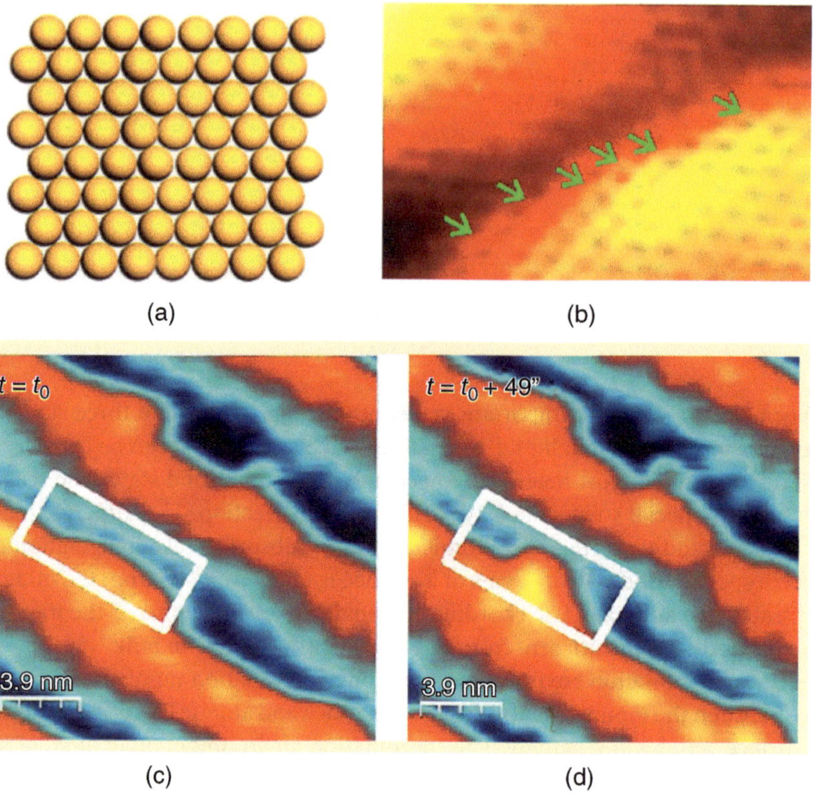

Figure 1.1 Surface structural model and scanning tunneling microscopy (STM) images of a real surface. (a) Pt(111) surface; each Pt atom bonds with six atoms of the topmost layer and three of the next layer (not drawn). (b) STM image of irregular packing of atoms at edge of a terrace on Pt(111) in 1 Torr CO. (c and d) STM images of the same area of Pt(111) single crystal model catalyst in 10 Torr CO collected at $t = t_0$ and $t = t_0 + 49$ seconds, respectively. The two images showing an evolution of the edge of a terrace within 49 seconds through repacking atoms at the edge of a terrace in 0.1 Torr CO at room temperature; the two images were taken at the same location on Pt(111). *Source:* Reproduced with permission from Nguyen et al.[1]/American Chemical Society.

It is not inappropriate to imagine a (111) surface drawn on a paper is perfect and all atoms are coordinated with nine Pt atoms. However, a real surface is way more complicated than the drawn. As shown in Figure 1.1b, even the simplest surface of Pt(111) single crystal could have certain portion of Pt atoms at step edge where a Pt atom coordinates with less than nine Pt atoms. These undercoordinated Pt atoms at step edge can be readily restructured. As seen in Figure 1.1c and d, the Pt—Pt bonds are readily broken at room temperature to form a peninsula-like cluster while a Pt(111) single crystal model catalyst is in a CO gas at a pressure as low as 0.1 Torr.[1] Clearly, we can simply claim the surface of a catalyst is full of defects. These defects make the surface act as an active reagent to participate in a chemical reaction of gaseous or/and liquid reactants, but the solid surface in terms of the catalyst is not consumed.

Pt(557) is relatively more complicated than Pt(111). As shown in Figure 1.2a, a Pt(557) surface has at least three different types of Pt atoms that coordinated with 7, 9, or 10 atoms,

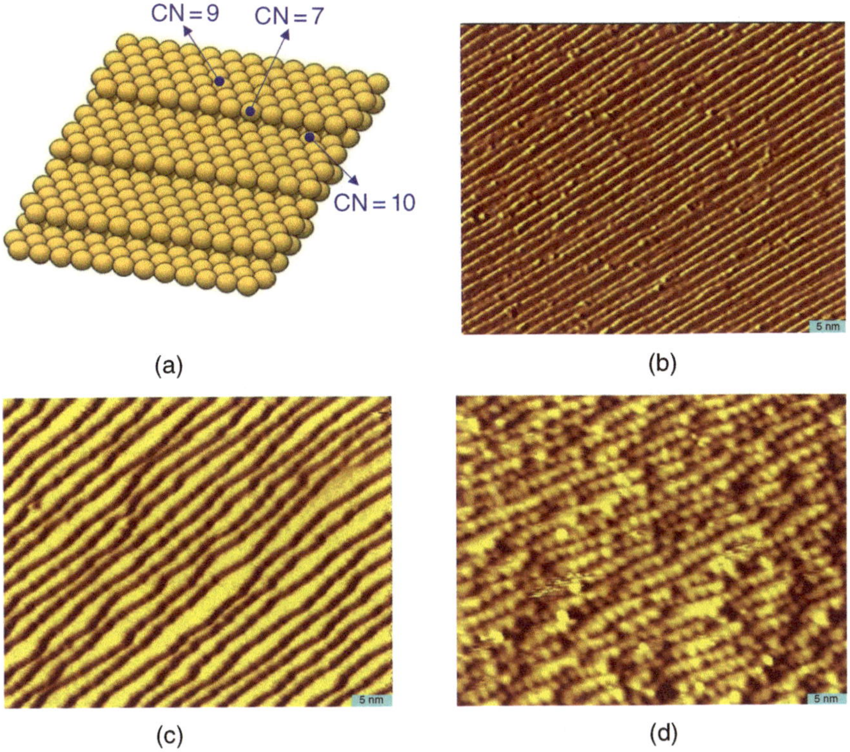

Figure 1.2 Structural model of a stepped model catalyst Pt(557) and evolution of its surface structure along increase of pressure of CO gas surrounding the model catalyst. (a) Structural model of Pt(557). *Source:* Adapted from Tao et al.[5] (b) STM image of clean Pt(557) at room temperature in UHV with a base pressure lower than 3×10^{-10} Torr. (c) STM image of Pt(557) at room temperature in 5×10^{-8} Torr CO. (d) STM image of Pt(557) at room temperature in 1 Torr CO. *Source:* Reproduced with permission from Tao et al.[5], Reprinted from AAAS.

respectively. The atoms within each terrace are just like atoms on Pt(111), coordinating with nine atoms. These atoms at the edge of a terrace or called a step have coordination number of 7. In addition, Pt atoms immediately under the step edge are partially exposed, coordinating with 10 atoms (Figure 1.2a). Notably, if one or two Pt atoms at step edge are removed, a kink site is created (Figure 1.3c). Typically, a Pt atom of a kink site coordinated with six Pt atoms or less. As these step-edge atoms have a low coordination number, 7, the Pt—Pt bonds at the step edge can be readily broken, resulting in a restructuring of the step edge in CO at a pressure as low as 1×10^{-8}Torr CO (Figure 1.2c). As the width of each step is narrow, a continuous break of Pt—Pt bonds and the resulting relocations of more Pt atoms make the step broken down.[1,2] Through forming an intermediate surface phase in terms of a double width terrace, the surface is broken into triangular nanoclusters with a width of 2.2 nm or so (Figure 1.2d).[5]

The surface of a catalyst nanoparticle is much more complicated. Figure 1.3c and d schematically presents the coexistence of various types of metal atoms and defects. The coexistence of various types of catalytic sites could create different reaction channels leading to different products. The heterogeneity of active sites on a catalyst surface is a factor

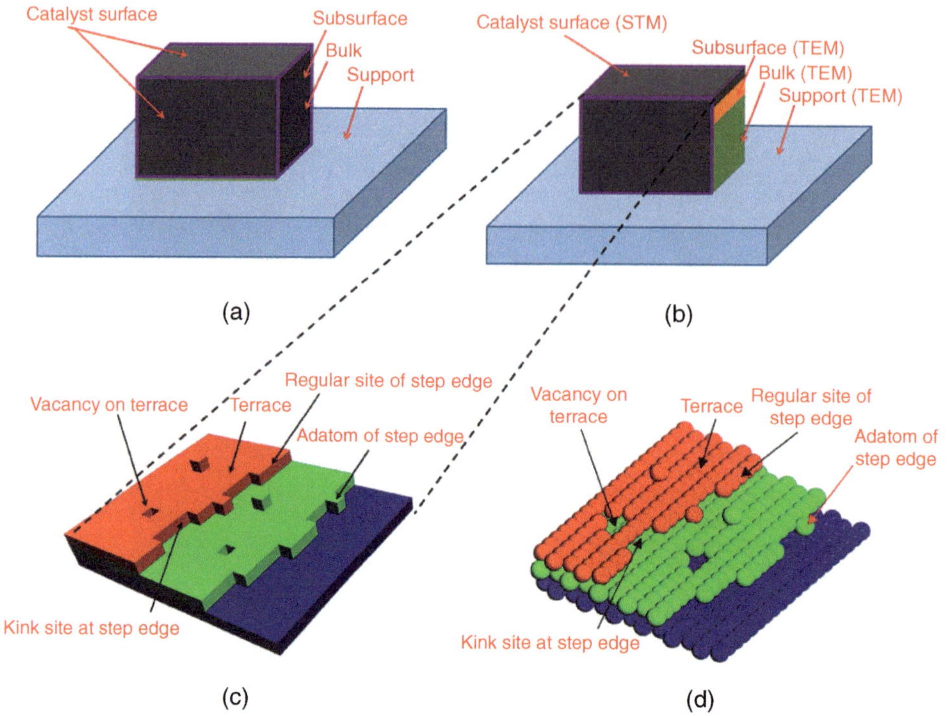

Figure 1.3 Schematics of the structure of a catalyst nanoparticle. (a) Catalyst nanoparticle (gray box) loaded on a support (light blue plate). (b) Surface (gray), subsurface (yellow), and bulk (green) of a catalyst nanoparticle supported on a support; the microscopic characterization techniques of structures of different sections are marked on the schematic. (c) Schematic of a topmost surface of a catalyst nanoparticle consisting of different types of catalytic sites. (d) Schematic of arrangements of catalyst atoms of the topmost surface of a nanoparticle. *Source:* Reproduced with permission from Tao and Crozier[6]/American Chemical Society.

contributing to a relatively low catalytic selectivity for producing an ideal product. Compared to a heterogeneous catalyst, a homogeneous catalyst also known as a molecular catalyst typically exhibits high selectivity for a product since each catalytic site of a molecular catalyst is nearly identical at the atomic level.

The complexity of a catalyst surface is largely enhanced upon considering the factors such as the existence of the second metal, the size-dependent molar fraction of atoms at edge or corner, and the coexistence of cations or anions with different valence states. Each of these factors adds at least one variable, elevating the complexity of a catalyst surface. By adding these variables, a surface has become quite complicated even if the surface is at room temperature in ultrahigh vacuum (UHV). In terms of characterizations of a catalyst surface in UHV, a great number of surface analytic techniques were developed in the past several decades for characterizing a catalyst in UHV environment. X-ray photoelectron spectroscopy (XPS) has been the significantly valuable spectroscopy to identify constituting elements of surface, measure surface composition, and investigate chemical environment of surface atoms. For instance, high-resolution XPS spectrum allows for distinguishing metal atoms with different coordination environments. In addition, scanning tunneling microscopy (STM) has been the most powerful microscopy to visualize atomic packing of surface at atomic scale. Most of these studies are performed in UHV at cryogenic or room temperature. Obviously, keeping a surface in UHV allows probing a surface readily without any perturbation from an external environment. A significant amount of information on the surface structure, adsorption, desorption, and diffusion was reported in literature in the past decades, which built up the traditional surface science.[2,3] It provides the base for exploring the surface in a complex environment.

Catalysis is performed on a catalyst surface at certain temperature in gas or liquid phase of reactants. We can imagine we have at least two more factors determining the structure of a catalyst surface during catalysis. They are temperature and pressure. Temperature is a necessary factor for most reactions. One reason is that a reaction rate is exponentially increased along the increase of catalysis temperature based on Arrhenius equation regardless of the endothermic or exothermic nature of this reaction; for an endothermic reaction, a high reaction temperature is favorable for a high conversion of reactants or even it is requested for spontaneity of an endothermic reaction. However, the catalysis temperature is a challenging factor for characterizations of a catalyst surface. Unfortunately, with the temperature factor, a catalyst surface at high temperature could be quite different from that at room temperature even if environment of the catalyst is still UHV. For instance, a Rh–Pt alloy surface can restructure to form a Pt-rich surface at high temperature in UHV since Pt atoms have lower surface energy than Rh. Obviously, the temperature factor has driven the segregation of Pt atoms in subsurface or bulk to the surface region since a high temperature can offer Pt atoms in subsurface and bulk enough energy to cross the diffusion barrier of Pt to emigrate and then segregate to the surface of the bimetallic nanoparticle.[7,8]

Another factor resulting in the complexity is the pressure of reactant gas(es). Most of the catalytic reactions are performed in gas phase at ambient pressure or high pressure or even in liquid phase. In such an environment, the molecular density of gas is much higher than that in UHV by at least 13 orders of magnitude. The entropy contribution of the surrounding gas or liquid phase potentially changes the surface structure at an atomic scale. The extent of changing surface structure of the catalyst by the gas or liquid phase is dependent on the original chemical environment of a catalytic site and the pressure, type, and composition of the gas around the catalyst surface while the catalyst is at a constant temperature.

To catalyze a reaction, typically a catalyst must be at room temperature or above in a gas phase at 1 atm or higher or in a liquid phase. Coexistence of the factors of temperature and pressure could largely change the surface structure in an unknown but convoluted manner. This complexity has driven people to study dynamics of a catalyst surface. The formation of such a dynamic surface is typically driven by the catalysis temperature or/and relatively high pressure of reactants; more importantly, in many cases the dynamic surface is maintained by catalysis temperature, type of gas, gas pressure, and composition of the constituting gases of the gaseous environment. In terms of catalysis on catalyst nanoparticles buried in a liquid phase, an expected dynamic surface is a convoluted consequence of catalysis temperature and density, type, and composition of the liquid surrounding the catalyst nanoparticles. Thus, the authentic surface of a catalyst during catalysis needs to be studied in the environment including pressure and temperature at which the catalyst works. The difference between a catalyst surface during catalysis and the catalyst at room temperature in UHV or air is analogous to the difference between on-site track of the living behavior of a deep ocean fish such as a viperfish and the off-site observation of a dead viperfish on a table.

From the catalyst development point of view, the complexity of the environment-induced dynamic surface of a catalyst has raised a key question, what is the base for the traditional approach of trial-and-error in development of a catalyst. In fact, the complexity of the environment-induced and maintained authentic surface of a catalyst during catalysis undermines the base of trial-and-error approach since the information of catalyst structure a trial-and-error approach typically relies on is mainly obtained through ex situ studies. The high complexity of the catalyst surface under a catalytic condition underlines that development of catalysts needs to rely on environment-induced dynamic surface instead of the surface of an as-prepared catalyst or surface of a used catalyst. A dynamic surface needs to be characterized with in situ spectroscopy or microscopy. One of the powerful in situ spectroscopies is ambient pressure X-ray photoelectron spectroscopy (AP-XPS). It is a surface analytical technique developed from high vacuum X-ray photoelectron spectroscopy (XPS).

References

1 Nguyen, L., Cheng, F., Zhang, S. *et al.* 2013. "Visualization of surfaces of Pt and Ni model catalysts in reactive environments using ambient pressure high temperature scanning tunneling microscopy and understanding the restructurings of surfaces of model metal catalysts under reaction conditions at near ambient pressure." *J. Phys. Chem. C* 117, 971–977.

2 Ertl, G., Knozinger, H., Schuth, F. *et al.* 2008. *Handbook of Heterogeneous Catalysis*. 2nd edn, Wiley-VCH.

3 Somorjai, G. A. and Li, Y. 2011. *Introduction to Surface Chemistry and Catalysis*. 2nd edn, Wiley-VCH.

4 Tang, Y., Li, Y., and Tao, F. 2022. "Activation and catalytic transformation of methane under mild conditions." *Chem. Soc. Rev.* 51, 376–423.

5 Tao, F., Dag, S., Wang, L. W. *et al.* 2010. "Break-up of stepped platinum catalyst surfaces by high CO coverage." *Science* 327, 850–853.

6 Tao, F. and Crozier, P. A. 2016. "Atomic-scale observations of catalyst structures under reaction conditions and during catalysis." *Chem. Rev.* 116, 3487–3539.

7 Tao, F., Grass, M. E., Zhang, Y. *et al.* 2008. "Reaction-driven restructuring of Rh-Pd and Pt-Pd core-shell nanoparticles." *Science* 322, 932–934.

8 Tao, F., Grass, M. E., Zhang, Y. *et al.* 2010. "Evolution of structure and chemistry of bimetallic nanoparticle catalysts under reaction conditions." *J. Am. Chem. Soc.* 132, 8697–8703.

2

Application of XPS: from Surface in UHV to Surface in Gas or Liquid Phase

2.1 Origin of X-ray Photoelectron Spectroscopy

X-ray photoelectron spectroscopy (XPS) is one of most widely used spectroscopies in catalysis, surface science, materials science, and energy science. Together with Auger electron spectroscopy (AES), it is traditionally called electron spectroscopy for chemical analysis (ESCA) since it identifies constituting elements, measures the composition of a solid surface, a liquid phase, or even a gas phase, and analyzes chemical environment of a type of atoms of a sample.

This spectroscopy is closely related to or even is originated at the very early discovery of the photoionization effect through irradiating surface of a sample by Heinrich Hertz in 1887.[1] At the beginning of twentieth century, Philippe Lenard defined the light-induced generation of electrons from a solid surface.[2,3] Lenard found the kinetic energy of electrons from a given emitter in terms of a sample does not exceed a maximum value related to the sample. In the nearly same period, Albert Einstein explained the generation of electrons from a metal surface through photoionization with a language of quantum theory.[4] Ten years later, Rutherford, Robinson, and Rawlinson were aware of the fact that the kinetic energy of generated electrons from a metal surface is the difference between photon energy and binding energy of electrons in their atoms.[5–7] Later Robinson and Young found that a different chemical bonding of atoms can shift kinetic energy of electrons generated from a solid surface, driving potential applications of the photoelectric phenomenon to understanding of chemical environment of a specific type of atoms on surface or in subsurface region of a solid.[5,8] The documentation of the development of X-ray photoelectron spectroscopy before 1960 can be found in literature.[13] Since 1950s,[9] pioneering efforts of Siegbahn et al. gave very early high-resolution XPS spectra.[10–13] Kai Siegbahn developed methods and instruments for achieving highly accurate measurements of kinetic energy of photoelectrons generated from atoms by irradiating them with X-ray.[10] Siegbahn published the first books of electron spectroscopy in 1960s.[11,12] He was honored with Nobel Prize in physics in 1981 for his seminal contribution to the development of high-resolution electron spectroscopy and spectrometer.[10]

Application of Ambient Pressure X-ray Photoelectron Spectroscopy to Catalysis, First Edition. Franklin Tao.

2.2 Applications of XPS to Study Surface in High Vacuum

Numerous excellent books of surface science or surface characterization have described the applications of XPS in high vacuum.[14–19] In brief, it can qualitatively identify constituting elements of a sample, interpret chemical environment of atoms, and quantitatively analyze its composition. It has been used in every corner of science, technology and engineering sectors of our society. Here it is not necessary to repeat its wide applications and significance.

When the analyte is deep layers away from the sample surface, a buried region, or even bulk of a sample, a sputtering-assisted depth profile analysis is necessary. Alternatively, hard X-ray XPS can provide important information of such an analyte. Hard X-ray generates photoelectrons with a higher energy that have longer inelastic mean free path, λ, in contrast to soft X-ray or tender X-ray. Thus, the information volume of hard X-ray XPS is much larger; in other words, much deeper region can also contribute to the intensity of an XPS peak if a hard X-ray XPS is used. The hard X-ray XPS has been used in studies of solid state science and functional devices such as development of battery materials, semiconducting devices, and superconducting materials.[20,21]

2.3 Applications of XPS to Study Sample in Gas Phase

Over three quarters of heterogeneous catalysis reactions are performed at solid–gas interface. For those reactions, a catalyst is retained in gas phase of reactants. As discussed in Chapter 1, surface of a catalyst likely restructures in gaseous environment of reactant(s) driven by both pressure and temperature factors. More importantly, in many cases these structures are dynamic, and they are maintained by the gaseous environment and temperature. When the gaseous environment or/and the temperature factor is withdrawn, the dynamics structure likely changes largely. To establish an intrinsic correlation between a catalyst surface and its corresponding catalytic performance, it is necessary to track the authentic surface formed at catalysis temperatures in gas phase and maintained under a catalytic condition. Ambient pressure X-ray photoelectron spectroscopy (AP-XPS) is the appropriate analytic technique for surface of a catalyst at catalysis temperature in gas phase. Chapters 3 and 4 present data analysis and instrumentation of XPS in general. Significance and challenges of studying surface of a catalyst in gaseous phase are discussed in Chapter 5. Instrumentation of ambient pressure X-ray photoelectron spectrometer is introduced in Chapter 6. Chapter 7 describes experimental methods of AP-XPS studies. Data analysis of AP-XPS was introduced in Chapter 8. Chapter 9 briefs the significance of using AP-XPS in studies of catalysis. Chapters 10–21 are AP-XPS studies of different types of catalytic reactions.

2.4 Applications of XPS to Study Sample in Liquid Phase

A high vacuum XPS can be used in fundamental studies of surface of nanoparticles in liquid phase. This is driven by the fact that nanoparticles function in liquid environment such as catalysis at a liquid-solid interface, electrocatalysis on anode or cathode in liquid, and molecular catalysis in liquid. Figure 2.1d schematically illustrates that the high density of

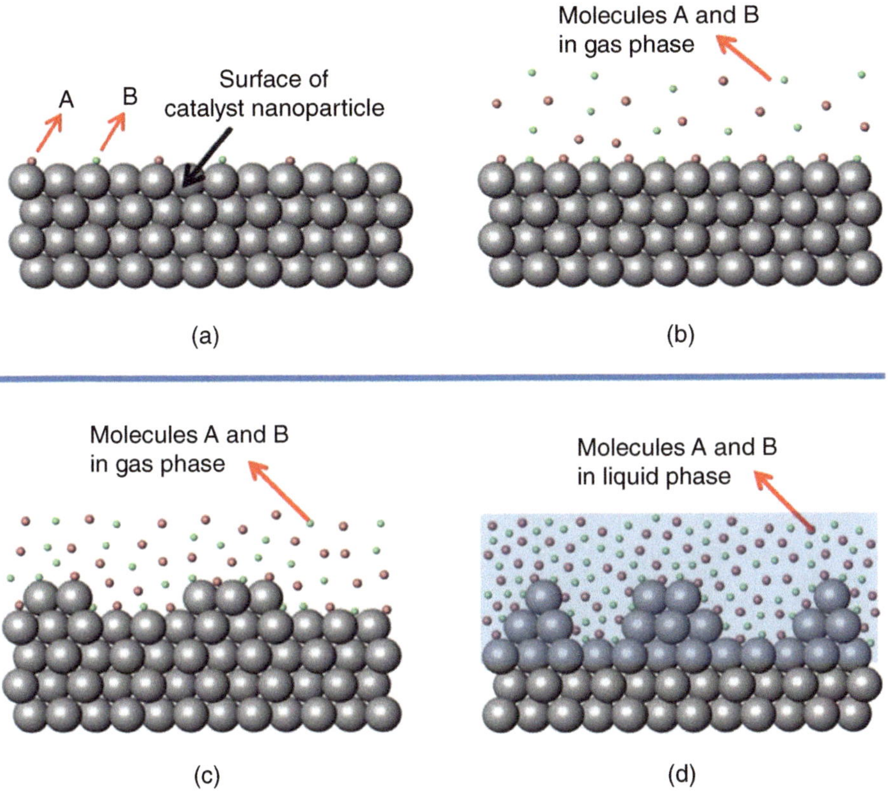

Figure 2.1 Schematics showing the potential impact of gas or liquid environment on surface of catalyst nanoparticles. (a) Surface of a catalyst nanoparticle with adsorbed molecules of reactants A and B in ultrahigh vacuum (UHV) environment. (b) Surface of a catalyst nanoparticle in a gas of reactants at a low pressure, p_1. (c) Surface of a catalyst nanoparticle in a gas of reactants at a relatively high pressure, p_2. (d) Surface of a catalyst nanoparticle surrounded by solvent and solute molecules of liquid. *Source:* Reproduced with permission from Nguyen et al.[22]/American Chemical Society.

molecules in liquid surrounding a nanoparticle could restructure surface of the nanoparticle.[22] From fundamental science point of view, it is necessary to explore surface of these nanoparticles functioning in liquid. An easy approach to study these nanoparticles is to filter these nanoparticles after catalysis, dry them, and then place them in a high vacuum environment of an XPS system for analysis. However, the separation of functional nanoparticles from its working environment could make us incapable of knowing its authentic surface chemistry under working condition. To uncover surface of a functional material such as a nanoparticle catalyst in liquid environment, XPS studies of these nanoparticles in static liquid for flowing liquids were demonstrated recently.[22–24]

2.4.1 XPS Studies of Surface of Nanoparticle Catalyst in Static Liquid

Figures 2.2 and 2.3 present how XPS was used to study surface of nanoparticles dispersed in liquid phase.[22,23] For this purpose, a specific cell is needed for retaining the sample consisting of nanoparticles and liquid. Simply speaking, the sample is a liquid containing

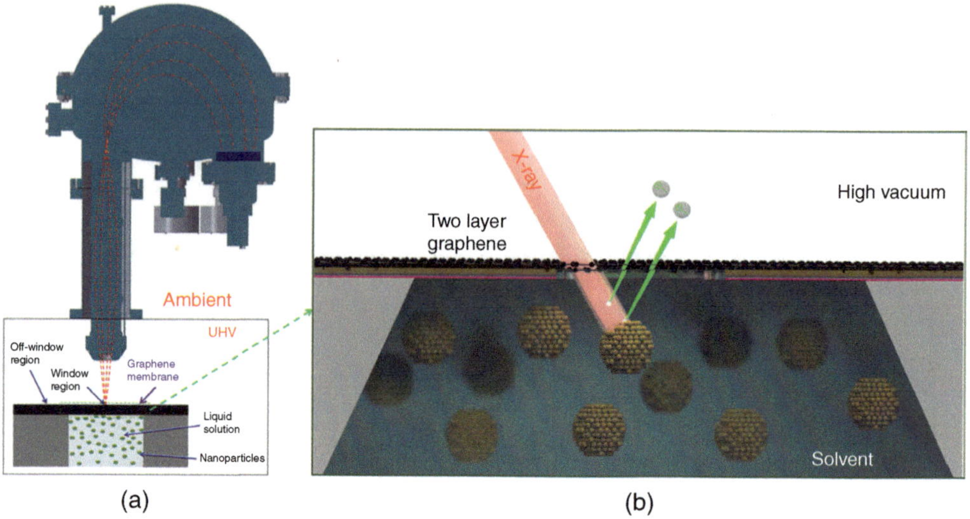

Figure 2.2 Schematic showing study of nanoparticles in a static liquid with XPS. (a) Schematic showing the integration of a liquid cell filled with a nanoparticle-containing liquid. (b) Schematic showing the interactions *between* X-ray photons *and* solvent and solute molecules and nanoparticles in the liquid, as well as the interactions *between* the generated photoelectrons *and* the solution between the nanoparticle surface and the two-layer graphene. *Source:* Reproduced with permission from Nguyen et al.[22]/American Chemical Society.

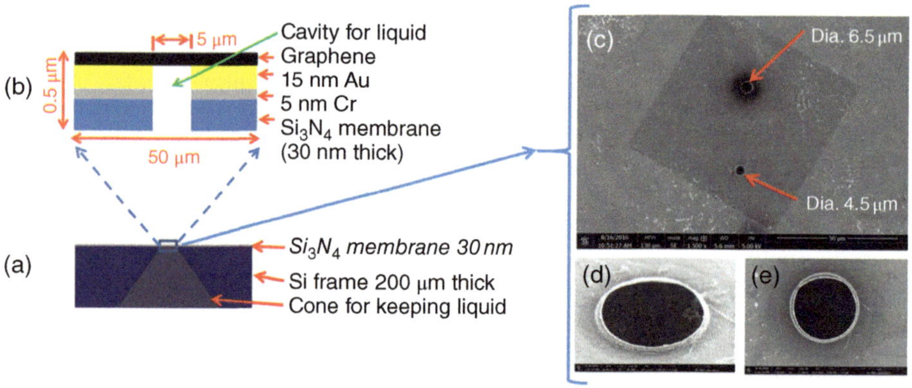

Figure 2.3 Schematics and scanning electron microscopy (SEM) images of the Si_3N_4 window for a liquid cell. (a) Schematic of a miniature silicon wafer with an ultrathin Si_3N_4 membrane at its center. (b) Cross-sectional schematic of the Si_3N_4 membrane located at the center of the wafer. (c) SEM image of pores bored in the Si_3N_4 membrane by SEM ion milling. Cr/Au layers were deposited on the Si_3N_4 window near to the pore to enhance the adhesion of the graphene layers. (d) SEM image of a graphene window covering the pore of the Si_3N_4 window. (e) SEM image of the same graphene window as (e) but taken with the sample tilted 50° from normal. *Source:* Reproduced with permission from Nguyen et al.[22]/ American Chemical Society.

well-dispersed nanoparticles as schematically shown in Figure 2.2b. Generally, there is no limit for size of these nanoparticles to be studied with this method.[22] The liquid can be a mixture of any solvent and solute. The solvent can be water or organics. In the following example, the solute and solvent are phenol and tripropylene glycol monomethyl ether, respectively. The mixture of phenol and tripropylene glycol monomethyl ether is the organic solution wherein Ag nanoparticles were dispersed. Figure 2.3a shows the cell structure. A two-layer graphene membrane is the wall to isolate the liquid containing well-dispersed nanoparticles from the high vacuum environment of XPS chamber (Figure 2.2b). Notably, the two-layer graphene membrane is transparent for X-ray photons and photoelectrons because they are single-atom thick graphene layers and there are sub-nanometer benzene-like holes in each single-atom layer. X-ray can irradiate nanoparticles in liquid. More importantly, this cell allows a part of the photoelectrons generated from the nanoparticles to transmit through the liquid between catalyst surface and then the holes in the graphene layers to enter the high vacuum environment of XPS immediately before they are analyzed by an energy analyzer. In principle, a high vacuum XPS can perform an analysis for surface of nanoparticles dispersed in a liquid if a lab-based X-ray beam can be strong enough to generate enough photoelectrons for being collected by an energy analyzer and then detected by an detector although a synchrotron beam having high X-ray flux density is preferred. As seen from the scanning electron microscopy (SEM) image in Figure 2.4c, the liquid containing well-dispersed Ag nanoparticles was placed in the cavity of the cell. The cavity of the cell is for holding liquid and is schematically shown in Figure 2.3b. As the graphene window well separated the liquid environment of the catalyst from the high vacuum environment of XPS, both the liquid of the cell and the high vacuum environment of XPS survive during XPS studies of nanoparticles in the liquid in the cell. Figure 2.4g is the XPS spectrum of Ag 3d photoelectrons generated from Ag nanoparticles dispersed in liquid.

Only these nanoparticles in liquid near the graphene window could contribute photoelectrons for analysis by the energy analyzer due to the limited mean free path of these generated photoelectrons in liquid. Notably, the nanoparticles studied in reference 22–23 were dispersed in liquid instead of being deposited on the graphene window. In general, nanoparticles in liquid with a distance of <10 nm from the graphene window can contribute to the intensity of XPS peak. This information depth is largely determined by (1) the original kinetic energy of photoelectrons of an interested element and the density of the interested elements in the liquid and (2) the molecular density and constituting elements of solute and solvent of the liquid. To increase the information depth, a hard X-ray can be used to generate photoelectrons with higher kinetic energies. Then, surface of nanoparticles farther from the graphene window can also contribute to the XPS peak. Thus, such a liquid cell-based XPS study of a nanoparticle catalyst in liquid using hard X-ray can provide more information of nanoparticles in a deeper region of the liquid although photoionization cross section of a specific subshell such as Ag 3d decreases along increase of X-ray energy. It is expected such a well-designed reaction cell can be used in studies of homogeneous catalysts in terms of molecular catalysts in solution or even under a catalytic condition.

2.4.2 XPS Studies of Surface of Nanoparticle Catalyst in Flowing Liquid

The dispersion of nanoparticles in a static liquid at the region near the graphene window may not reflect the dispersion of nanoparticles in a flowing liquid due to gravitation or surface tension of liquid. The catalyst nanoparticles or even reactant molecules may not be distributed in the solution homogeneously. In addition, the accumulation of product molecules on or near the catalyst nanoparticle surface could prevent reactant molecules or intermediates from accessing catalytic sites on the nanoparticle surface. Driven by these

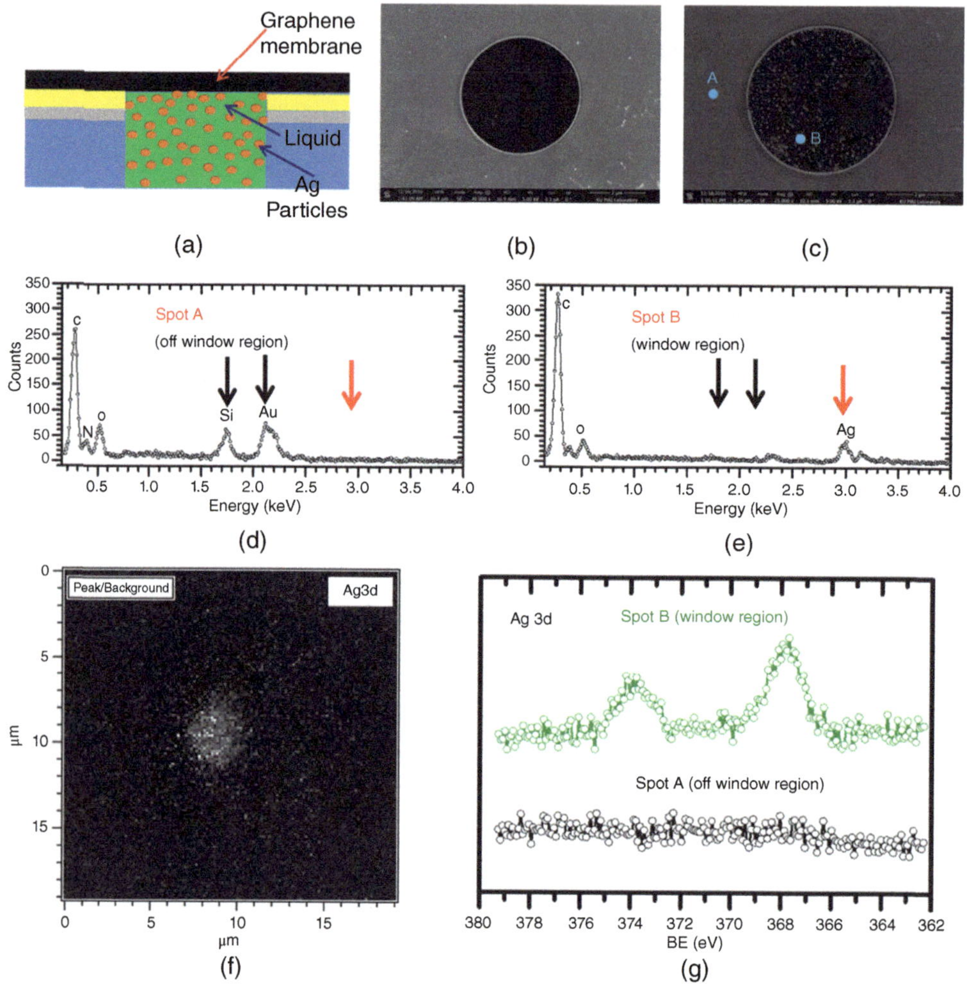

Figure 2.4 XPS studies of Ag nanoparticles dispersed in a mixture of tripropylene glycol monomethyl ether and phenol sealed in a cell of static liquid. (a) Schematic of the micro-opening region of the cell covered by the graphene membrane, which separates the liquid from the UHV environment of XPS. (b, c) SEM images of a graphene liquid cell (b) without liquid and (c) filled with liquid containing tripropylene glycol methyl ether (TPM) (solvent), phenol (reactant), and Ag nanoparticles; the concentration of Ag nanoparticles in solution is 10 wt% or 1.08 vol%; the bright spots are Ag nanoparticles dispersed in the liquid. (d, e) EDX spectra acquired at spot A (off graphene window region) of (c) and spot B (graphene window region) of the liquid cell in (c). (f) Chemical mapping of the distribution of Ag 3d photoelectrons generated from a 20 μm × 20 μm region of the cell in (c); the bright region of the mapping is a Ag nanoparticle dispersed in the liquid sealed in the cell. (g) XPS spectra of Ag 3d of spot A (black) and spot B (green); spot B are Ag nanoparticles dispersed in the liquid sealed in the cell. *Source:* Reproduced with permission from Nguyen et al.[22]/American Chemical Society.

concerns, a flowing reaction cell for XPS studies of a catalyst in liquid was designed and successfully fabricated by Tao group as shown in Figures 2.5–2.8. With this flowing cell, the nanoparticles in a flowing liquid can be characterized with a high vacuum XPS as well. In other words, the nanoparticles in liquid are flowing through the cell when they are being characterized by a high vacuum XPS.[23]

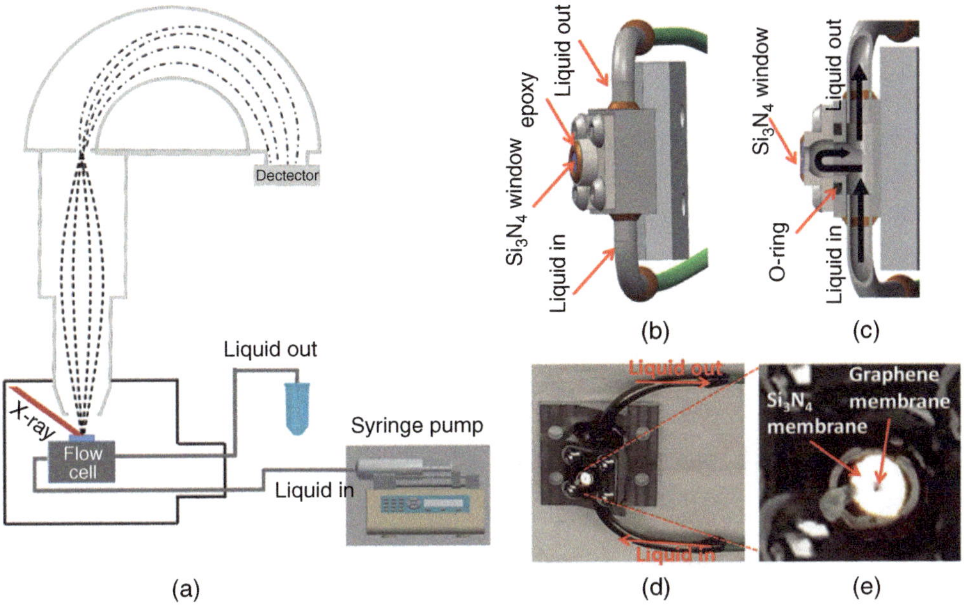

Figure 2.5 Reactor system designed for studies of the surface of catalyst nanoparticles in a flowing liquid. (a) Schematic of the whole system. (b) and (c) External and internal views of the Si_3N_4 cell; the brown ring shows epoxy instead of an O-ring (epoxy was used to seal the two parts externally). (d) Photograph of the reaction cell. (e) Enlargement of the graphene membrane on a Si_3N_4 membrane. *Source:* Reproduced with permission from Nguyen et al.[23]/Royal Society of Chemistry.

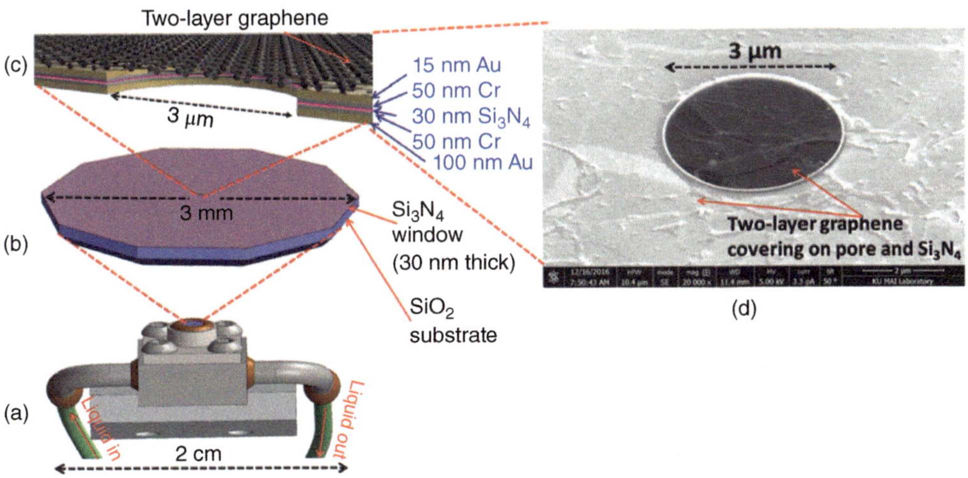

Figure 2.6 Structure of a Si_3N_4 membrane cell with a window covered with a two-layer graphene membrane for studying nanoparticles dispersed in a flowing liquid and for studying a flowing gas at a pressure of 1 atm or higher pressure. (a) Reaction cell of flowing liquid. (b) A Si_3N_4 membrane grown on a silicon wafer; there is only a Si_3N_4 membrane (30 nm thick) in the middle of the wafer. (c) Cross-sectional view of a pore of Si_3N_4 and the supported graphene membrane. (d) SEM image of a two-layer graphene membrane covering the window of the Si_3N_4 membrane. *Source:* Reproduced with permission from Nguyen et al.[23]/Royal Society of Chemistry.

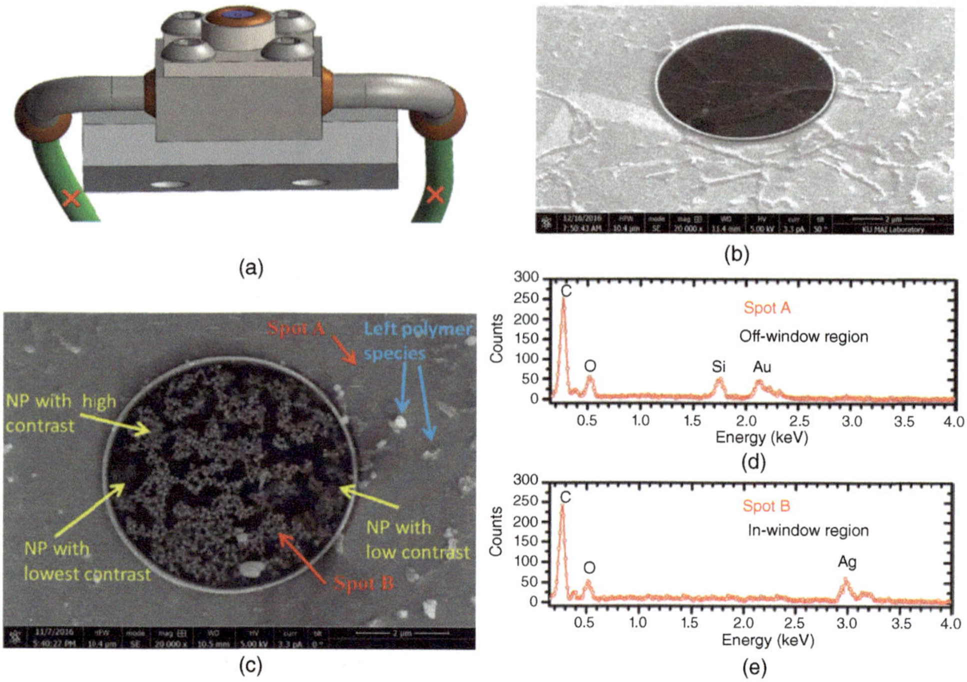

Figure 2.7 SEM studies of the reaction cell filled with a solution of Ag nanoparticles. (a) The designed reaction system (schematic). (b) SEM image of the window of the Si_3N_4 membrane cell covered with a two-layer graphene membrane without a solution of Ag nanoparticles. (c) SEM image of the Si_3N_4 cell covered with a two-layer graphene membrane filled with a solution of Ag nanoparticles. (d) SEM-EDX spectrum at spot A (Si_3N_4 region in terms of the off-window region) of the Si_3N_4 cell. (e) SEM-EDX spectrum at spot B (in-window region) of the Si_3N_4 cell. *Source:* Reproduced with permission from Nguyen et al.[23]/Royal Society of Chemistry.

The schematic showing the setup for studying nanoparticles dispersed in a flowing liquid is presented in Figure 2.5. Basically, a high-performance liquid chromatography (HPLC) syringe pump was used to drive liquid to circulate in a loop consisting of a reaction cell made with the method described in literature.[23] The structure of this flowing cell was shown in detail in Figures 2.5–2.7. Different from the static cell (Figure 2.4a), the liquid in Figure 2.5 must flow through the cell (Figure 2.5c) while the liquid is well separated from the high vacuum senvironment of XPS. Figure 2.7c is the SEM image showing that Ag nanoparticles were dispersed in the liquid. The chemical identity of the element of the metal nanoparticles dispersed in the flowing liquid cell which was originally located in a syringe pump was confirmed by scanning electron microscopy-energy dispersive spectroscopy (SEM-EDS) spectrum of the cell. As shown in Figure 2.7e, SEM-EDX identified the Ag element, suggesting the liquid containing Ag nanoparticle catalyst was successfully flown from its original location, the syringe pump to the cell. X-ray source irradiated the Ag nanoparticles for exciting photoelectrons escaping from surface of Ag nanoparticles in the flowing liquid. Similar to the batch reaction cell, the generated Ag 3d photoelectrons first escaped the liquid layers between

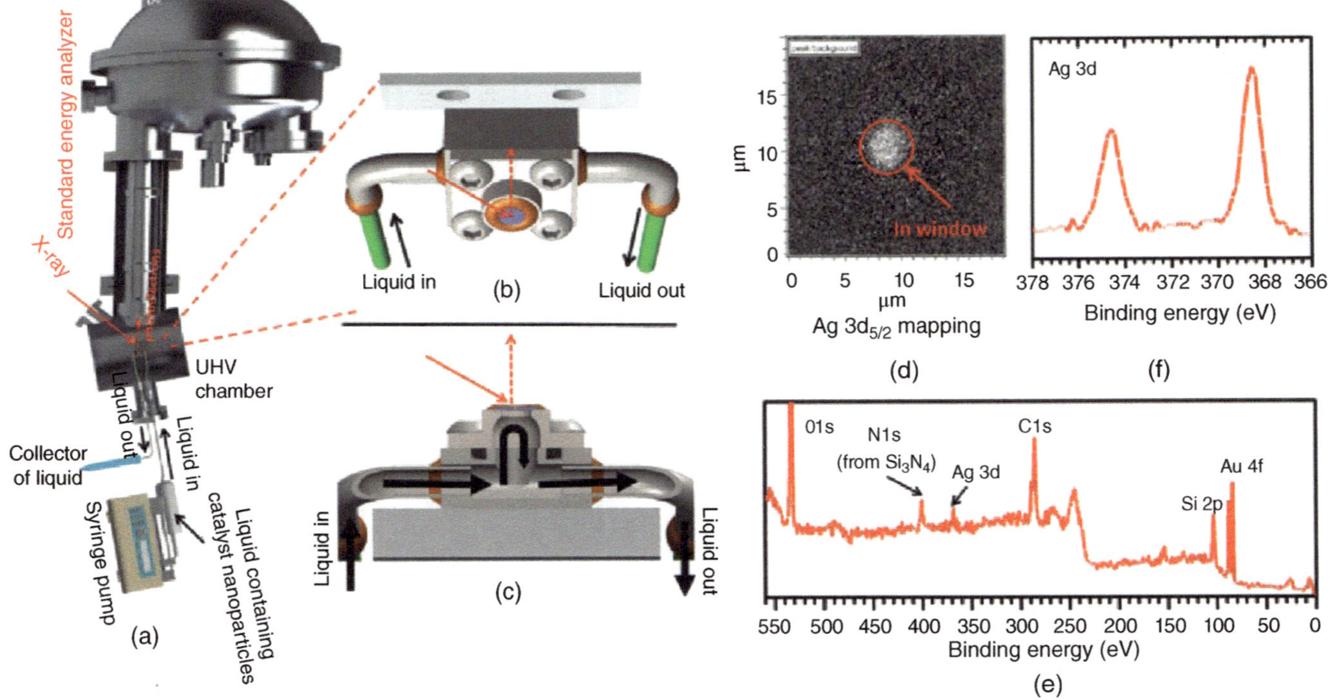

Figure 2.8 XPS studies of a solution of Ag nanoparticles flowing through a window of the Si_3N_4 membrane covered with a two-layer graphene membrane. The solution containing a solvent (a mixture of tripropylene glycol methyl ether and phenol) and Ag nanoparticles dispersed in the solution was pumped by a syringe pump. The flow rate of the tested liquid is 0.5 ml/min. (a) The entire reactor system of flowing liquid. (b and c) External and internal structure of the Si_3N_4 membrane cell with a window covered with the graphene membrane. (d) Mapping of Ag $3d_{5/2}$ photoelectrons. (e) XPS survey of the window region (covered with a graphene membrane) in which the solution of Ag nanoparticles was flowing through the surface of the graphene membrane at a slow rate of 0.5 ml/min. (f) XPS spectrum of Ag 3d photoelectrons of Ag nanoparticles dispersed in liquid that is flowing through the reaction cell. *Source:* Reproduced with permission from Nguyen et al.[23]/Royal Society of Chemistry.

their home nanoparticles and the graphene window and then entered the high vacuum environment of XPS. The XPS spectrum of Ag 3d was collected from the Ag nanoparticles in the liquid flowing through the window. The flowing direction of the liquid is schematically shown in Figure 2.8c. The successful demonstration of measuring metal nanoparticles in a flowing liquid was reported in 2018.[23] It demonstrated the capability of examining surface of nanoparticles dispersed in a *flowing* liquid. It offers the capability of studying a functional material such as functional particles dispersed in a flowing liquid when the material is functioning in liquid. In terms of catalysis, it suggests the capability of studying catalyst nanoparticles or a molecular catalyst catalyzing a chemical synthesis in a flowing liquid while the catalysis is being performed in the flowing liquid. It is expected that such an XPS study of using flowing liquid cells can be used in exploring the electronic state of a catalyst surface or a molecular catalyst during catalysis in many fields.

2.4.3 XPS Study of Flowing Gas with a Pressure of 1 atm or Higher

The well developed flowing cell[23,24] can be used to study flowing gas at a pressure of 1 atm or even higher by Scholgl group and Tao group. Air was used as a probe gas to demonstrate how the well-designed flowing cell (Figures 2.5–2.8) can be used for studying a catalyst surface in a flowing gas with a gas pressure of 1 atm or even higher. As shown in Figure 2.9, air

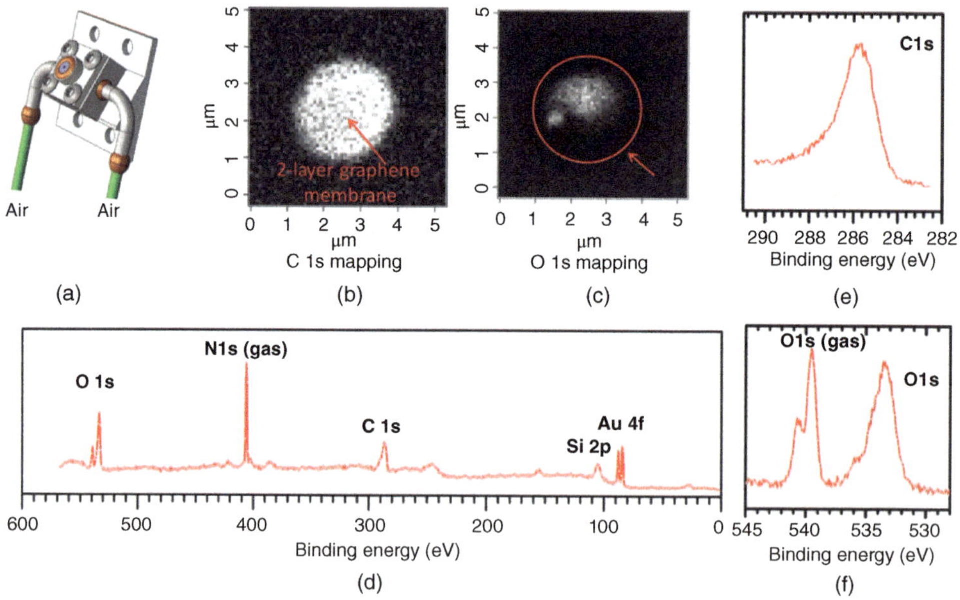

Figure 2.9 XPS studies of air flowing through the reaction cell. (a) Air flowing through the reaction cell. (b) Mapping of C 1s photoelectron intensity of the two-layer graphene membrane. (c) Mapping of O 1s of O_2 in air. (d) XPS of the graphene window region of the Si_3N_4 membrane of the reaction cell; both O 1s of O_2 and N 1s of N_2 of the flowing air were observed. (e) C 1s peak of the graphene membrane. (f) O 1s peaks of O_2 of air in the graphene membrane during preparation; the doublet of the O 1s peaks at 540 eV or so is due to the paramagnetic property of molecular O_2 since there are two unpaired single electrons in π^*_{2p}. *Source:* Reproduced with permission from Nguyen et al.[23]/Royal Society of Chemistry.

was flowing through the cell (Figure 2.9a) since the air flowing through the cell was well separated from the high vacuum environment of XPS chamber by the two-layer graphene membrane (Figure 2.8c). With this method, the flowing air at 1 atm was studied with the high vacuum XPS. The doublet peak of O 1s of gas phase O_2 of air at 540 eV shown in Figure 2.9d and f is due to the paramagnetic properties of molecular O_2 that has two unpaired electrons in π^*_{2p}. As shown in Figure 2.9d and f, both XPS peaks of N 1s of N_2 of air (Figure 2.9d) and O 1s of O_2 of air (Figure 2.9f) were simultaneously observed from the flowing air, showing the flowing air was analyzed by the high vacuum XPS. This study confirmed that the XPS characterization of a catalyst in the gas phase at a pressure of 1 atm or even higher with this flowing cell is durable. It is expected the flowing liquid cell approach can be widely used in various fields such as catalysis, environmental science, and even the biomedical field.

References

1 Hertz, H. 1887. "XIV. Ueber einen Einfluss des ultravioletten Lichtes auf die electrische Entladung." *Ann. Phys.* 267, 983–1000.

2 Lenard, P. 1902. "Energy and mass in relativity theory." *Ann. Phys.* 8, 169.

3 Lenard, P. and Wolf, M. 1889. "Zerstäuben der Körper durch das ultraviolette Licht." *Ann. Phys.* 37, 443–456.

4 Einstein, A. 1905. "Über einen die Erzeugung und Verwandlung des Lichtes betreffenden heuristischen Gesichtspunkt." *Ann. Phys.* 17, 132.

5 Robinson, H. 1923. "The secondary corpuscular rays produced by homogeneous X-rays." *Proc. R. Soc. A* 104, 455–479.

6 Rutherford, E. 1914. "XXXVII. The connexion between the β and γ-ray spectra." *London, Edinburgh Dublin Philos. Mag. J. Sci.* 28, 305–319.

7 Robinson, H. and Rawlinson, W. F. 1914. "XXXIII. The magnetic spectrum of the β rays excited in metals by soft X rays." *London, Edinburgh Dublin Philos. Mag. J. Sci.* 28, 277–281.

8 Robinson, H. R. and Young, C. L. 1930. "The influence of chemical state on critical X-ray absorption frequencies." *London, Edinburgh Dublin Philos. Mag. J. Sci.* 10, 71–75.

9 Steinhardt, R. G. and Serfass, E. J. 1951. "X-ray photoelectron spectrometer for chemical analysis." *Anal. Chem.* 23, 1585–1590.

10 Hollander, J. M. and Shirley, D. A. 1981. "The 1981 Nobel prize in physics." *Science* 214, 629–631.

11 Siegbahn, K. 1967. *Atomic, Molecular, and Solid State Structure Studied by Means of Electron Spectroscopy. ESCA.* Almqvist & Wiksell.

12 Siegbahn, K. 1969. *Applied to Free Molecules. ESCA.* North-Holland.

13 Jenkin, J. G., Leckey, R. C. G., and Liesegang, J. 1977. "The development of X-ray photoelectron spectroscopy: 1900–1960." *J. Electron Spectrosc. Relat. Phenom.* 12, 1–35.

14 Vickerman, J. C. and Gilmore, I. S. 1977. *Surface Analysis – The Principle Techniques.* Wiley.

15 Briggs, D. and Grant, J. T. 2003. *Surface Analysis by Auger and X-Ray Photoelectron Spectroscopy.* IM Publisher.

16 O'Connor, D. J., Sexton, B. A., and Smart, R. S. C. 2003. *Surface Analysis Methods in Materials Science.* Springer-Verlag.

17 Hofmann, S. 2012. *Auger- and X-Ray Photoelectron Spectroscopy in Materials Science: A User-Oriented Guide*. Springer, Heidelberg.

18 van der Heide, P. 2012. *X-Ray Photoelectron Spectroscopy: An Introduction to Principles and Practices*, Wiley.

19 Watts, J. F. and Wolstenholme, J. 2020. *An Introduction to Surface Analysis by XPS and AES*. Wiley.

20 Kalha, C., Fernando, N. K., Bhatt, P. *et al.* 2021. "Hard x-ray photoelectron spectroscopy: a snapshot of the state-of-the-art in 2020." *J. Phys. Condens. Matter* 33, 233001.

21 Weiland, C., Rumaiz, A. K., Pianetta, P. *et al.* 2016. "Recent applications of hard X-ray photoelectron spectroscopy." *J. Vac. Sci. Technol.* 34, 030801.

22 Nguyen, L., Tao, P., Liu, H. *et al.* 2018. "X-ray photoelectron spectroscopy studies of nanoparticles dispersed in static liquid." *Langmuir* 34, 9906–9916.

23 Nguyen, L., Tao, P., Liu, H. *et al.* 2018. "Studies of surface of metal nanoparticles in a flowing liquid with XPS." *Chem. Commun. (Camb.)* 54, 9981–9984.

24 Velasco-Velez, J. J., Pfeifer, V., Hävecker, M. et al. 2015. "Photoelectron spectroscopy at the graphene–liquid interface reveals the electronic structure of an electrodeposited cobalt/graphene electrocatalyst." *Angew. Chem. Int. Ed.* 54, 14554–14558.

3

Fundamentals of X-ray Photoelectron Spectroscopy

3.1 Principle of XPS

X-ray photoelectron spectroscopy (XPS) is one of the main analytical techniques in catalysis, surface science, materials science, and energy science. Majority of these elements constituting heterogeneous catalysts are transition metals. About half of these transition metals used in heterogeneous catalysts have quite low abundances in the earth as shown in Figures 3.1 and 3.2. Efficient use of these transition metal elements, particularly these low abundance metal elements in catalysis significantly relies on qualitative identification and quantitative analyses of these elements on a catalyst surface using XPS. XPS analyzes binding energy (BE) of photoelectrons of subshell of atoms upon irradiation of a sample with X-ray of a specific energy ($h\nu$) through an accurate measurement of kinetic energy of the electrons emitted from these subshells. Sometimes the electrons in subshell are called core electrons as they are electrons in a core shell in contrast to electrons in valence shell. In other words, an electron of an atom is either a core electron or a valence shell electron. Binding energy is calculated through the equation, $BE = h\nu - KE - \phi$; here BE, KE, and ϕ are binding energy of a photoelectron, kinetic energy of the photoelectron and work function of the XPS spectrometer, respectively. It can analyze most elements of the periodic table.

Figure 3.3a lists binding energies of the most intense or/and most used sub-shells. Once an element is suspected, its most intense or most used subshell and the corresponding binding energy can be found from the reference periodic table in Figure 3.3a. With this reference table (Figure 3.3a), the constituting elements of the catalyst surface can be identified. The XPS peak intensity of a subshell such as Pt 4f is proportional to the number of Pt atoms and the photoionization cross section of Pt 4f at a specific radiation such as Al Kα. The photo-ionization cross sections of most elements excited with a radiation at an energy of 1486.6 eV (Al Kα), 1253.6 eV (Mg Kα), 1041.0, 800.0, 600.0, 300.0, or 200.0 eV[1,2] are listed in Figure 3.3b, c, d, e, f, g, h, and i, respectively. For instance, by dividing XPS peak intensity of a subshell such as Pt 4f with the photo-ionization cross section corresponding to the radiation used to excite the subshell electrons available in one of the reference tables in Figure 3.3b-i, the number of Pt atoms in surface region can be evaluated. With the same method, the number of atoms of another constituting element such as Ce can be evaluated. Due to the difference in inelastic mean free paths of the two subshells of the two elements,

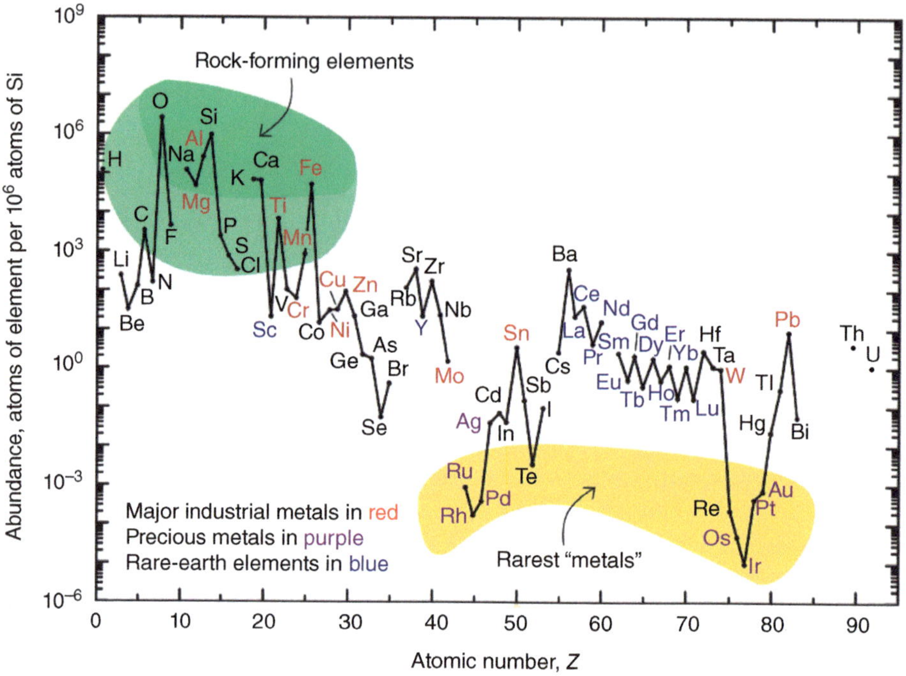

Figure 3.1 Abundances of elements on Earth shown in molar ratio. The value of *Y*-axis is the ratio of the number of atoms of a specific element to 10^{-6} times of all Si atoms on Earth. The molar ratio of an element (such as Rh) to Si on Earth (about 1×10^{-10} – 10^{-9}) is obtained by dividing the reading of an element (such as Rh) found from the *Y*-axis (about 10^{-4} – 10^{-3}) by 10^{6}. *Source:* Gordon B. Haxel, Sara Boore, and Susan Mayfield from USGS; vectorized by User: michbich - http://pubs.usgs.gov/fs/2002/fs087-02/, Public Domain, https://commons.wikimedia.org/w/index.php?curid=11215468.

the atomic ratio of the two elements in the same region such as the top 1 nm region can be found out upon calibration with their inelastic mean free paths. A significant application of these photoionization cross sections of different radiations to most elements represented in Figure 3.3b–i is to accurately, readily measure the atomic ratio of two elements in the surface region of a catalyst if synchrotron radiations with variable energies can be accessed. For instance, both Pt and Ce elements coexist in a catalyst. For example, to measure the atomic ratio of Pt/Ce, we can choose 1041 and 200 eV radiations to excite Ce 3d and Pt 4f, respectively to generate Ce 3d and Pt 4f photoelectrons with energies in the range of 120–160 eV. The photoionization cross sections of Ce 3d excited with radiation of 1041 eV and Pt 4f with radiation of 200 eV can be readily found from Figure 3.3e and i. They are 1.714 for Ce 3d and 3.826 for Pt 4f, respectively. By dividing the peak intensities of Ce 3d and Pt 4f with 1.714 and 3.826, respectively, the atomic ratio of Pt to Ce is readily found out.

It can identify constituting elements and quantify the elemental concentration of surface region of a sample and deduce chemical environment of an interested type of atoms. Most applications of XPS go to solid samples although it can study both gas and liquid phase. With modern energy analyzers having high transmission of photoelectrons and high magnification of intensity, the lower limit of measurable concentration of an element can go down to 1 at% or even lower.

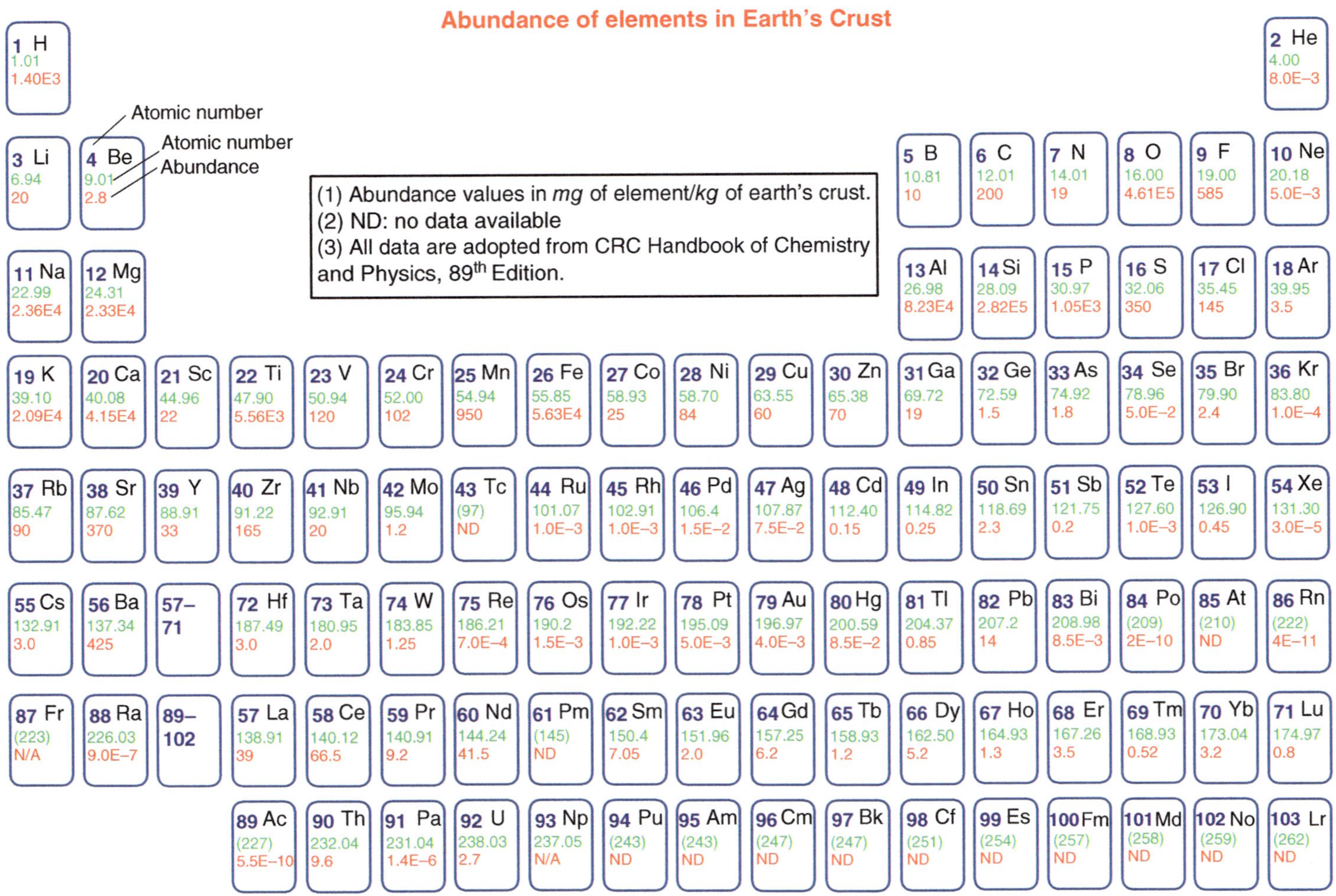

Figure 3.2 Periodic table listing abundance of elements on Earth shown in mass ratio. All data are adopted from CRC Handbook of Chemistry and Physics, 89th Edition. *Source:* Adapted from Science Notes and Projects.

Most intense/most used subshells and their binding energies

Key (example cell 4 Be): Atomic number (4); Atomic mass (9.01); The most intense/most used subshell of the element (1s); Binding energy (eV) (111.8)

(1) All binding energies (BE) except values marked with * are values experimentally measured and adopted from ref. 1.
(2) Majority of BEs are BEs of elements except those marked with triangles (Li 1s of LiF,N1 s of BN,O is of Al_2O_3, F 1s in LiF, Cl 2p3/2 in KCl, Ne 1s, Ar 2p3/2, Kr 3d5/2, or Xe 3d5/2 of inert gas implanted in graphite, Br in KBr, I in KI, Hg in HgS).
(3) Computational values of binding energies of elements adopted from ref. 2 are marked with * due to the lack of experimental data.
(4) The Computational values could be different from measured values (if available).

1 H 1.01 N/A 13.6*																	2 He 4.00 N/A 23.4*
3 Li 6.94 1s Δ 55.6	4 Be 9.01 1s 111.8											5 B 10.81 1s 189.4	6 C 12.01 1s 284.5	7 N 14.01 1s Δ 398.1	8 O 16.00 1s Δ 531.0	9 F 19.00 1s Δ 684.9	10 Ne 20.18 1s Δ 863.1
11 Na 22.99 1s 1071.8	12 Mg 24.31 2p 49.8											13 Al 26.98 2p 72.9	14 Si 28.09 2p 99.3	15 P 30.97 2p3/2 129.9	16 S 32.06 2p3/2 164.0	17 Cl 35.45 2p3/2 Δ 198.5	18 Ar 39.95 2p3/2 Δ 241.9
19 K 39.10 2p3/2 294.4	20 Ca 40.08 2p3/2 346.7	21 Sc 44.96 2p3/2 398.6	22 Ti 47.90 2p3/2 454.1	23 V 50.94 2p3/2 512.2	24 Cr 52.00 2p3/2 574.4	25 Mn 54.94 2p3/2 639.0	26 Fe 55.85 2p3/2 707.0	27 Co 58.93 2p3/2 778.3	28 Ni 58.70 2p3/2 852.7	29 Cu 63.55 2p3/2 932.7	30 Zn 65.38 2p3/2 1021.8	31 Ga 69.72 2p3/2 1116.7	32 Ge 72.59 3d 29.4	33 As 74.92 3d5/2 41.6	34 Se 78.96 3d5/2 55.6	35 Br 79.90 3d5/2 Δ 68.8	36 Kr 83.80 3d5/2 Δ 87.0
37 Rb 85.47 3d5/2 111.5	38 Sr 87.62 3d5/2 134.3	39 Y 88.91 3d5/2 156.0	40 Zr 91.22 3d5/2 178.9	41 Nb 92.91 3d5/2 202.4	42 Mo 95.94 3d5/2 228.0	43 Tc (97) 3d 273.0*	44 Ru 101.07 3d5/2 280.1	45 Rh 102.91 3d5/2 307.2	46 Pd 106.4 3d5/2 335.1	47 Ag 107.87 3d5/2 368.3	48 Cd 112.40 3d5/2 405.1	49 In 114.82 3d5/2 443.9	50 Sn 118.69 3d5/2 485.0	51 Sb 121.75 3d5/2 528.3	52 Te 127.60 3d5/2 573.1	53 I 126.90 3d5/2 Δ 619.3	54 Xe 131.30 3d5/2 Δ 669.7
55 Cs 132.91 3d5/2 726.4	56 Ba 137.34 3d5/2 780.6	57–71	72 Hf 187.49 4f7/2 14.3	73 Ta 180.95 4f7/2 21.9	74 W 183.85 4f7/2 31.4	75 Re 186.21 4f7/2 40.3	76 Os 190.2 4f7/2 50.7	77 Ir 192.22 4f7/2 60.9	78 Pt 195.09 4f7/2 71.2	79 Au 196.97 4f7/2 84.0	80 Hg 200.59 4f7/2 Δ 101.0	81 Tl 204.37 4f7/2 117.7	82 Pb 207.2 4f7/2 136.9	83 Bi 208.98 4f7/2 157.0	84 Po (209) 4f 216.2*	85 At (210) 4f 240.5*	86 Rn (222) 4f 265.6*
87 Fr (223) 4f 296.6*	88 Ra 226.03 4f 328.6*	89–102	57 La 138.91 3d5/2 835.8	58 Ce 140.12 3d5/2 883.8	59 Pr 140.91 3d5/2 931.8	60 Nd 144.24 3d5/2 980.8	61 Pm (145) 3d 1033*	62 Sm 150.4 3d5/2 1081.1	63 Eu 151.96 3d5/2 1125.6	64 Gd 157.25 4d 140.4	65 Tb 158.93 4d 146.0	66 Dy 162.50 4d 152.4	67 Ho 164.93 4d 159.6	68 Er 167.26 4d 167.3	69 Tm 168.93 4d 175.4	70 Yb 173.04 4d 182.4	71 Lu 174.97 4f7/2 7.3
			89 Ac (227) 4f 359.2*	90 Th 232.04 4f7/2 333.2	91 Pa 231.04 4f 409.3*	92 U 238.03 4f7/2 377.3	93 Np 237.05 4f 460.4*	94 Pu (243) 4f 479.8*	95 Am (243) 4f 505.9*	96 Cm (247) 4f 539.2*	97 Bk (247) 4f 566.2*	98 Cf (251) 4f 593.5*	99 Es (254) 4f 621.2*	100 Fm (257) 4f 649.2*	101 Md (258) 4f 677.7*	102 No (255) 4f 706.5*	103 Lr (262) 4f 735.7*

Figure 3.3a Periodic table listing binding energy of the most intense or most used core levels of elements. *Source:* Adapted from M.C. Biesinger (www.xpsfitting.com).

Photoionization cross section of most intense/most used subshells excited by Cu Kα

Atomic number / Most intense/most used subshell / Binding energy of the most intense/most used subshell / Photoionization cross section of the most intense/most used subshell (example: 4 Be, 1s, 111.8, .13E–4)

(1) The datum at the bottom of each box is the cross-section of photoionization under irradiation of Cu Kα (hν = 8047.8 eV). NE stands for lack of photoionization as the photon energy is close to or lower than the binding energy of the subshell electrons.
(2) For a *p*, *d*, or *f* subshell, its photoionization cross section listed in a box is the total of the two split peaks although only BE of one of the two split peaks is listed in the box.
(3) All binding energies except values marked with * are values experimentally measured and adopted from ref. 1.
(4) Computational values of binding energies adopted from ref. 2 are marked with * due to the lack of experimental data. The Computational values could be different from measured values (if available).

Atomic number	Element	Subshell	Binding energy	Cross section
1	H	1s	13.6*	.21E–6
2	He	1s	23.4*	.24E–6
3	Li	1s	55.6	.41E–5
4	Be	1s	111.8	.13E–4
5	B	1s	189.4	.36E–4
6	C	1s	284.5	.80E–4
7	N	1s	398.1	.15E–3
8	O	1s	531.0	.27E–3
9	F	1s	684.9	.45E–3
10	Ne	1s	863.1	.70E–3
11	Na	1s	1071.8	.10E–2
12	Mg	2p	49.8	.13E–4
13	Al	2p	72.9	.17E–4
14	Si	2p	99.3	.30E–4
15	P	2p3/2	129.9	.47E–4
16	S	2p3/2	164.0	.66E–4
17	Cl	2p3/2	198.5	.97E–4
18	Ar	2p3/2	241.9	.13E–3
19	K	2p3/2	294.4	.19E–3
20	Ca	2p3/2	346.7	.26E–3
21	Sc	2p3/2	398.6	.35E–3
22	Ti	2p3/2	454.1	.47E–3
23	V	2p3/2	512.2	.59E–3
24	Cr	2p3/2	574.4	.75E–3
25	Mn	2p3/2	639.0	.95E–3
26	Fe	2p3/2	707.0	.11E–2
27	Co	2p3/2	778.3	.14E–2
28	Ni	2p3/2	852.7	.17E–2
29	Cu	2p3/2	932.7	.20E–2
30	Zn	2p3/2	1021.8	.24E–2
31	Ga	2p3/2	1116.7	.29E–2
32	Ge	3d	29.4	.28E–4
33	As	3d5/2	41.6	.37E–4
34	Se	3d5/2	55.6	.49E–4
35	Br	3d5/2	68.8	.73E–4
36	Kr	3d5/2	87.0	.82E–4
37	Rb	3d5/2	111.5	.12E–3
38	Sr	3d5/2	134.3	.14E–3
39	Y	3d5/2	156.0	.17E–3
40	Zr	3d5/2	178.9	.23E–3
41	Nb	3d5/2	202.4	.28E–3
42	Mo	3d5/2	228.0	.33E–3
43	Tc	3d	273.0*	.40E–3
44	Ru	3d5/2	280.1	.46E–3
45	Rh	3d5/2	307.2	.55E–3
46	Pd	3d5/2	335.1	.67E–3
47	Ag	3d5/2	368.3	.78E–3
48	Cd	3d5/2	405.1	.93E–3
49	In	3d5/2	443.9	.10E–2
50	Sn	3d5/2	485.0	.12E–2
51	Sb	3d5/2	528.3	.14E–2
52	Te	3d5/2	573.1	.16E–2
53	I	3d5/2	619.3	.18E–2
54	Xe	3d5/2	669.7	.20E–3
55	Cs	3d5/2	726.4	.23E–2
56	Ba	3d5/2	780.6	.27E–2
57–71				
72	Hf	4f7/2	14.3	.16E–3
73	Ta	4f7/2	21.9	.19E–3
74	W	4f7/2	31.4	.24E–3
75	Re	4f7/2	40.3	.29E–3
76	Os	4f7/2	50.7	.35E–3
77	Ir	4f7/2	60.9	.39E–3
78	Pt	4f7/2	71.2	.46E–3
79	Au	4f7/2	84.0	.52E–3
80	Hg	4f7/2	101.0	.58E–3
81	Tl	4f7/2	117.7	.67E–3
82	Pb	4f7/2	136.9	.76E–3
83	Bi	4f7/2	157.0	.85E–3
84	Po	4f	216.2*	.97E–3
85	At	4f	240.5*	.10E–2
86	Rn	4f	265.6*	.12E–2
87	Fr	4f	296.6*	.13E–2
88	Ra	4f	328.6*	.15E–2
89–102				
57	La	3d5/2	835.8	.30E–2
58	Ce	3d5/2	883.8	.34E–2
59	Pr	3d5/2	931.8	.38E–2
60	Nd	3d5/2	980.8	.42E–2
61	Pm	3d	1033*	.47E–2
62	Sm	3d5/2	1081.1	.52E–2
63	Eu	3d5/2	1125.6	.58E–2
64	Gd	4d	140.4	.12E–2
65	Tb	4d	146.0	.13E–2
66	Dy	4d	152.4	.14E–2
67	Ho	4d	159.6	.16E–2
68	Er	4d	167.3	.17E–2
69	Tm	4d	175.4	.19E–2
70	Yb	4d	182.4	.20E–2
71	Lu	4f7/2	7.3	.15E–3
89	Ac	4f	359.2*	.17E–2
90	Th	4f7/2	333.2	.19E–2
91	Pa	4f	409.3*	.21E–2
92	U	4f7/2	377.3	.23E–2
93	Np	4f	460.4*	.25E–2
94	Pu	4f	479.8*	.28E–2
95	Am	4f	505.9*	.31E–2
96	Cm	4f	539.2*	.33E–2
97	Bk	4f	566.2*	.37E–2
98	Cf	4f	593.5*	.40E–2
99	Es	4f	621.2*	.43E–2
100	Fm	4f	649.2*	.47E–2
101	Md	4f	677.7*	.50E–2
102	No	4f	706.5*	.55E–2
103	Lr	4f	735.7*	.60E–2

Figure 3.3b List of atomic subshell photoionization cross sections of the most intense or most used subshells of elements $1 \leq Z \leq 103$ while Cu Kα (hυ=8047.8 eV) was used. Cross sections were adopted from ref. 2.

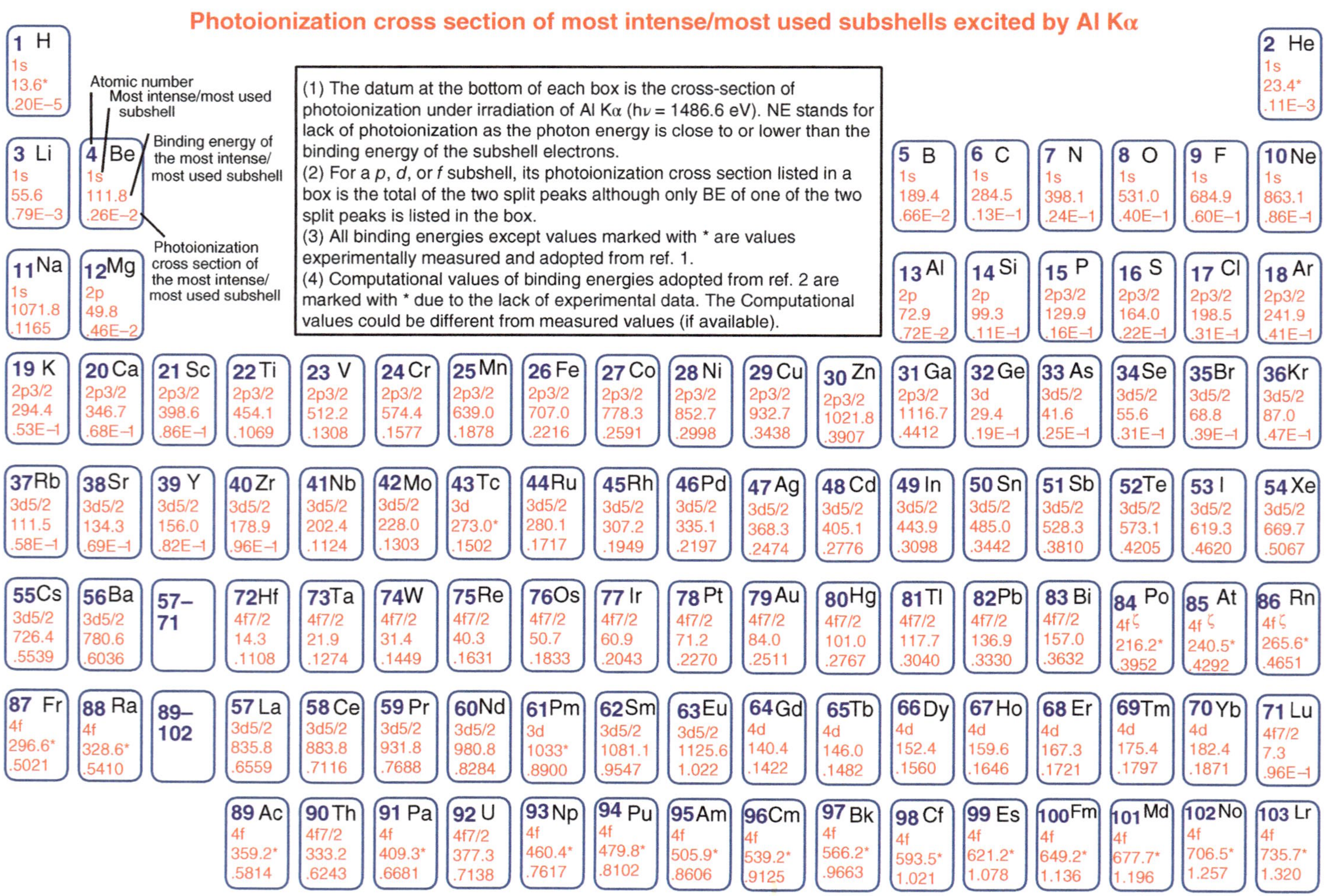

Figure 3.3c List of atomic subshell photoionization cross sections of the most intense or most used subshells of elements $1 \le Z \le 103$ while Al Kα (hυ=1486.6 eV) was used. Cross sections were adopted from ref. 2.

Photoionization cross section of most intense/most used subshells excited by Mg Kα

(1) The datum at the bottom of each box is the cross section of photoionization under irradiation of Mg Kα (hν = 1253.6 eV). NE stands for lack of photoionization as the photon energy is close to or lower than the binding energy of the subshell electrons.
(2) For a *p*, *d*, or *f* subshell, its photoionization cross section listed in a box is the total of the two split peaks although only BE of one of the two split peaks is listed in the box.
(3) All binding energies except values marked with * are values experimentally measured and adopted from ref. 1.
(4) Computational values of binding energies adopted from ref. 2 are marked with * due to the lack of experimental data. The Computational values could be different from measured values (if available).

Key (example box, 4 Be): Atomic number; Most intense/most used subshell; Binding energy of the most intense/most used subshell; Photoionization cross section of the most intense/most used subshell.

Atomic number	Element	Most intense/most used subshell	Binding energy	Photoionization cross section
1	H	1s	13.6*	.46E–5
2	He	1s	23.4*	.18E–3
3	Li	1s	55.6	.13E–2
4	Be	1s	111.8	.44E–2
5	B	1s	189.4	.10E–1
6	C	1s	284.5	.22E–1
7	N	1s	398.1	.39E–1
8	O	1s	531.0	.63E–1
9	F	1s	684.9	.94E–1
10	Ne	1s	863.1	.1328
11	Na	1s	1071.8	.1781
12	Mg	2p	49.8	.77E–2
13	Al	2p	72.9	.12E–1
14	Si	2p	99.3	.19E–1
15	P	2p3/2	129.9	.27E–1
16	S	2p3/2	164.0	.38E–1
17	Cl	2p3/2	198.5	.52E–1
18	Ar	2p3/2	241.9	.69E–1
19	K	2p3/2	294.4	.89E–1
20	Ca	2p3/2	346.7	.1132
21	Sc	2p3/2	398.6	.1419
22	Ti	2p3/2	454.1	.1746
23	V	2p3/2	512.2	.2117
24	Cr	2p3/2	574.4	.2540
25	Mn	2p3/2	639.0	.3011
26	Fe	2p3/2	707.0	.3529
27	Co	2p3/2	778.3	.4090
28	Ni	2p3/2	852.7	.4691
29	Cu	2p3/2	932.7	.5345
30	Zn	2p3/2	1021.8	.6057
31	Ga	2p3/2	1116.7	.6851
32	Ge	3d	29.4	.34E–1
33	As	3d5/2	41.6	.44E–1
34	Se	3d5/2	55.6	.55E–1
35	Br	3d5/2	68.8	.67E–1
36	Kr	3d5/2	87.0	.83E–1
37	Rb	3d5/2	111.5	.99E–1
38	Sr	3d5/2	134.3	.1186
39	Y	3d5/2	156.0	.1400
40	Zr	3d5/2	178.9	.1633
41	Nb	3d5/2	202.4	.1900
42	Mo	3d5/2	228.0	.2186
43	Tc	3d	273.0*	.2500
44	Ru	3d5/2	280.1	.2836
45	Rh	3d5/2	307.2	.3210
46	Pd	3d5/2	335.1	.3617
47	Ag	3d5/2	368.3	.4052
48	Cd	3d5/2	405.1	.4519
49	In	3d5/2	443.9	.5017
50	Sn	3d5/2	485.0	.5551
51	Sb	3d5/2	528.3	.6116
52	Te	3d5/2	573.1	.6720
53	I	3d5/2	619.3	.7359
54	Xe	3d5/2	669.7	.8023
55	Cs	3d5/2	726.4	.8717
56	Ba	3d5/2	780.6	.9444
57–71				
72	Hf	4f7/2	14.3	.1922
73	Ta	4f7/2	21.9	.2194
74	W	4f7/2	31.4	.2463
75	Re	4f7/2	40.3	.2776
76	Os	4f7/2	50.7	.3095
77	Ir	4f7/2	60.9	.3438
78	Pt	4f7/2	71.2	.3792
79	Au	4f7/2	84.0	.4176
80	Hg	4f7/2	101.0	.4584
81	Tl	4f7/2	107.7	.5008
82	Pb	4f7/2	136.9	.5452
83	Bi	4f7/2	157.0	.5929
84	Po	4f	216.2*	.6431
85	At	4f	240.5*	.6948
86	Rn	4f	265.6*	.7477
87	Fr	4f	296.6*	.8048
88	Ra	4f	328.6*	.8640
89–102				
57	La	3d5/2	835.8	1.020
58	Ce	3d5/2	883.8	1.100
59	Pr	3d5/2	931.8	1.183
60	Nd	3d5/2	980.8	1.269
61	Pm	3d	1033*	1.359
62	Sm	3d5/2	1081.1	1.456
63	Eu	3d5/2	1125.6	1.562
64	Gd	4d	140.4	.2054
65	Tb	4d	146.0	.2128
66	Dy	4d	152.4	.2224
67	Ho	4d	159.6	.2310
68	Er	4d	167.3	.2404
69	Tm	4d	175.4	.2497
70	Yb	4d	182.4	.2587
71	Lu	4f7/2	7.3	.1670
89	Ac	4f	359.2*	.9260
90	Th	4f7/2	333.2	.9895
91	Pa	4f	409.3*	1.054
92	U	4f7/2	377.3	1.121
93	Np	4f	460.4*	1.188
94	Pu	4f	479.8*	1.259
95	Am	4f	505.9*	1.331
96	Cm	4f	539.2*	1.405
97	Bk	4f	566.2*	1.481
98	Cf	4f	593.5*	1.558
99	Es	4f	621.2*	1.637
100	Fm	4f	649.2*	1.716
101	Md	4f	677.7*	1.797
102	No	4f	706.5*	1.879
103	Lr	4f	735.7*	1.963

Figure 3.3d List of atomic subshell photoionization cross sections of the most intense or most used subshells of elements $1 \leq Z \leq 103$ while Mg Kα (hυ=1253.6 eV) was used. Cross sections were adopted from ref. 2.

Photoionization cross section of most intense/most used subshells excited by X-ray (1041.0 eV)

Key (example box 4 Be): Atomic number; Most intense/most used subshell; Binding energy of the most intense/most used subshell; Photoionization cross section of the most intense/most used subshell.

(1) The datum at the bottom of each box is the cross-section of photoionization under irradiation of energy (hν = 1041.0 eV). NE stands for lack of photoionization as the photon energy is close to or lower than the binding energy of the subshell electrons.
(2) For a *p*, *d*, or *f* subshell, its photoionization cross section listed in a box is the total of the two split peaks although only BE of one of the two split peaks is listed in the box.
(3) All binding energies except values marked with * are values experimentally measured and adopted from ref. 1.
(4) Computational values of binding energies adopted from ref. 2 are marked with * due to the lack of experimental data. The Computational values could be different from measured values (if available).

1 H 1s 13.6* .13E–4																	2 He 1s 23.4* .33E–3
3 Li 1s 55.6 .23E–2	4 Be 1s 111.8 .77E–2											5 B 1s 189.4 .18E–1	6 C 1s 284.5 .37E–1	7 N 1s 398.1 .65E–1	8 O 1s 531.0 .1042	9 F 1s 684.9 .1525	10 Ne 1s 863.1 .2093
11 Na 1s 1071.8 NE	12 Mg 2p 49.8 .14E–1											13 Al 2p 72.9 .22E–1	14 Si 2p 99.3 .34E–1	15 P 2p3/2 129.9 .49E–1	16 S 2p3/2 164.0 .67E–1	17 Cl 2p3/2 198.5 .91E–1	18 Ar 2p3/2 241.9 .1193
19 K 2p3/2 294.4 .1534	20 Ca 2p3/2 346.7 .1933	21 Sc 2p3/2 398.6 .2396	22 Ti 2p3/2 454.1 .2930	23 V 2p3/2 512.2 .3535	24 Cr 2p3/2 574.4 .4205	25 Mn 2p3/2 639.0 .4936	26 Fe 2p3/2 707.0 .5723	27 Co 2p3/2 778.3 .6569	28 Ni 2p3/2 852.7 .7513	29 Cu 2p3/2 932.7 .8603	30 Zn 2p3/2 1021.8 .7746	31 Ga 2p3/2 1116.7 NE	32 Ge 3d 29.4 .64E–1	33 As 3d5/2 41.6 .80E–1	34 Se 3d5/2 55.6 .99E–1	35 Br 3d5/2 68.8 .1230	36 Kr 3d5/2 87.0 .1475
37 Rb 3d5/2 111.5 .1770	38 Sr 3d5/2 134.3 .2090	39 Y 3d5/2 156.0 .2447	40 Zr 3d5/2 178.9 .2849	41 Nb 3d5/2 202.4 .3275	42 Mo 3d5/2 228.0 .3748	43 Tc 3d 273.0* .4263	44 Ru 3d5/2 280.1 .4827	45 Rh 3d5/2 307.2 .5429	46 Pd 3d5/2 335.1 .6073	47 Ag 3d5/2 368.3 .6762	48 Cd 3d5/2 405.1 .7495	49 In 3d5/2 443.9 .8280	50 Sn 3d5/2 485.0 .9108	51 Sb 3d5/2 528.3 .9981	52 Te 3d5/2 573.1 1.088	53 I 3d5/2 619.3 1.182	54 Xe 3d5/2 669.7 1.279
55 Cs 3d5/2 726.4 1.382	56 Ba 3d5/2 780.6 1.487	57–71	72 Hf 4f7/2 14.3 .3396	73 Ta 4f7/2 21.9 .3824	74 W 4f7/2 31.4 .4294	75 Re 4f7/2 40.3 .4781	76 Os 4f7/2 50.7 .5299	77 Ir 4f7/2 60.9 .5852	78 Pt 4f7/2 71.2 .6439	79 Au 4f7/2 84.0 .7041	80 Hg 4f7/2 101.0 .7678	81 Tl 4f7/2 117.7 .8352	82 Pb 4f7/2 136.9 .9064	83 Bi 4f7/2 157.0 .9791	84 Po 4f 216.2* 1.054	85 At 4f 240 .5* 1.133	86 Rn 4f 265.6* 1.217
87 Fr 4f 296.6* 1.301	88 Ra 4f 328.6* 1.388	89–102	57 La 3d5/2 835.8 1.592	58 Ce 3d5/2 883.8 1.714	59 Pr 3d5/2 931.8 1.850	60 Nd 3d5/2 980.8 2.014	61 Pm 3d 1033* NE	62 Sm 3d5/2 1081.1 NE	63 Eu 3d5/2 1125.6 NE	64 Gd 4d 140.4 .2944	65 Tb 4d 146.0 .3052	66 Dy 4d 152.4 .3141	67 Ho 4d 159.6 .3263	68 Er 4d 167.3 .3369	69 Tm 4d 175.4 .3470	70 Yb 4d 182.4 .3565	71 Lu 4f7/2 7.3 .2961
			89 Ac 4f 359.2* 1.477	90 Th 4f7/2 333.2 1.568	91 Pa 4f 409.3* 1.659	92 U 4f7/2 377.3 1.754	93 Np 4f 460.4* 1. 850	94 Pu 4f 479.8* 1.946	95 Am 4f 505.9* 2.044	96 Cm 4f 539.2* 2.143	97 Bk 4f 566.2* 2.242	98 Cf 4f 593.5* 2.342	99 Es 4f 621.2* 2.443	100 Fm 4f 649.2* 2.546	101 Md 4f 677.7* 2.655	102 No 4f 706.5* 2.772	103 Lr 4f 735.7* 2.900

Figure 3.3e List of atomic subshell photoionization cross sections of the most intense or most used subshells of elements $1 \leq Z \leq 103$ while radiation with an energy of 1041.0 eV was used. Cross sections were adopted from ref. 2.

Photoionization cross section of most intense/most used subshells excited by X-ray (800.0 eV)

Key (4 Be box): Atomic number; Most intense/most used subshell; Binding energy of the most intense/most used subshell; Photoionization cross section of the most intense/most used subshell

(1) The datum at the bottom of each box is the cross-section of photoionization under irradiation of energy ($h\nu = 800.0$ eV). NE stands for lack of photoionization as the photon energy is close to or lower than the binding energy of the subshell electrons.
(2) For a *p*, *d*, or *f* subshell, its photoionization cross section listed in a box is the total of the two split peaks although only BE of one of the two split peaks is listed in the box.
(3) All binding energies except values marked with * are values experimentally measured and adopted from ref 1.
(4) Computational values of binding energies adopted from ref. 2 are marked with * due to the lack of experimental data. The Computational values could be different from measured values (if available).

1 H 1s 13.6* .29E–4																	2 He 1s 23.4* .86E–3
3 Li 1s 55.6 .52E–2	4 Be 1s 111.8 .16E–1											5 B 1s 189.4 .39E–1	6 C 1s 284.5 .77E–1	7 N 1s 398.1 .1329	8 O 1s 531.0 .2044	9 F 1s 684.9 .2923	10 Ne 1s 863.1 NE
11 Na 1s 1071.8 NE	12 Mg 2p 49.8 .34E–1											13 Al 2p 72.9 .51E–1	14 Si 2p 99.3 .76E–1	15 P 2p3/2 129.9 .1081	16 S 2p3/2 164.0 .1475	17 Cl 2p3/2 198.5 .1951	18 Ar 2p3/2 241.9 .2533
19 K 2p3/2 294.4 .3207	20 Ca 2p3/2 346.7 .3999	21 Sc 2p3/2 398.6 .4913	22 Ti 2p3/2 454.1 .5931	23 V 2p3/2 512.2 .7034	24 Cr 2p3/2 574.4 .8235	25 Mn 2p3/2 639.0 .9533	26 Fe 2p3/2 707.0 1.107	27 Co 2p3/2 778.3 .8899	28 Ni 2p3/2 852.7 NE	29 Cu 2p3/2 932.7 NE	30 Zn 2p3/2 1021.8 NE	31 Ga 2p3/2 1116.7 NE	32 Ge 3d 29.4 .1465	33 As 3d5/2 41.6 .1827	34 Se 3d5/2 55.6 .2239	35 Br 3d5/2 68.8 .2715	36 Kr 3d5/2 87.0 .3234
37 Rb 3d5/2 111.5 .3826	38 Sr 3d5/2 134.3 .4478	39 Y 3d5/2 156.0 .5186	40 Zr 3d5/2 178.9 .5960	41 Nb 3d5/2 202.4 .6815	42 Mo 3d5/2 228.0 .7731	43 Tc 3d 273.0* .8718	44 Ru 3d5/2 280.1 .9758	45 Rh 3d5/2 307.2 1.087	46 Pd 3d5/2 335.1 1.205	47 Ag 3d5/2 368.3 1.329	48 Cd 3d5/2 405.1 1.457	49 In 3d5/2 443.9 1.588	50 Sn 3d5/2 485.0 1.723	51 Sb 3d5/2 528.3 1.866	52 Te 3d5/2 573.1 2.013	53 I 3d5/2 619.3 2.177	54 Xe 3d5/2 669.7 2.356
55 Cs 3d5/2 726.4 2.661	56 Ba 3d5/2 780.6 NE	57–71	72 Hf 4f7/2 14.3 .7139	73 Ta 4f7/2 21.9 .7959	74 W 4f7/2 31.4 .8859	75 Re 4f7/2 40.3 .9763	76 Os 4f7/2 50.7 1.075	77 Ir 4f7/2 60.9 1.174	78 Pt 4f7/2 71.2 1.276	79 Au 4f7/2 84.0 1.387	80 Hg 4f7/2 101.0 1.497	81 Tl 4f7/2 117.7 1.612	82 Pb 4f7/2 136.9 1.731	83 Bi 4f7/2 157.0 1.852	84 Po 4f 216.2* 1.977	85 At 4f 240.5* 2.102	86 Rn 4f 265.6* 2.226
87 Fr 4f 296.6* 2.353	88 Ra 4f 328.6* 2.481	89–102	57 La 3d5/2 835.8 NE	58 Ce 3d5/2 883.8 NE	59 Pr 3d5/2 931.8 NE	60 Nd 3d5/2 980.8 NE	61 Pm 3d 1033* NE	62 Sm 3d5/2 1081.1 NE	63 Eu 3d5/2 1125.6 NE	64 Gd 4d 140.4 .4613	65 Tb 4d 146.0 .4676	66 Dy 4d 152.4 .4793	67 Ho 4d 159.6 .4889	68 Er 4d 167.3 .4988	69 Tm 4d 175.4 .5076	70 Yb 4d 182.4 .5154	71 Lu 4f7/2 7.3 .6311
			89 Ac 4f 359.2* 2.609	90 Th 4f7/2 333.2 2.734	91 Pa 4f 409.3* 2.856	92 U 4f7/2 377.3 2.974	93 Np 4f 460.4* 3.089	94 Pu 4f 479.8* 3.205	95 Am 4f 505.9* 3.318	96 Cm 4f 539.2* 3.428	97 Bk 4f 566.2* 3.545	98 Cf 4f 593.5* 3.665	99 Es 4f 621.2* 3.786	100 Fm 4f 649.2* 3.833	101 Md 4f 677.7* 3.547	102 No 4f 706.5* 2.274	103 Lr 4f 735.7* .6392

Figure 3.3f List of atomic subshell photoionization cross sections of the most intense or most used subshells of elements $1 \leq Z \leq 103$ while radiation with an energy of 800.0 eV was used. Cross sections were adopted from ref. 2.

Photoionization cross section of most intense/most used subshells excited by X-ray (600.0 eV)

Atomic number
Most intense/most used subshell
Binding energy of the most intense/most used subshell
Photoionization cross section of the most intense/most used subshell

(1) The datum at the bottom of each box is the cross-section of photoionization under irradiation of energy (hν = 600.0 eV). NE stands for lack of photoionization as the photon energy is close to or lower than the binding energy of the subshell electrons.
(2) For a *p*, *d*, or *f* subshell, its photoionization cross section listed in a box is the total of the two split peaks although only BE of one of the two split peaks is listed in the box.
(3) All binding energies except values marked with * are values experimentally measured and adopted from ref. 1.
(4) Computational values of binding energies adopted from ref. 2 are marked with * due to the lack of experimental data. The Computational values could be different from measured values (if available).

1 H 1s 13.6* .71E−4																	2 He 1s 23.4* .20E−2
3 Li 1s 55.6 .12E−1	4 Be 1s 111.8 .38E−1											5 B 1s 189.4 .88E−1	6 C 1s 284.5 .1680	7 N 1s 398.1 .2763	8 O 1s 531.0 .4119	9 F 1s 684.9 NE	10 Ne 1s 863.1 NE
11 Na 1s 1071.8 NE	12 Mg 2p 49.8 .81E−1											13 Al 2p 72.9 .1230	14 Si 2p 99.3 .1776	15 P 2p3/2 129.9 .2460	16 S 2p3/2 164.0 .3315	17 Cl 2p3/2 198.5 .4330	18 Ar 2p3/2 241.9 .5526
19 K 2p3/2 294.4 .6920	20 Ca 2p3/2 346.7 .8482	21 Sc 2p3/2 398.6 1.019	22 Ti 2p3/2 454.1 1.202	23 V 2p3/2 512.2 1.410	24 Cr 2p3/2 574.4 1.645	25 Mn 2p3/2 639.0 NE	26 Fe 2p3/2 707.0 NE	27 Co 2p3/2 778.3 NE	28 Ni 2p3/2 852.7 NE	29 Cu 2p3/2 932.7 NE	30 Zn 2p3/2 1021.8 NE	31 Ga 2p3/2 1116.7 NE	32 Ge 3d 29.4 .3438	33 As 3d5/2 41.6 .4227	34 Se 3d5/2 55.6 .5117	35 Br 3d5/2 68.8 .6076	36 Kr 3d5/2 87.0 .7182
37 Rb 3d5/2 111.5 .8343	38 Sr 3d5/2 134.3 .9554	39 Y 3d5/2 156.0 1.108	40 Zr 3d5/2 178.9 1.259	41 Nb 3d5/2 202.4 1.418	42 Mo 3d5/2 228.0 1.588	43 Tc 3d 273.0* 1.766	44 Ru 3d5/2 280.1 1.948	45 Rh 3d5/2 307.2 2.133	46 Pd 3d5/2 335.1 2.314	47 Ag 3d5/2 368.3 2.503	48 Cd 3d5/2 405.1 2.700	49 In 3d5/2 443.9 2.925	50 Sn 3d5/2 485.0 3.238	51 Sb 3d5/2 528.3 4.016	52 Te 3d5/2 573.1 .1578	53 I 3d5/2 619.3 NE	54 Xe 3d5/2 669.7 NE
55 Cs 3d5/2 726.4 NE	56 Ba 3d5/2 780.6 NE	57–71	72 Hf 4f7/2 14.3 1.470	73 Ta 4f7/2 21.9 1.617	74 W 4f7/2 31.4 1.778	75 Re 4f7/2 40.3 1.935	76 Os 4f7/2 50.7 2.101	77 Ir 4f7/2 60.9 2.267	78 Pt 4f7/2 71.2 2.432	79 Au 4f7/2 84.0 2.599	80 Hg 4f7/2 101.0 2.771	81 Tl 4f7/2 117.7 2.937	82 Pb 4f7/2 136.9 3.098	83 Bi 4f7/2 157.0 3.256	84 Po 4f 216.2* 3.407	85 At 4f 240.5* 3.549	86 Rn 4f 265.6* 3.684
87 Fr 4f 296.6* 3.798	88 Ra 4f 328.6* 3.895	89–102	57 La 3d5/2 835.8 NE	58 Ce 3d5/2 883.8 NE	59 Pr 3d5/2 931.8 NE	60 Nd 3d5/2 980.8 NE	61 Pm 3d 1033* NE	62 Sm 3d5/2 1081.1 NE	63 Eu 3d5/2 1125.6 NE	64 Gd 4d 140.4 .6719	65 Tb 4d 146.0 .6682	66 Dy 4d 152.4 .6736	67 Ho 4d 159.6 .6773	68 Er 4d 167.3 .6781	69 Tm 4d 175.4 .6769	70 Yb 4d 182.4 .6737	71 Lu 4f7/2 7.3 1.321
			89 Ac 4f 359.2* 3.972	90 Th 4f7/2 333.2 4.036	91 Pa 4f 409.3* 4.076	92 U 4f7/2 377.3 4.077	93 Np 4f 460.4* 3.980	94 Pu 4f 479.8* 3.633	95 Am 4f 505.9* 2.763	96 Cm 4f 539.2* 1.251	97 Bk 4f 566.2* .2186	98 Cf 4f 593.5* .92E−1	99 Es 4f 621.2* NE	100 Fm 4f 649.2* NE	101 Md 4f 677.7* NE	102 No 4f 706.5* NE	103 Lr 4f 735.7* NE

Figure 3.3g List of atomic subshell photoionization cross sections of the most intense or most used subshells of elements $1 \leq Z \leq 103$ while radiation with an energy of 600.0 eV was used. Cross sections were adopted from ref. 2.

Photoionization cross section of most intense/most used subshells excited by X-ray (300.0 eV)

Legend (example box 4 Be): Atomic number; Most intense/most used subshell; Binding energy of the most intense/ most used subshell; Photoionization cross section of the most intense/ most used subshell

(1) The datum at the bottom of each box is the cross-section of photoionization under irradiation of energy ($h\nu = 300.0$ eV). NE stands for lack of photoionization as the photon energy is close to or lower than the binding energy of the subshell electrons.
(2) For a *p*, *d*, or *f* subshell, its photoionization cross section listed in a box is the total of the two split peaks although only BE of one of the two split peaks is listed in the box.
(3) All binding energies except values marked with * are values experimentally measured and adopted from ref. 1.
(4) Computational values of binding energies adopted from ref. 2 are marked with * due to the lack of experimental data. The Computational values could be different from measured values (if available).

1 H 1s 13.6* .61E–3																	2 He 1s 23.4* .17E–1
3 Li 1s 55.6 .91E–1	4 Be 1s 111.8 .2590											5 B 1s 189.4 .5403	6 C 1s 284.5 .8591	7 N 1s 398.1 NE	8 O 1s 531.0 NE	9 F 1s 684.9 NE	10 Ne 1s 863.1 NE
11 Na 1s 1071.8 NE	12 Mg 2p 49.8 .5675											13 Al 2p 72.9 .8226	14 Si 2p 99.3 1.132	15 P 2p3/2 129.9 1.501	16 S 2p3/2 164.0 1.920	17 Cl 2p3/2 198.5 2.321	18 Ar 2p3/2 241.9 2.752
19 K 2p3/2 294.4 3.177	20 Ca 2p3/2 346.7 NE	21 Sc 2p3/2 398.6 NE	22 Ti 2p3/2 454.1 NE	23 V 2p3/2 512.2 NE	24 Cr 2p3/2 574.4 NE	25 Mn 2p3/2 639.0 NE	26 Fe 2p3/2 707.0 NE	27 Co 2p3/2 778.3 NE	28 Ni 2p3/2 852.7 NE	29 Cu 2p3/2 932.7 NE	30 Zn 2p3/2 1021.8 NE	31 Ga 2p3/2 1116.7 NE	32 Ge 3d 29.4 2.006	33 As 3d5/2 41.6 2.357	34 Se 3d5/2 55.6 2.718	35 Br 3d5/2 68.8 3.095	36 Kr 3d5/2 87.0 3.468
37 Rb 3d5/2 111.5 3.816	38 Sr 3d5/2 134.3 4.129	39 Y 3d5/2 156.0 4.382	40 Zr 3d5/2 178.9 4.525	41 Nb 3d5/2 202.4 4.551	42 Mo 3d5/2 228.0 4.361	43 Tc 3d 273.0* 2.827	44 Ru 3d5/2 280.1 .1898	45 Rh 3d5/2 307.2 NE	46 Pd 3d5/2 335.1 NE	47 Ag 3d5/2 368.3 NE	48 Cd 3d5/2 405.1 NE	49 In 3d5/2 443.9 NE	50 Sn 3d5/2 485.0 NE	51 Sb 3d5/2 528.3 NE	52 Te 3d5/2 573.1 NE	53 I 3d5/2 619.3 NE	54 Xe 3d5/2 669.7 NE
55 Cs 3d5/2 726.4 NE	56 Ba 3d5/2 780.6 NE	57–71	72 Hf 4f7/2 14.3 4.979	73 Ta 4f7/2 21.9 5.202	74 W 4f7/2 31.4 5.397	75 Re 4f7/2 40.3 5.542	76 Os 4f7/2 50.7 5.606	77 Ir 4f7/2 60.9 5.594	78 Pt 4f7/2 71.2 5.497	79 Au 4f7/2 84.0 5.274	80 Hg 4f7/2 101.0 4.922	81 Tl 4f7/2 117.7 4.421	82 Pb 4f7/2 136.9 3.761	83 Bi 4f7/2 157.0 2.918	84 Po 4f 216.2* 1.933	85 At 4f 240.5* .9412	86 Rn 4f 265.6* .3292
87 Fr 4f 296.6* .2213	88 Ra 4f 328.6* NE	89–102	57 La 3d5/2 835.8 NE	58 Ce 3d5/2 883.8 NE	59 Pr 3d5/2 931.8 NE	60 Nd 3d5/2 980.8 NE	61 Pm 3d 1033* NE	62 Sm 3d5/2 1081.1 NE	63 Eu 3d5/2 1125.6 NE	64 Gd 4d 140.4 .6946	65 Tb 4d 146.0 .6186	66 Dy 4d 152.4 .5636	67 Ho 4d 159.6 .5089	68 Er 4d 167.3 .4569	69 Tm 4d 175.4 .4083	70 Yb 4d 182.4 .3642	71 Lu 4f7/2 7.3 4.703
			89 Ac 4f 359.2* NE	90 Th 4f7/2 333.2 NE	91 Pa 4f 409.3* NE	92 U 4f7/2 377.3 NE	93 Np 4f 460.4* NE	94 Pu 4f 479.8* NE	95 Am 4f 505.9* NE	96 Cm 4f 539.2* NE	97 Bk 4f 566.2* NE	98 Cf 4f 593.5* NE	99 Es 4f 621.2* NE	100 Fm 4f 649.2* NE	101 Md 4f 677.7* NE	102 No 4f 706.5* NE	103 Lr 4f 735.7* NE

Figure 3.3h List of atomic subshell photoionization cross sections of the most intense or most used subshells of elements $1 \leq Z \leq 103$ while radiation with an energy of 300.0 eV was used. Cross sections were adopted from ref. 2.

Photoionization cross section of most intense/most used subshells excited by X-ray (200.0 eV)

Key (example box 4 Be / 1s / 111.8 / .7359): Atomic number; Most intense/most used subshell; Binding energy of the most intense/most used subshell; Photoionization cross section of the most intense/most used subshell.

(1) The datum at the bottom of each box is the cross-section of photoionization under irradiation of energy ($h\nu$ = 200.0 eV). NE: lack of photoionization as the photon energy is close to or lower than the binding energy of the subshell electrons. ND: No data available.
(2) For a *p*, *d*, or *f* subshell, its photoionization cross section listed in a box is the total of the two split peaks although only BE of one of the two split peaks is listed in the box.
(3) All binding energies except values marked with * are values experimentally measured and adopted from ref. 1.
(4) Computational values of binding energies adopted from ref. 2 are marked with * due to the lack of experimental data. The Computational values could be different from measured values (if available).

1 H 1s 13.6* .21E−2																	2 He 1s 23.4* .55E−1
3 Li 1s 55.6 ND	4 Be 1s 111.8 .7359											5 B 1s 189.4 1.335	6 C 1s 284.5 NE	7 N 1s 398.1 NE	8 O 1s 531.0 NE	9 F 1s 684.9 NE	10 Ne 1s 863.1 NE
11 Na 1s 1071.8 NE	12 Mg 2p 49.8 1.568											13 Al 2p 72.9 2.176	14 Si 2p 99.3 2.857	15 P 2p3/2 129.9 3.489	16 S 2p3/2 164.0 3.799	17 Cl 2p3/2 198.5 NE	18 Ar 2p3/2 241.9 NE
19 K 2p3/2 294.4 NE	20 Ca 2p3/2 346.7 NE	21 Sc 2p3/2 398.6 NE	22 Ti 2p3/2 454.1 NE	23 V 2p3/2 512.2 NE	24 Cr 2p3/2 574.4 NE	25 Mn 2p3/2 639.0 NE	26 Fe 2p3/2 707.0 NE	27 Co 2p3/2 778.3 NE	28 Ni 2p3/2 852.7 NE	29 Cu 2p3/2 932.7 NE	30 Zn 2p3/2 1021.8 NE	31 Ga 2p3/2 1116.7 NE	32 Ge 3d 29.4 4.343	33 As 3d5/2 41.6 4.871	34 Se 3d5/2 55.6 5.282	35 Br 3d5/2 68.8 5.576	36 Kr 3d5/2 87.0 5.691
37 Rb 3d5/2 111.5 5.469	38 Sr 3d5/2 134.3 4.856	39 Y 3d5/2 156.0 3.522	40 Zr 3d5/2 178.9 .3690	41 Nb 3d5/2 202.4 NE	42 Mo 3d5/2 228.0 NE	43 Tc 3d 273.0* NE	44 Ru 3d5/2 280.1 NE	45 Rh 3d5/2 307.2 NE	46 Pd 3d5/2 335.1 NE	47 Ag 3d5/2 368.3 NE	48 Cd 3d5/2 405.1 NE	49 In 3d5/2 443.9 NE	50 Sn 3d5/2 485.0 NE	51 Sb 3d5/2 528.3 NE	52 Te 3d5/2 573.1 NE	53 I 3d5/2 619.3 NE	54 Xe 3d5/2 669.7 NE
55 Cs 3d5/2 726.4 NE	56 Ba 3d5/2 780.6 NE	57–71	72 Hf 4f7/2 14.3 6.668	73 Ta 4f7/2 21.9 6.537	74 W 4f7/2 31.4 6.258	75 Re 4f7/2 40.3 5.797	76 Os 4f7/2 50.7 5.127	77 Ir 4f7/2 60.9 4.281	78 Pt 4f7/2 71.2 3.286	79 Au 4f7/2 84.0 2.274	80 Hg 4f7/2 101.0 1.381	81 Tl 4f7/2 117.7 .7155	82 Pb 4f7/2 136.9 .3744	83 Bi 4f7/2 157.0 .3397	84 Po 4f 216.2* NE	85 At 4f 240.5* NE	86 Rn 4f 265.6* NE
87 Fr 4f 296.6* NE	88 Ra 4f 328.6* NE	89–102	57 La 3d5/2 835.8 NE	58 Ce 3d5/2 883.8 NE	59 Pr 3d5/2 931.8 NE	60 Nd 3d5/2 980.8 NE	61 Pm 3d 1033* NE	62 Sm 3d5/2 1081.1 NE	63 Eu 3d5/2 1125.6 NE	64 Gd 4d 140.4 .4471	65 Tb 4d 146.0 .5146	66 Dy 4d 152.4 .6172	67 Ho 4d 159.6 .7348	68 Er 4d 167.3 .8424	69 Tm 4d 175.4 .8837	70 Yb 4d 182.4 .7572	71 Lu 4f7/2 7.3 6.656
			89 Ac 4f 359.2* NE	90 Th 4f7/2 333.2 NE	91 Pa 4f 409.3* NE	92 U 4f7/2 377.3 NE	93 Np 4f 460.4* NE	94 Pu 4f 479.8* NE	95 Am 4f 505.9* NE	96 Cm 4f 539.2* NE	97 Bk 4f 566.2* NE	98 Cf 4f 593.5* NE	99 Es 4f 621.2* NE	100 Fm 4f 649.2* NE	101 Md 4f 677.7* NE	102 No 4f 706.5* NE	103 Lr 4f 735.7* NE

Figure 3.3i List of atomic subshell photoionization cross sections of the most intense or most used subshells of elements $1 \leq Z \leq 103$ while radiation with an energy of 200.0 eV was used. Cross sections were adopted from ref. 2.

One of the main functions of XPS is identification of chemical environments of constituting elements in surface region of a sample. The XPS spectral feature of core-level electrons represents the electron state of a specific element. By scanning a sample, scanning an anode by a focused electron beam, or scanning lens of an energy analyzer, the spatial distribution of a constituting element and spatial mapping of a specific oxidation state of an element can be obtained. With a depth-profile analysis, the depth-dependent concentration and depth-dependent chemical environment of a constituting element can be revealed. With XPS, the band structure of a specific molecule can be derived or deduced. To study valence band or molecular orbitals of a sample, an X-ray source is replaced with ultraviolet radiation; the photoelectron spectroscopy using ultraviolet radiation to excite electrons in the valence band of a sample is called ultraviolet photoelectron spectroscopy (UPS). UPS has a working principle similar to XPS.

In literature, both spectroscopic and X-ray notations are used in the description of electrons in atoms or ions. The spectroscopic notations are widely used in chemistry and physics. Main quantum numbers (n), 1, 2, 3, 4, 5, 6, and 7 label the core level of electrons around a nucleus. s subshell corresponds to electrons with angular momentum quantum number $l = 0$; here, *s* stands for sharp. *p* subshell corresponds to electrons with angular momentum quantum number $l = 1$; here, *p* stands for principle. *d* and *f* subshells correspond to electrons with angular momentum quantum numbers $l = 2$ and 3, respectively; here, *d* and *f* stand for diffuse and fundamental, respectively.

X-ray notations are mainly used in physics. K, L, M, N, O, and so on represent the core shell $n = 1, 2, 3, 4, 5$, respectively. Considering *J–J* coupling, electrons in subshells are categorized as s, $p_{1/2}$, $p_{3/2}$, $d_{3/2}$, $d_{5/2}$, $f_{5/2}$, and $f_{7/2}$ along the order of decrease of BE. In the same shell, different subshells are distinguished through a subscript 1, 2, 3, ... For instance, electrons 3s in 3s subshell and electrons $3p_{1/2}$ and $3p_{3/2}$ in 3p subshell of the third core level (M) are denoted with M_1, M_2, and M_3, respectively. Table 3.1 lists the symbols of energy levels of electrons in atoms denoted by the two notation systems.

Table 3.1 List of symbols of energy levels of electrons in atoms.

	Quantum number				
Entry	***n***	m_l	m_s	**X-ray notation**	**XPS notation**
1	1	0	±1/2	K	1s
3		0	±1/2	L_1	2s
3	2	−1	±1/2	L_2	$2p_{1/2}$
4		1	±1/2	L_3	$2p_{3/2}$
5		0	±1/2	M_1	3s
6		−1	±1/2	M_2	$3p_{1/2}$
7	3	1	±1/2	M_3	$3p_{3/2}$
8		−2	±1/2	M_4	$3d_{3/2}$
9		2	±1/2	M_5	$3d_{5/2}$

These notations are used in description of charge transfer occurred upon photoionization in XPS. For instance, upon a core-hole state is created on Ni^{2+} due to the X-ray photoionization of Ni^{2+}, there is a charge transfer from O^{2-} to Ni^{2+}. In this charge transfer, the core hole of Ni^{2+} received an electron from its neighboring O^{2-} atom. The electron configuration of Ni^{2+} before and after emitting a photoelectron of Ni 2p subshell are $2s^22p^63s^23p^63d^8$ and $2s^22p^53s^23p^63d^8$. Alternatively, an electron of a ligand, here an oxygen atom can transfer to 3d orbital of Ni^{2+} of NiO; this transfer can be written as $L2p^53d^8 \rightarrow L^{-1}2p^53d^9$; L stands for ligand; the O atom losing one electron to the core hole of Ni^{2+} is denoted with L^{-1}. Sometimes, the core hole is designed to c^{-1} as "c" stands for "core" and "−1" means loss of one electron. Thus, this electron transfer can be written as $Lc^{-1} \rightarrow L^{-1}2p^53d^9$ as well.

3.2 Generation of X-ray

X-ray is produced through bombardment of surface of a metallic solid or a liquid metal in some unusual case with high-energy electron beam. The metallic solid or liquid is called X-ray anode and the electron gun or electron beam is the cathode. In most XPS systems, the anode is a metallic solid. Al Kα is the mostly used X-ray source on lab-based XPS systems. Al Kα is produced through the following process (Figure 3.4). Electrons with high energy bombard the surface of a chosen metal solid; here it is metallic Al thin film with a thickness of about 10 μm deposited on a Cu block that is cooled through a circulating water with certain flow rate. Upon the bombardment, part of the energy of the impinging high-energy electrons is spent for exciting Al 1s electrons. Al 1s electrons are emitted upon this scattering, generating a core hole on Al 1s subshell. The core hole is filled with electrons from Al $2p_{1/2}$, Al $2p_{3/2}$, Al 3s, Al $3p_{1/2}$, or Al $3p_{3/2}$. Upon the core hole in the low-energy Al 1s subshell is filled by electrons originating at Al $2p_{1/2}$, Al $2p_{3/2}$, Al 3s, Al $3p_{1/2}$, or Al $3p_{3/2}$, the extra energies of these electrons are released in the form of fluorescence radiations or Auger electrons. The released fluorescence radiations are the generated X-rays that are Al $K\alpha_1$ released upon transition of an electron from Al $2p_{3/2}$ state to Al 1s, Al $K\alpha_2$ upon a transition from Al $2p_{3/2}$ to Al 1s, and Al K_β upon a transition from Al 3s to Al 1s. Figure 3.4a is the schematic showing the generation of X-rays. As shown in Figure 3.4b and Table 3.2, Al $K\alpha_{1,2}$ is the main portion of the X-rays produced from an aluminum anode, being 90% of the intensity of all X-rays of Al X-ray anode. It is the major component of Al Kα. The minor difference in energy between Al $K\alpha_1$ (1486.295 eV) and Al $K\alpha_2$ (1486.708 eV) is only ~0.4 eV. Other radiations of Al atoms are called satellite X-rays which also excite sub-shell electrons to form peaks in a spectrum as shown in Figure 3.4b. Al $K\alpha_{3,4}$, Al K_β, and Bremsstrahlung radiation are the main satellite lines, circled in Figure 3.4b. Al $K\alpha_{3,4}$ X-rays are generated from transition of electrons at Al $2p_{1/2}$ and Al $2p_{3/2}$ states of Al^+ cations to Al 1s subshell. Al K_β was produced from the process of filling the hole on Al 1s subshell with an Al 3s electron. Bremsstrahlung radiation appears as a broad background of X-ray spectrum of a metal; they are photons covering continuous energy range; it is generated from the deceleration of a charged particle by deflection of electrons. The relative position and intensity of these satellite peaks are listed in Table 3.2. These satellite X-rays largely contribute to the width of X-rays of Al anode if

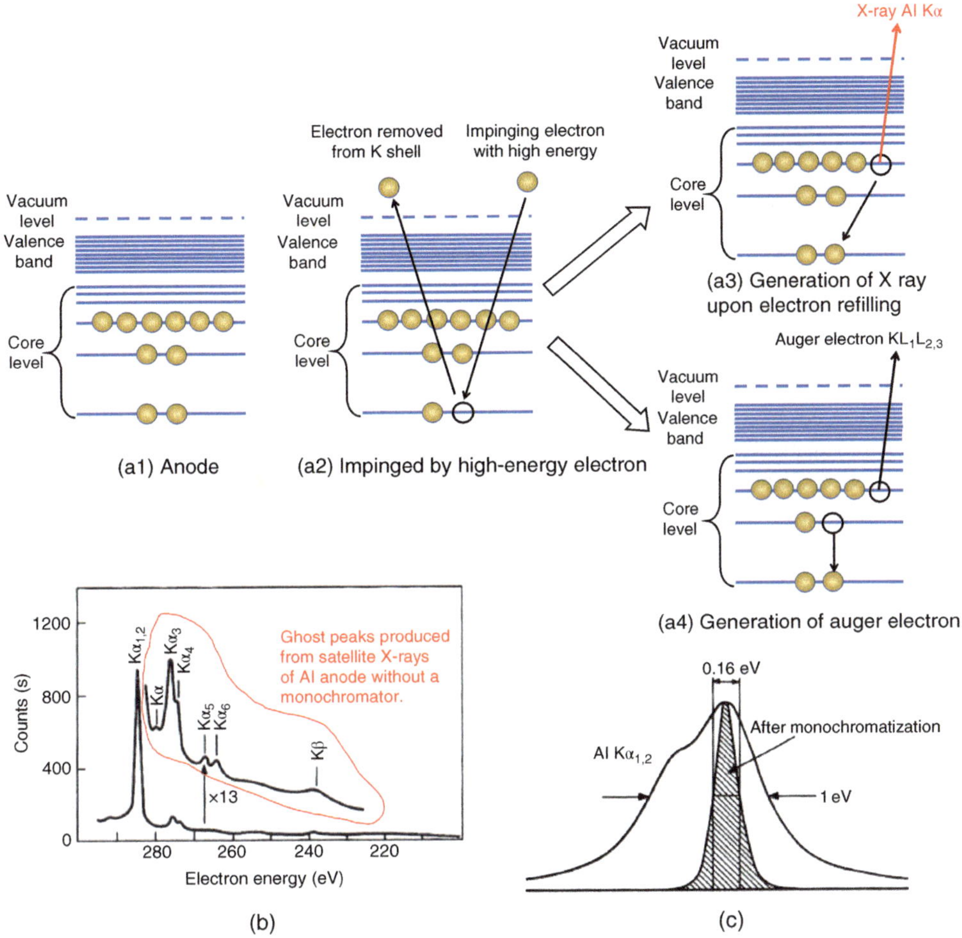

Figure 3.4 Schematic showing the processes of the generation of X-ray (a1–a4), C 1s XPS peak upon X-rays (Al $K\alpha_{1,2}$, Al $K\alpha_{3,4}$, Al $K\alpha_{5,6}$, and $K\beta$ generated through an Al X-ray source) (b), and Al $K\alpha_{1,2}$ without monochromator and with monochromator (c). *Source:* (b, c) Reproduced with permission from Cardona and Ley[3]/Springer.

Table 3.2 Distribution of intensity and energy of X-rays generated from Al and Mg anode.

Anode	Energy displacement (eV) by comparing to Al$K_{\alpha_{1,2}}$ and relative intensity by defining Al$K_{\alpha_{1,2}}$ to 100					
Radiation	$\alpha_{1,2}$	α_3	α_4	α_5	α_6	β
Al	0 eV, 100	9.8 eV, 6.4	11.8 eV, 3.2	20.1 eV, 0.4	23.4 eV, 0.3	69.7 eV, 0.55
Mg	0 eV, 100	8.4 eV, 8.0	10.2 eV, 4.1	17.5 eV, 0.55	20.0 eV, 0.45	48.5 eV, 0.5

Source: Adapted from Cardona and Ley.[3]

a monochromator is not used. Consequently, these satellite X-rays largely broaden the XPS peak. For instance, the X-rays of Al anode without monochromator give a C 1s spectrum across an energy window of nearly 50 eV (Figure 3.4b).[3] The large full width at half maximum (FWHM) of C 1s results from the coexisting Al Kα_1, Al Kα_1, Al Kβ, Cu Lα, and Cu Kα and Bremsstrahlung radiation. If we only count Al K$\alpha_{1,2}$ and no monochromator is used, the width of the energy window or energy resolution of the X-rays of Al anode can be represented with FWHM of the generated X-rays and it is 0.8 eV. By using a monochromator, the FWHM of the X-rays of Al anode can decrease to 0.3 eV or lower (Figure 3.4c). As the sharpness of X-ray in terms of the width of energy window of the X-rays generated from Al anode is one of the major factors determining the resolution of an XPS peak of an element to be analyzed, a monochromator of an Al Kα X-ray source can significantly improve resolution of an XPS peak.

Other than Al, many metals such as Mg, Ti, Ag, or Cu can be used as an X-ray source. Choice of an X-ray source depends on requirements of energy, resolution, and flux of an X-ray radiation. Table 3.3 lists energies and FWHMs of X-rays generated from elements at their stable states in high vacuum. Their intensities were not listed as they are varied significantly by voltage of electron beam and flux density of electron beam bombarding the metal surface and other parameters of the X-ray source. Energies of characteristic X-rays of more elements can be found from Figure 3.5. These characteristic energies of metal elements listed in Figure 3.5 are highly valuable in performing X-ray absorption spectroscopy (XAS), including X-ray absorption near edge structure (XANES) and extended X-ray absorption fine structure (EXAFS) spectroscopies since they are important parameters in

Table 3.3 Energy and line width of X-rays generated from elements at their stable states.

Entry	Radiation lines	Energy (eV)	Line width (eV)
1	Na Kα (1)	1041.0	0.70
2	Mg Kα (2)	1253.6	0.70
3	Al Kα (3)	1486.6	0.85
4	Si Kα (4)	1739.5	1.0
5	Ti Kα (5)	4510.0	2.0
6	Cu Kα (6)	8048	2.6
7	Y $M\zeta$	132.3	0.47
8	Zr $L\alpha$	2042.4	1.7
9	Zr $M\zeta$	151.4	0.77
10	Cr Kα	5417.0	2.1
11	Ni $L\alpha$	851.5	2.5
12	Y $L\alpha$	1922.6	1.5
13	Ti $L\alpha$	395.3	3.0
14	Nb $M\zeta$	171.4	1.21
15	Mo $M\zeta$	192.3	1.53

Notes: (1), (4), and (5): highly sensitive to environment; (2), (3), and (6): most commonly used.

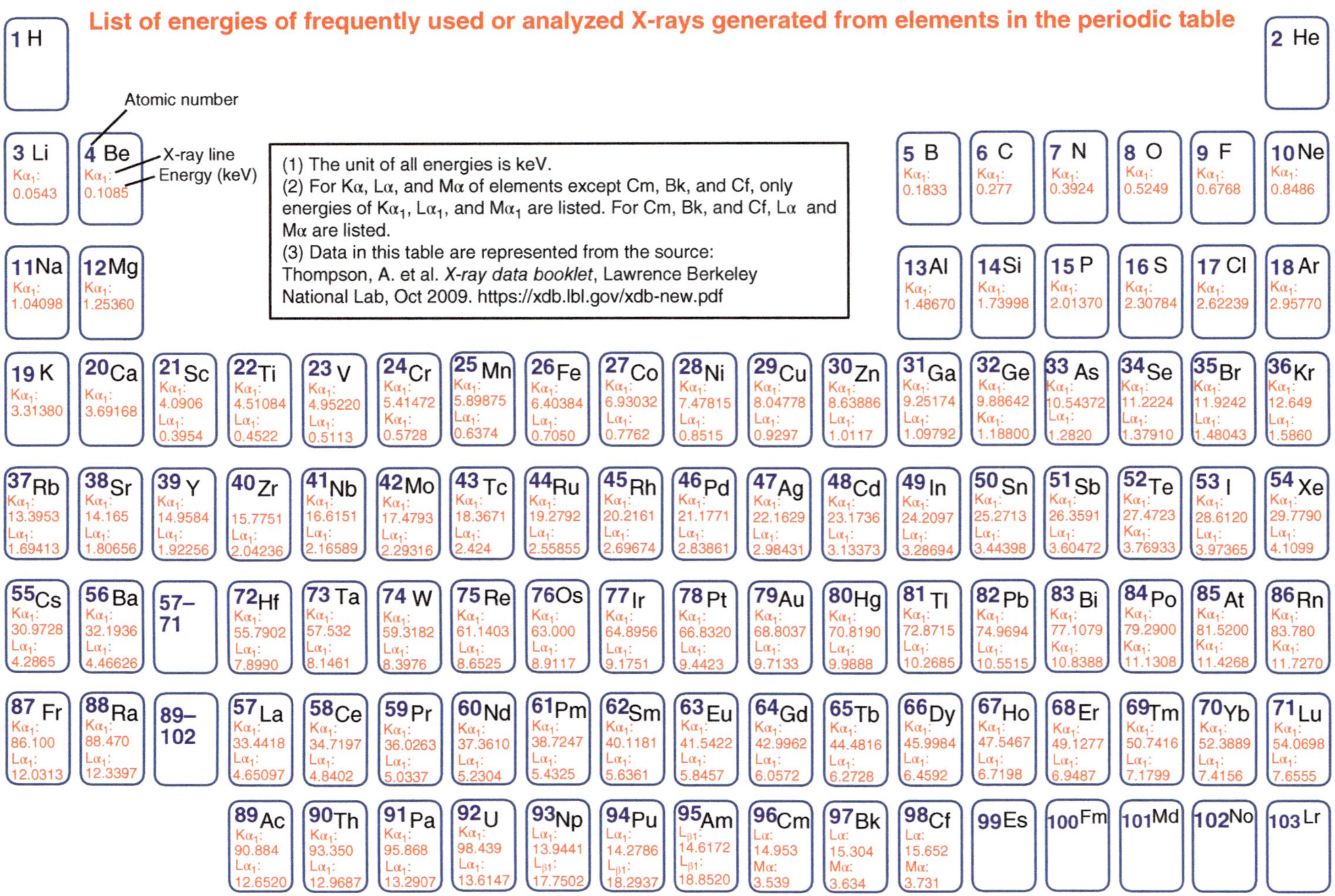

Figure 3.5 Periodic table listing energies of frequently used or analyzed X-rays generated from elements. All Data are adopted from Thompson, A. et al. X-ray data booklet, Lawrence Berkeley National Lab. https://xdb.lbl.gov/xdb-new.pdf.

determining a specific type of reaction cell of XAS studies to be used and whether an absorption edge of a metal element could be observed if an absorption edge of another element could overlap it.

3.3 Excitation of Photoelectron and Chemical Shift

In XPS, X-ray directly emitted from an anode or monochromator irradiates a sample surface. Al Kα radiation transits in a solid as deep as a few micrometers in general. Upon receiving energy of X-ray photons, an electron in a subshell of an atom of a sample can be excited to leave its home atom by overcoming its binding to the nucleus of its home atom. By collecting the emitted photoelectrons and measuring their kinetic energies, binding energy (BE) of these photoelectrons is then known. BE of electrons in a specific subshell is a fingerprint of an element although it is variable by up to a few electron Volts (eV) due to a different chemical environment. Thus, the BE read from an XPS spectrum allows for identifying the elements of the sample. The measured BE of electrons of a specific subshell of atoms also reflects the specific chemical environment of these atoms. The high resolution of energy analyzer of an XPS instrument allows an accurate measurement of BE (<0.1 eV) and distinguishing different chemical environments.

Understanding of the specific energy position of an XPS peak in terms of the specific BE of nucleus to an electron in a specific chemical environment is the key of XPS studies. In general, BE of subshell photoelectrons is basically determined by the effective nuclear charge, Z^*, a value generated by subtracting the shielding effect from the apparent charge Z. This definition does not distinguish the measurable difference in energy resulting from different binding environments in terms of chemical coordination of the same element. The measurable difference in BE between the different chemical environments of an interested subshell such as C 1s or Co $2p_{3/2}$, often called a chemical shift, is a summation of initial state and final state effects.

Interpretation of the chemical shift must be based on a profound understanding of initial state and final state effects. Typically, an initial state effect of BE of electrons on a subshell of an atom is attributed to the bonding between this atom and its surrounding atoms. As these bonding electrons are localized at ionic bonds of an ionic solid such as NaCl, covalent bonds of a covalent network solid such as SiO_2, or covalent bonds in a molecule of a molecular solid such as an ester, the difference in a coordination or bonding environment of a home atom can vary the attraction of a nucleus to its surrounding core electrons and valence electrons to some extent. Such differences can be reflected by variation of BE of core electrons when the home atom of the core electrons bonds with different atoms in terms of different chemical environments. The contribution of initial effects to the chemical shift constitutes one of the major applications of XPS, which is the interpretation of chemical environment of an element on the surface of a sample. Different from the initial state effect, a final state effect is due to the relaxation of electronic structure immediately upon photoionization.

In brief, there are two different types of initial state effects. One can be understood as bonding environment-induced polarization; it is a type of external initial state effect. The other is spin orbital splitting; it is a type of internal initial state effect, independent of bonding environment in general. It originates at a coupling of the spin and orbital angular momentum. It is noted that a charge-potential model in Koopmans' theorem assumes that spin-orbital splitting remains no variation for atoms of the same element in different chemical environments in general. In Koopmans' theorem,[4] it is assumed that the first ionization energy of a molecule is equal to the negative eigenvalue of the highest occupied molecular orbital, provided the distributions of the remaining electrons do not change. A detailed description is given in the following Section 3.3.1.

Final state effects are related to secondary excitation of electrons or rearrangement of core/valence electrons. Notably, the BE predicted from Koopmans' theorem does not include the final state effect as the Koopmans' theorem is a type of frozen orbital approach.[4] A strong final state effect occurs when lifetime of a core hole generated from photoionization is shorter since a shorter lifetime of core hole leaves a longer time for relaxation in terms of additional excitation or rearrangement of valence shell of the home atom of photoelectrons. In addition, the strength of a final state effect is related to the initial state effect of the home atom of the photoelectrons to be photoionized. For example, an obvious Russell–Saunders effect could result in a strong final state effect.[5]

3.3.1 Initial State Effect

Bonding environment-induced polarization is one type of initial effects. It can be understood through a charge density model. In a charge density model, the BE of an electron in a subshell originates at the Coulombic electrostatic attraction between the electron to be emitted and the nucleus. Notably, other than the attraction by nucleus, an electron in a specific subshell is certainly repelled by other electrons of the atom.

Bonding environment-induced polarization can be evaluated with a point charge model. In this model, the home atom of a photoelectron is considered as a hollow sphere. This model assumes that there is a sphere with a radius of r_v consisting of valence electrons and the center of the sphere which is the nucleus of the atom. This model ignores core electrons as these core electrons do not participate in bonding of this atom to its surrounding atoms. In addition, this model assumes that variation of valence shell should change the Coulombic attraction *between* electrons on subshells and the remaining valence electrons *and* the nucleus. Thus, variation of valence shell of the home atom resulting from a different bonding environment of the home atom should alter BEs of electrons of all subshell of the home atom to it nucleus. Based on this mode, chemical shift, ΔBE, created by the change of bonding environment of the home atom, can be evaluated with the following equation[6]:

$$\Delta BE = K\left(k\Delta q_i + \Delta\sum_{i\neq j}\frac{q_j}{d_{i-j}}\right) \tag{3.1}$$

here, K: constant to correlate ΔBE and charge terms. k: coupling constant, approximately $\frac{1}{r_v}$. d_{i-j}: bond distance between atoms i and j. $k\Delta q_i$ refers to all intra-atomic effects; for atoms of the same element, $k\Delta q_i$ is obviously different since their chemical environment or bonding atoms, j, are different. $\Delta\sum_{i\neq j}\frac{q_j}{d_{i-j}}$ describes the contribution of all atoms, j, surrounding to the home atom i; q_j is the charge of a surrounding atom (j) adjacent to atom i. q_j can be

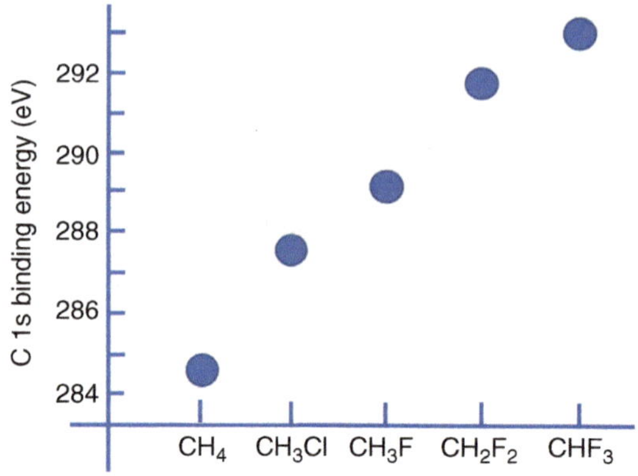

Figure 3.6 Examples showing correlation between EN of the element of atom *j* surrounding the home atom *i* and the BE of the C 1s of the home atom *i* (here C) in CH_4, CH_3F, CH_2F_2, and CHF_3.[1] Here *j* is F and H atoms of CH_4, CH_3F, CH_2F_2, and CHF_3, respectively.

reflected by electronegativity (EN) of the element of atom *j*. In some cases, a straightforward correlation between ΔBE of a subshell of the home atom *i* and EN of the element of the surrounding atom *j* can be readily established (Figure 3.6). In general, an increase of EN of element of surrounding atom *j* makes ΔBE of the core electrons of the home atom *i* more positive or less negative. An empirical rule is that if the home atom *i* bonds to atom(s), *j*, of an element with EN higher than element of the home atom *i*, the BE of electrons of atom *i* is increased in contrast to BE of the pure element of atom *i*. For example, BEs of 2p subshells of transition metal elements in a metal oxide sample such as BE of Fe 2p of Fe_2O_3 are higher than Fe 2p of metal Fe since its surrounding O atoms have higher EN than Fe. Another example is the obvious increase of BE of C 1s of the central atom C along the increase of EN of its surrounding atoms (Figure 3.6). Figure 3.6 presents the increase of BEs of C 1s in CH_4, CH_3Cl, CH_3F, CH_2F_2, and CH_2F_3 along the increase of EN of element(s) of the surrounding atoms of the home atom, C. In summary, the simple point charge model can well rationalize the influence of EN of surrounding atoms *j* on BE of the central atom *i* in initial state effects.

3.3.2 Final State Effect

Final state effects include any effects resulting from relaxation of core hole state. The core hole state is the state formed immediately when a core electron is emitted from the home atom. Based on the consequences of different relaxations, it can be categorized into core hole-induced polarization effect and core hole-induced rearrangement effect.

3.3.2.1 Core Hole-Induced Polarization Final State Effect

The consequence of the core hole-induced polarization effect is the increase of attraction of nucleus to the remaining electrons since the loss of one electron with negative charge and a specific spin can be considered as an addition of a positive charge and opposing spin to the home atom. A simple model for this effect is that the loss of one electron increases the effective nucleus charge (Z^*) to electrons in the same or outer subshell of the leaving electron, and the increase of Z^* could increases BE of these electrons. In addition, the increase of effective charge results in decrease of average valence radius, which increases the shielding to leaving photoelectron; therefore, the increased shielding effect to a leaving photoelectron decreases kinetic energy of the leaving photoelectron, by which the BE is increased. The amount of increased BE is determined by the electronic configuration of the home atom and the core hole-induced polarization that the home atom experiences. The polarization is partially influenced by the chemical environment of the home atom. Such a shift of BE resulting from the core hole-induced polarization could be as large as a few eV.

The existence of core hole-induced polarization effect is evidenced by observed multiplet splitting in XPS spectral features of a metal atom having an unpaired valence electron(s). Multiplet splitting results from interactions between a core hole with the unpaired electron and any unpaired electron(s) in valence shell such as 3d subshell of transition metals of the fourth period. Similar to spin-orbital splitting, multiplet splitting forms from interactions of different magnetic fields of two or more localized spinning electrons, particularly the magnetic field formed from one or more unpaired electrons created due to the loss of a photoelectron and the field formed from an unpaired electron in the valence shell. The unpaired electrons of valence shell can be these originally existing before photoionization or newly formed in valence shell through relaxation of the core hole.

Different from spin-orbital splitting that occurs to p, d, and f orbitals instead of s orbital, multiplet splitting can modify BE of any electrons of any orbitals of the home atom such as 3s orbital as shown in Figure 3.7. For instance, m_s of the single electron left in the valence shell (3d subshell) of Cr_2O_3 can be $+\frac{1}{2}$ or $-\frac{1}{2}$; the magnetic field resulting from this spin (+1/2 or −1/2) interacts with the magnetic field induced by the unpaired electron(s) in valence shell ($3d^3$) of Cr^{3+}, resulting in the splitting of BE of 3s electrons of Cr atoms of Cr_2O_3. As shown in the middle of Figure 3.7a, the double peaks of Cr 3s of Cr_2O_3 were clearly observed. A great number of multiple splitting can be found in Figure 3.7.

Based on van der Heide's work,[7] the energy separation of multiplet splitting for 3s level of these transition metals (Sc, Ti, V, Cr, Mn, Fe, Co, Ni, Cu, Zn) in the fouth peroid can be evaluated with the following equation[7]:

$$\Delta E = K\frac{2(S_v + S_c)+1}{0.6BE^{1/9}} \tag{3.2}$$

In Eq. (3.2), K is the exchange integral of multiple splitting; S_v is the total 3d level spin of the atom; S_c is the core level spin (+1/2). van der Heider calculated the energy separation (ΔE) of multiplet splitting and found a linear increase in the energy separation, ΔE, as a function of S_v.[7]

The multiplet splitting in the home atom is sensitive to chemical environment of the home atom. This is because the constituting electrons of this splitting are the unpaired electrons in valence shell as mentioned earlier. As electrons in valence shell of a home atom participate in bonding with surrounding atoms, the splitting is obviously influenced by the surrounding atoms in terms of the chemical environment of the home atom. Thus, the multiplet splitting reflects the local chemical environment of the home atom.

As we know the electron rearrangement of the core hole state can make valence shell generate unpaired electron(s); both a diamagnetic ion originally without an unpaired electron and paramagnetic ion originally having unpaired electron(s) could exhibit multiplet splitting based on the mechanism of forming a multiplet splitting described in the second paragraph in Section 3.3.2.1. For instance, a very weak multiple splitting was observed in Cu 2p of CuO since electron rearrangement makes Cu^{2+} of CuO eligible for a multiplet splitting. As valence shells of Cu atoms of Cu metal and Cu_2O do not have an unpaired

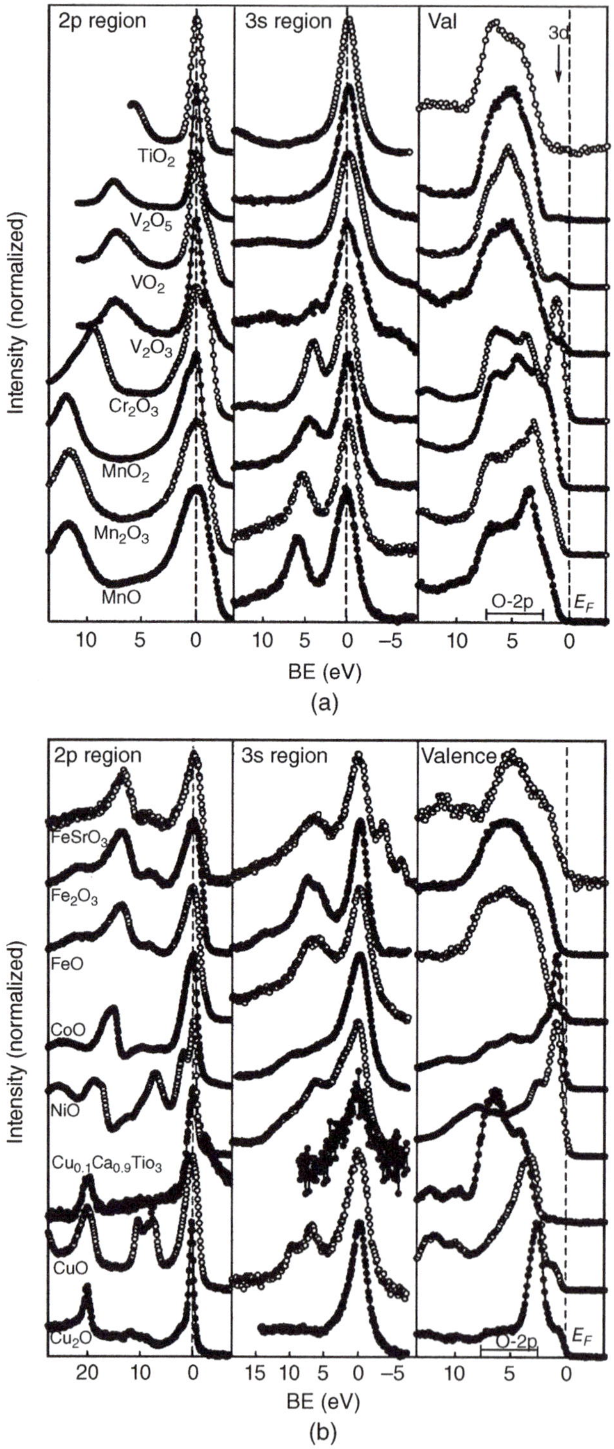

Figure 3.7 XPS spectra of the 2p, 3s and valence regions of transition metal oxides including TiO_2, V_2O_5, VO_2, V_2O_3, Cr_2O_3, MnO_2, Mn_2O_3, and MnO (a) and $FeSrO_3$, Fe_2O_3, FeO, CoO, NiO, $Cu_{0.1}Ca_{0.9}TiO_2$, CuO, and Cu_2O (b). For the sake of clarity, all intensity values are normalized to unity and all 2p and 3d BEs are plotted on a relative BE scale. *Source:* Reproduced with permission from van der Heide[7]/Elsevier.

electron in valence shell ($3d^{10}$), no multiplet splitting was observed from Cu_2O in Figure 3.7b. However, the multiplet splitting is also responsive for the electron rearrangement induced by core hole as discussed in Paragraph 2 in Section 3.3.2.1. Since the multiplet splitting is applicable to rearrange a core hole state, a peak next to the Cu $2p_{3/2}$ satellite peak (formed from shake up in terms of a type of electron rearrangement of a core hole state) is observed in Cu 2p of CuO. This additional peak is located at 2–3 eV above the satellite peak (7–8 eV above the Cu $2p_{3/2}$ main peak) as shown in Figure 3.7b.

Multiplet splitting, being one type of core hole-induced polarization, can vary the energy splitting of the pair of XPS peak of spin-orbital splitting such as BEs of Co $2p_{1/2}$ and Co $2p_{3/2}$. For instance, as shown in Figure 3.8 the energy split of Co $2p_{3/2}$ and Co $2p_{1/2}$ for metal Co and that for oxide CoO are 14.97 and 15.2 eV, respectively, based on the reference.[1] The difference in energy split (Δ) of Co $2p_{1/2}$ and Co $2p_{3/2}$ between metal Co and oxide CoO are given in Figure 3.8a and b, respectively. In addition, the split of Co $2p_{1/2}$ and Co $2p_{3/2}$ of Co_2O_3 is 16.0 eV, further different from that of metal Co.[8] Therefore, these differences in energy split among Co, CoO and Co_2O_3 support the feature that the multiplet splitting resulting from core-hole induced polarization can vary the energy split of spin-orbital coupling.

3.3.2.2 Core Hole-Induced Rearrangement Final State Effect

Relaxation of a core hole state can result in excitation of electrons in valence shell of the home atom, which is called rearrangement of valence electrons. The consequence of a rearrangement of valence electrons is a variation of the kinetic energy of the leaving core electron when it departs the atom. Here the remaining core electrons and the leaving core electron are interchangeable. Here using the term "core electron" is to make readers realize that the cause is the rearrangement of valence electrons and the consequence is the variation of kinetic energy of core electrons. If the valence electron is excited to leave its original

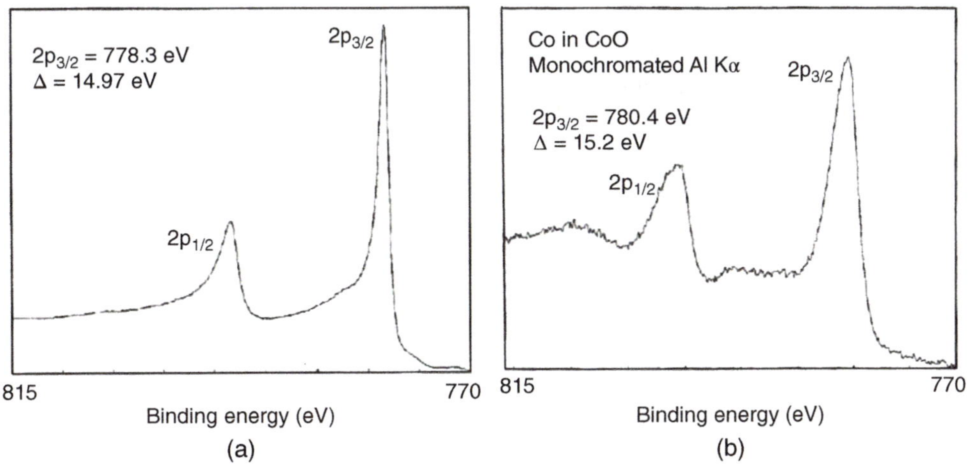

Figure 3.8 XPS spectra of Co 2p of Co metal and CoO. The spin-orbital split of Co 2p of Co and CoO is different. (a) Co 2p of metallic Co. (b) Co 2p of CoO. *Source:* Reproduced with permission from Moulder et al.[1]/Perkin-Elmer Corporation.

valence level before the photoelectron leaves the atom, the KE of the leaving photoelectron is decreased. Thus, its peak appears at the side of higher BE. There are three types of valence electron rearrangements that are shake-up, shake-off, and plasmon generation. Figure 3.9 schematically shows the shake-up and shake-off processes. However, if the excited valence electron returns to its valence shell or even a lower energy level, the energy released in this returning could transfer to the leaving photoelectron; thus, it could even increase KE of the leaving photoelectron; in other words, it could make the peak of the photoelectrons appear at the side of lower BE.

Shake-up processes are the transitions of valence electrons to a discrete or non-discrete valence level, a higher energy level that is closer to the Fermi level, as shown in Figure 3.9a, leading to their a separate peak with measurable intensity at the higher BE side of the main XPS peak or an additional tail at the higher BE side of the main peak. In other words, a shake-up process typically decrease kinetic energy of leaving core electrons and thus their contributions appear at higher BE side of the main XPS peak in the form of a separate satellite peak (Figure 3.10a) or a tail (Figure 3.11). The feature of a tail is not so obvious as a separate satellite peak. The addition of a tail resulting from the shake-up process to the high BE side of a main XPS peak results in asymmetry of the main XPS peaks of subshells of transition metal elements. For instance, XPS peaks of Ti $2p_{3/2}$ of TiO_2 and Cr $2p_{3/2}$ of Cr_2O_3 are obviously asymmetric in Figure 3.11.[1]

The shake-up processes were seen in CuO but neither Cu metal nor Cu_2O (Figure 3.11a).[1] Based on Kotani–Toyozawa model,[9] the satellite peak of CuO was formed from the electron transition from filled ligand state (O^{2-} atom bonded to Cu of Cu^{2+}) to (4s + 4p) band of Cu^{2+}. If c^- refers to the core hole due to photoionization and L^{-1} a ligand losing an electron, this electron transfer of shake-up process can be denoted as $Lc^{-1}3d^9(4s+4p)^0 \rightarrow L^{-1}c^{-1}3d^9\,(4s+4p)^1$; in this process, the excited ligand electron transits to the (4s + 4p) band. It forms the satellite peaks as marked in Figure 3.10a. Compared to CuO,[10,11] metal Cu does not have a ligand oxygen atom; thus, no shake-up process $Lc^{-1}3d^9(4s+4p)^0 \rightarrow L^{-1}c^{-1}3d^9\,(4s+4p)^1$ occurs (Figure 3.10b). The explanation for the lack of satellite peak in Cu 2p of metal Cu is supported by the observation of satellite peaks from NiO ($3d^8$) marked as "S" in Figure 3.12b. In addition, shake-up process can well explain the observed satellite peaks in Ni $2p_{3/2}$ for Ni^{2+} of NiO (Figure 3.12b). Notably, the multiplet splitting of the main peak of Ni $2p_{3/2}$ was observed and marked with "M" in Ni 2p of NiO in Figure 3.12b as well.

Different from the shake-up process, shake-off process makes valence electron(s) move to unbound continuum states that are above vacuum level (Figure 3.9b). Discrete shake-off peaks are rarely identified. Similar to shake-up processes (Figure 3.12), shake-off processes decrease the KE of leaving core electron, by which their contribution to the XPS spectrum is at the higher BE side of the main XPS peak. The contribution of shake-off processes to the spectrum is typically a broad background at the higher BE side of the main photoelectron peak. In most cases, the contribution of shake-off process is not accountable or even not observable. The difference in physics between shake-off and shake-up processes is that former makes valence electron(s) leave the atom but the latter not (Figure 3.9).

Other than the shake-up and shake-off processes, another format of a core hole-induced rearrangement final state effect is plasmon loss. Plasmon is a collective oscillation in all electrons in energy levels near to Fermi level. It is produced while electrons in conduction band move to respond to a sudden change of charge density that occurs in the formation of

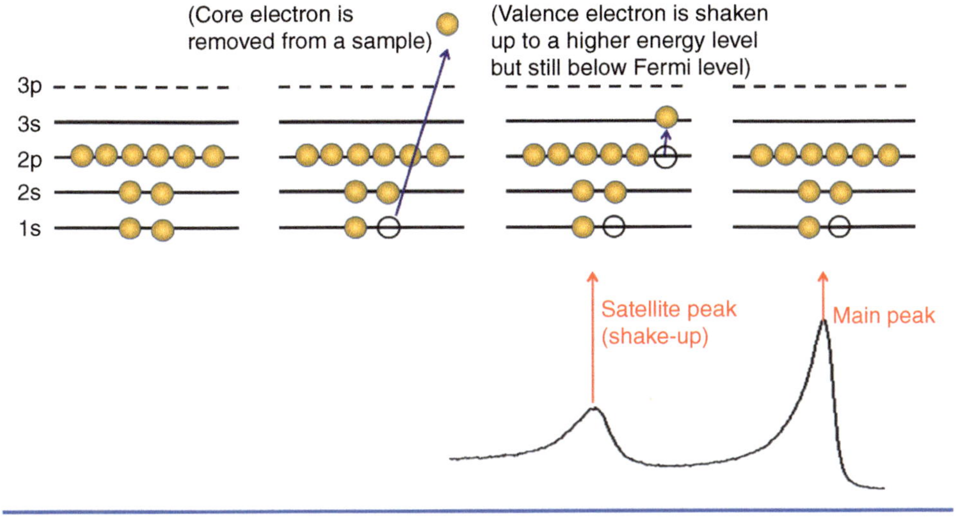

(a)

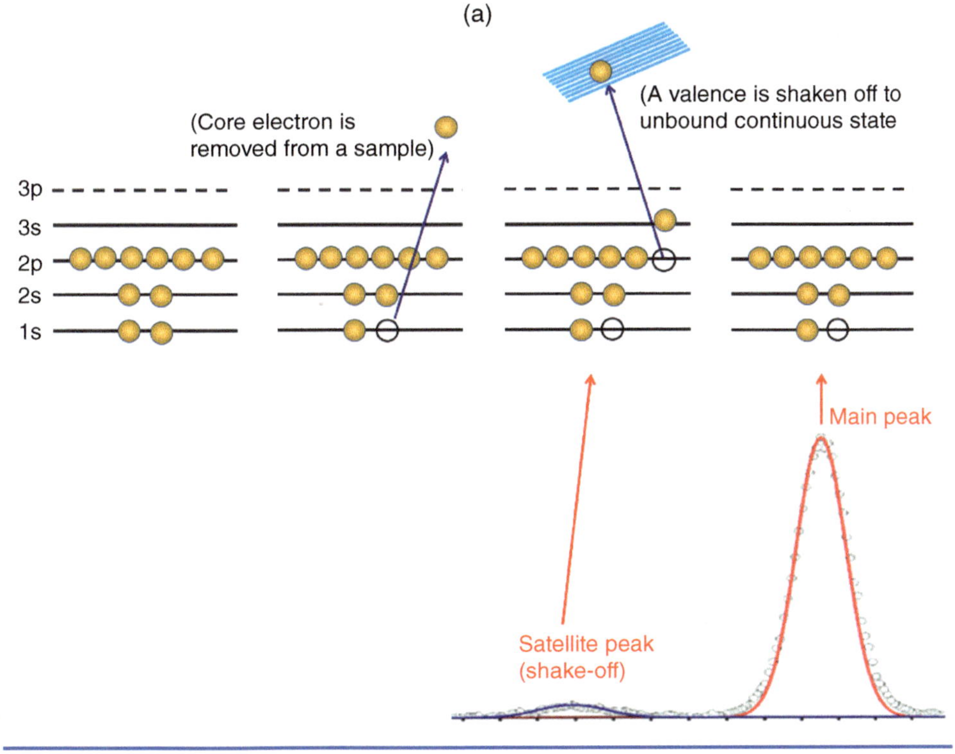

(b)

Figure 3.9 Schematics showing (a) shake-up effect process and a spectrum consisting of main peak, and a shake-up satellite peak and (b) shake-off effect process and a spectrum consisting of main XPS peak and a shake-off satellite peak.

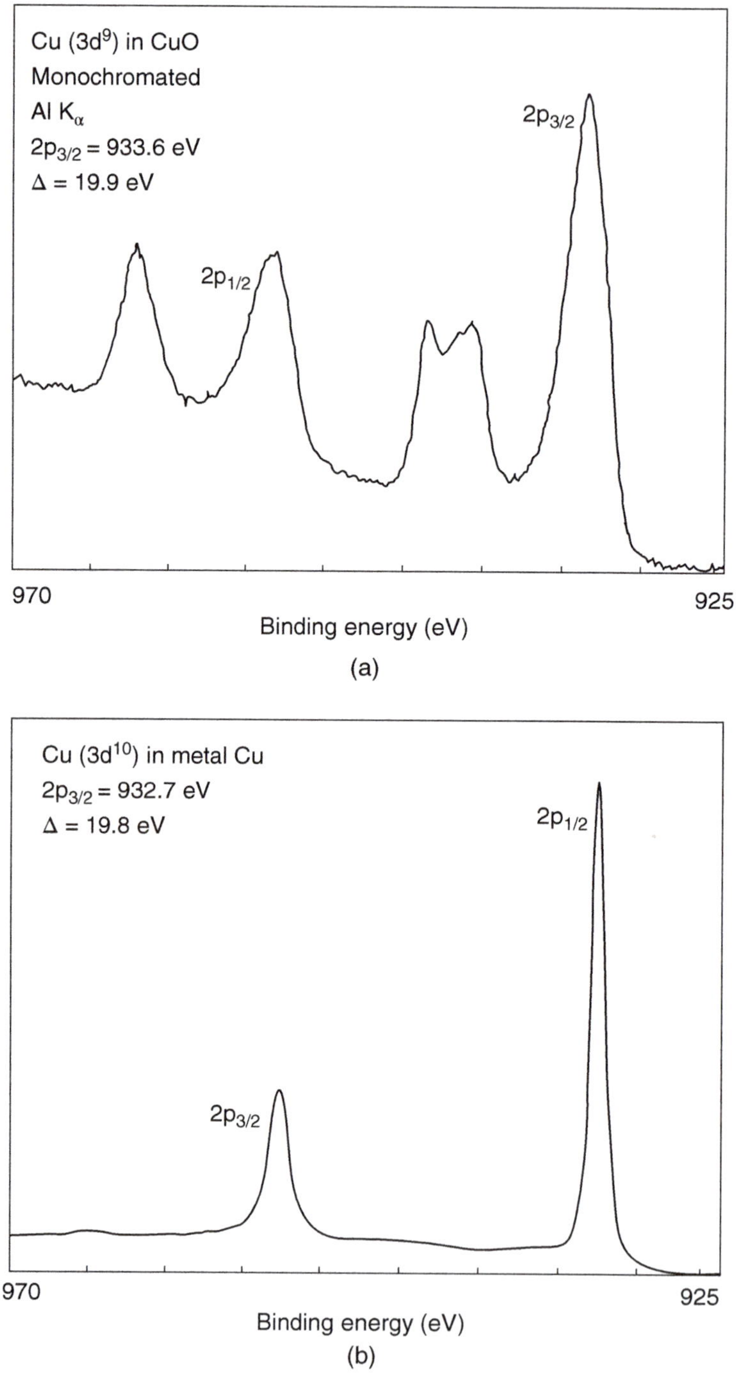

Figure 3.10 XPS peaks of Cu 2p of CuO (a) and metal Cu (b). *Source:* Reproduced with permission from Moulder et al.[1]/Perkin-Elmer Corporation.

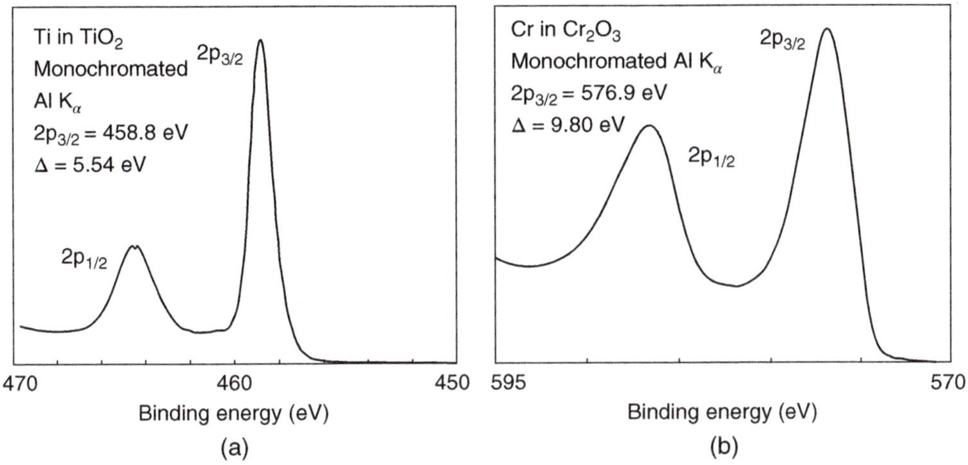

Figure 3.11 XPS peaks of Ti 2p and Cr 2p of Ti and Cr metals. *Source:* Reproduced with permission from Moulder et al.[1]/Perkin-Elmer Corporation.

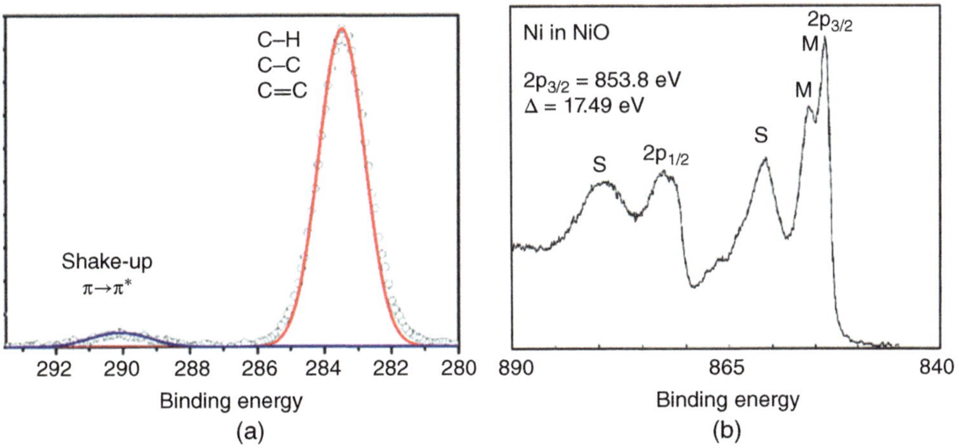

Figure 3.12 Examples of shake-up peaks. (a) C 1s of polystyrene and its shake-up satellite peak. (b) Ni $2p_{3/2}$ and Ni $2p_{1/2}$ and their shake-up satellite peaks (S) and the coexisting multiplet splitting peaks (M).

core hole. Such a movement of electrons initiates collective oscillation in the conduction band, producing plasmon. A plasmon can be parameterized with a specific plasmon frequency and plasmon energy. The frequency and energy of a plasmon of collective oscillation are determined by the density of free electrons in the metal. Be more specific, the bulk and surface plasmon energies of a metal can be evaluated with measurable terms including molar mass (M), density of the metal (ρ), and number of electrons in valence shell (N_v), through the equations $E_{\text{bulk plasmon}} = \frac{28.8}{\sqrt{N_v \rho / M}}$ and $E_{surface\ plasmon} = \frac{28.8}{\sqrt{2N_v \rho / M}}$.[12,13] The energy of surface plasmon of a metal is $\frac{1}{\sqrt{2}}$ of bulk plasmon of the same metal. In addition, the intensity of bulk plasmon of a metal is much higher than surface plasmon of the same sample. As the response of the rapid change of electron density occurring in the formation of core holes forms bulk and surface plasmons, the formation of these plasmons makes kinetic energy of leaving core electrons decrease by the amount of the energies of these plasmons. Thus, formation of these plasmons creates new peaks at higher BE side of the main XPS peak of a subshell. As shown in Figure 3.13, bulk plasmon peaks of silicon appear at the higher BE side of the main Si 2s peaks.[14] They are marked with B_1 and B_2 in Figure 3.13. Similarly, two bulk plasmon peaks (B_1 and B_2) are observed at the higher BE side of the Si 2p main peak. The peaks of the surface plasmons relevant to Si 2s and Si 2p are marked with S_1 in Figure 3.13. There are three notable features for the satellite peaks resulting from bulk and surface plasmon losses (Figure 3.13). One is that the energy separation, $\Delta_{bulk\ plasmon}$, defined to the difference between peak of bulk plasmon loss (B_1) and the main peak, Si 2s or Si 2p, is obviously larger than $\Delta_{surface\ plasmon}$, defined to the difference between peak of surface plasmon loss (S_1) and the main peak, Si 2s or Si 2p, as shown in Figure 3.13. The ratio of the two energy separations, $\frac{\Delta_{bulk\ plasmon}}{\Delta_{surface\ plasmon}}$, is consistent with the ratio of bulk plasmon energy to surface plasmon energy, $\sqrt{2}: 1$. In addition, the intensity of the peak arising from the bulk plasmon loss (B_1 in Figure 3.13) is much higher than that contributed from the surface plasmon loss (S_1 in Figure 3.13). The third feature is that the peaks of bulk and surface plasmon losses always appear at the higher BE sides of

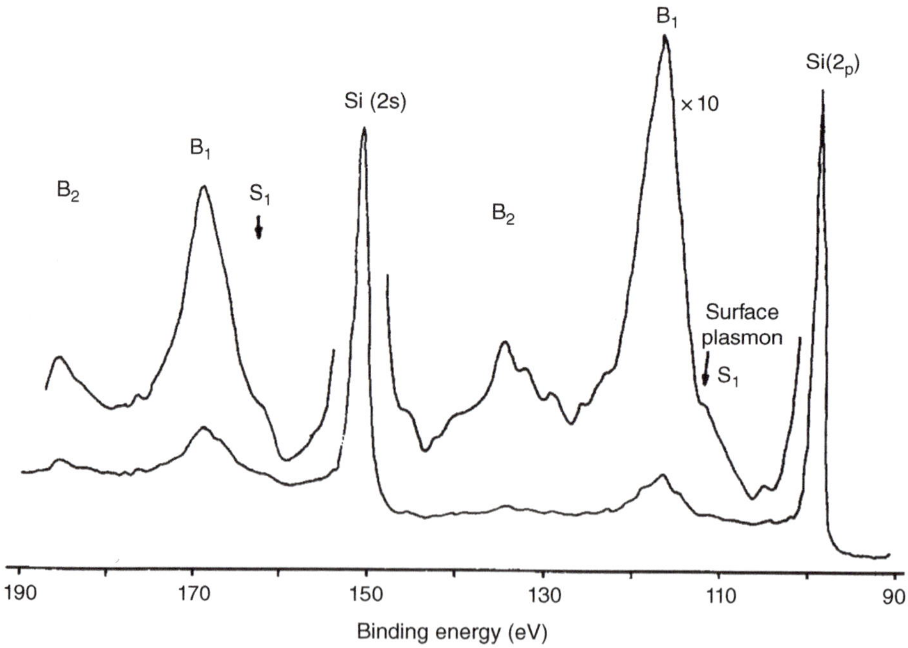

Figure 3.13 The observed satellite peaks due to surface (S) and bulk (B) plasmon loss of silicon that appear at the higher BE sides of the main XPS peaks of Si 2s and Si 2p. *Source:* Reproduced with permission from Barr[14]/Elsevier.

their main XPS peaks such as Si 2s and Si 2p (Figure 3.13). Satellite peaks arising from bulk and surface plasmons can be observed from Al, Si, Ti, and Ag.[15–17]

Other than the shake-up, shake-off, and generation of bulk and surface plasmon losses, the relaxation of a core hole state can be realized through releasing Auger electrons. Compared to peaks of plasmon losses only observable from a limited number of elements including Al, Si, Ti, and Ag, the release of Auger electrons in XPS process is very common for all elements in XPS. Release of Auger electron was first reported by Auger in 1925.[18] It involves three electrons in subshells (Figure 3.14). In terms of Auger electron released in the XPS process, the primary electron is the photoelectron. The other two electrons in the Auger process dissipate the energy associated with the core hole formed by X-ray excitation. The removed electron is labeled with X-ray notation. For instance, KL_1L_2 refers to a process in which a hole on K shell is filled by one electron of L_1 subshell and then an electron of L_2 subshell is emitted from the atom to vacuum by carrying certain KE as shown in Figure 3.14b. Since Auger electrons emitted in XPS process carries on specific KEs, their KEs can be measured. In general, they are collected by the XPS energy analyzer and thus appear as a "fake" XPS peak in the XPS spectrum, mixing up with the real XPS peaks. Whether a suspected peak is Auger or XPS peak can be identified by collecting another XPS spectrum in the same kinetics energy range from the same sample surface by using a different X-ray source. This can be readily done on a system with dual anodes such as Al Kα versus Mg Kα. As kinetic energy of the same Auger electrons is the same under the two experiments using different X-ray sources, their positions in the two XPS spectra plotting intensity versus BE must be different. However, the suspected peak has the same peak position in XPS spectra of Mg Kα spectra plotting intensity versus BE as that in the XPS spectra of Al Kα spectra, it is an XPS peak.

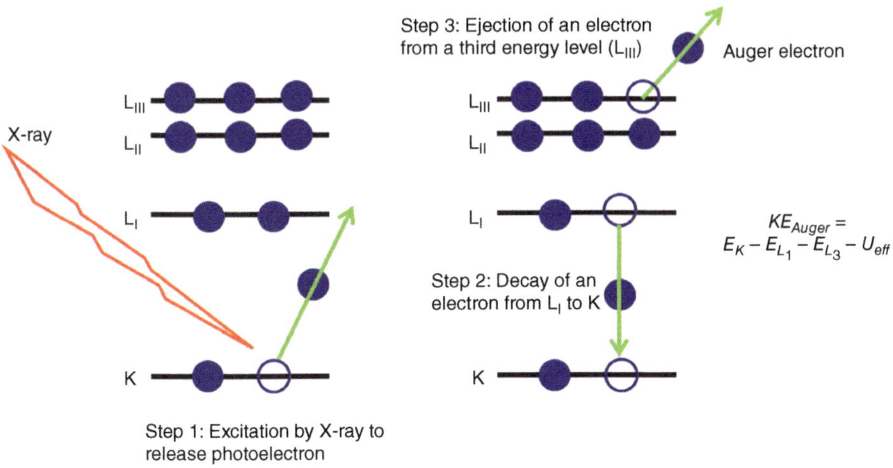

Figure 3.14 Schematic showing the formation of an Auger electron along with a XPS process. Ionization of a core hole resulting from X-ray irradiation (left panel) leads to a decay of an electron at a higher energy level (L_I). The energy released from this decay causes an ejection of an electron from L_{III}. The electron finally released from the atom is called Auger electron. Its kinetic energy (KE) while leaving the atom is calculated with the equation shown in the figure. U_{eff} is the extra energy needed to remove an electron from an ionized atom.

The observation of Auger peak in XPS spectrum can aid distinguishing metal elements at different oxidation states while their spectral features of XPS peaks are similar. For instance, Auger parameter was used in distinguishing Cu_2O and Cu. CuO can be readily distinguished from Cu_2O and metal Cu since the satellite peaks of shake-up process can be observed from CuO but not from Cu_2O and metal Cu. In general, Cu_2O and metallic Cu are differentiated with Auger parameters based on the Auger peak of Cu $L_3M_{4,5}M_{4,5}$.

3.4 Measurements of Energy of Photoelectrons

Accurate measurement of binding energies of photoelectrons generated from XPS process is crucial for interpretation of surface chemistry of a sample since BE of these photoelectrons is the X-axis in a XPS spectrum plotting intensity of photoelectrons as a function of BE. It relies on accurate and precise measurements of kinetic energy of photoelectrons. As photoelectrons from different subshells of different atoms of different elements in surface region of a sample are produced simultaneously, separation of these photoelectrons having different kinetic energies is necessary. Depending on the energy analyzer, the separation can be done in different ways. Concentric hemispherical analyzer (CHA) also called hemispherical sector analyzer (HSA) is the mostly used energy analyzer. It can run in a constant analyzer energy (CAE) or a fixed retard ratio (FRR) mode. In terms of CAE mode, photoelectrons with different kinetic energies have to be retarded or accelerated to a constant analyzer energy or pass energy (PE) such as 20.0 eV before electrons with same KE can pass through the HSA. Depending on KEs of photoelectrons, a special potential (V) needs to be applied to retard photoelectrons with a specific kinetic energy such as 1000.0 eV to reach the set pass energy before these photoelectrons can pass through the HSA. For instance, to allow photoelectrons with kinetic energy of 1000.0 eV to pass through the HSA to reach the detector at the exit of HSA for counting the number of these photoelectrons of 1000.0 eV, a retarded potential of 980.0 V must be applied to decrease their KE to 20.0 eV before entering the energy analyzer. Once these electrons pass through the HSA, intensity of these electrons with original KE of 1000.0 eV is measured by a detector attached to the exit of the energy analyzer. The retard potential V is set to continuously change as a function of time, $V = V_0 + Kt$. Here t is the time in unit of second. K is the constant; a larger K makes the change of retard potential faster. For instance, if K is 1.0 V/s and V_0 is 980.0 V, V is 979.0 V at $t_1 = 1.0$ second; at t_1, these photoelectrons with KE of 999.0 eV will be retarded to 20.0 eV by applying a retard potential of 979.0 V before entering the HSA to reach the detector of photoelectrons; the intensity of photoelectrons collected by the detector at t_1 is measured to be I_{t_1} by the detector. Then, at t_2=2.0 second, V is 978.0 V and thus only photoelectrons with original KE of 998.0 eV can be retarded to $KE = 20.0$ before passing through the HAS to reach the detector to count their intensity at t_2. Along the continuous change of retard potential as a function of time, photoelectrons will be separated as the time goes and the intensities of all photoelectrons with different original kinetic energies are measured. With this approach, the kinetic energy of photoelectrons is calculated from the retard potential V through $KE = V + PE + C$; here C is a constant. Then, BE of photoelectrons with a specific kinetic energy is calculated with the equation, $BE = h\nu - V - PE - C - \phi$. Here, ϕ is the work function of the spectrometer or called energy analyzer. By plotting intensity of these photoelectrons having a specific BE as a function of BE, an XPS spectrum is formed. Certainly, in some cases, the XPS spectrum is presented by plotting the intensity of photoelectrons having the specific kinetic energy as a function of kinetic energy for some specific purpose. Typically, the X-axis of a XPS spectrum presents the decreasing binding energy from left to right side, likely due to a used presentation of the X-axis in increasing kinetic energy in history. In order to retain the shape of an XPS peak of a subshell such as Cu 2p, the X-axis has to present decreasing binding energy so that the peak such as Cu 2p can be appropriately compared to another peak of the same subshell such as Cu 2p plotted with intensity versus increasing kinetic energy.

3.5 Measurements of Intensity of Photoelectrons

As described above, the accurate measurement of intensity of photoelectrons with a specific kinetic energy is as important as the measurement of KE of these photoelectrons. Obviously, a direct measurement of the number of photoelectrons before a magnification would give huge noise. Magnification of the number of these photoelectrons upon leaving the exit of energy analyzer is the key for obtaining a spectrum with high signal-to-noise ratio. To obtain a high sensitivity in measuring number of photoelectrons, a detector being able to record individual electrons is requested. It typically works in pulse counting mode. The sensitivity of XPS in measuring concentration of a constituting element of a sample surface is determined by lower limit of intensity of photoelectrons that can be measured by the detector. Essentially, it is determined by both pulse counting electronics and the magnification electronics. Sensitivity of a conventional pulse counting electronics is at the level of detection limit of 10^{-15} Ampere (A); in other words, it can detect the minimum charge of 10^{-15} Coulomb (C) per second. As charge of an electron is 1.60×10^{-19} C, the minimum number of electrons per second to be detected with this pulse counting electronics is at least 6240 within 1 second to reach the minimum charge detectable by the detector. In order to reach the sensitivity of recording an individual electron per second, a magnification electronics needs to magnify a single electron by at least 6240 times each second. To detect a smaller number of electrons per second, a pulse counting electronics with a lower detection limit (X) or a magnification electronics with a higher multiplication factor (Y) is needed. The minimum number of electrons the detector can measure is determined by the factor, R, which is calculated with the equation $R = \frac{1.6 \times 10^{-19} Y}{X}$. When R is larger, the minimum detectable concentration of a constituting element is lower, and the XPS instrument has a higher sensitivity in quantitative analysis. R has been increased significantly in the past two decades. The function and working principles of pulse counting electronics and electron magnification electronics will be described in a later section. Through further development of these electronics, the sensitivity of XPS in quantitative analysis is expected to be enhanced. That is a challenging task for physicist and microelectronic engineers but is significantly valuable. For instance, a trace amount of alkali metal such as lithium and precious metal such as platinum plays key roles in catalytic chemical transformation. Currently, it is quite challenging, if not possible, to identify the oxidation state and analyze the chemical environment of trace amounts of these metals in a great number of important catalysts even in UHV.

References

1 Moulder, J. F., Stickle, W. F., Sobol, P. E. *et al.* 1992. *Handbook of X-ray Photoelectron Spectroscopy: A Reference Book of Standard Spectra for Identification and Interpretation of XPS Data*. Perkin-Elmer Corporation.

2 Yeh, J. J. and Lindau, I. 1985. *Atomic Data and Nuclear Data Tables*, 32, 1.

3 Cardona, M. and Ley, L. 1978. *Photoemission in Solids*. Vol. 26. Springer.

4 Koopmans, T. 1933. "Ordering of wave functions and eigenenergies to the individual electrons of an atom." *Physica* 1, 104–113.

5 Bagus, P. S., Ilton, E., and Nelin, C. J. 2018. "Extracting chemical information from XPS spectra: a perspective." *Catal. Lett.* 148, 1785–1802.

6 van der Heide, P. A. W. 2006. "Photoelectron binding energy shifts observed during oxidation of group IIA, IIIA and IVA elemental surfaces." *J. Electron. Spectrosc. Relat. Phenom.* 151, 79–91.

7 van der Heide, P. A. W. 2008. "Multiplet splitting patterns exhibited by the first row transition metal oxides in X-ray photoelectron spectroscopy." *J. Electron. Spectrosc. Relat. Phenom.* 164, 8–18.

8 Briggs, D. and Seah, M. P. 1990. *Practical Surface Analysis, Volume 1: Auger and X – ray Photoelectron Spectroscopy*. John Wiley & Sons.

9 Kotani, A. and Toyozawa, Y. 1974. "Photoelectron spectra of core electrons in metals with an incomplete shell." *J. Phys. Soc. Jpn.* 37, 912–919.

10 Zaanen, J., Sawastsky, G. A., and Allen, J. W. 1985. "Band gaps and electronic structure of transition-metal compounds." *Phys. Rev. Lett.* 55, 418–421.

11 Imada, M., Fujimori, A., and Tokura, Y. 1998. "Metal-insulator transitions." *Rev. Mod. Phys.* 70, 1039–1263.

12 Bethe, H. 1930. "Zur Theorie des Durchgangs schneller Korpuskularstrahlen durch Materie." *Ann. Phys.* 397, 325–400.

13 Tanuma, S., Powell, C. J., and Penn, O. R. 1993. "Calculations of electron inelastic mean free paths (IMFPS). IV. Evaluation of calculated IMFPs and of the predictive IMFP formula TPP-2 for electron energies between 50 and 2000 eV." *Surf. Interface Anal.* 20, 77–89.

14 Barr, T. L. 1983. "An XPS study of Si as it occurs in adsorbents, catalysts, and thin films." *Appl. Surf. Sci.* 15, 1–35.

15 Barrie, A. 1973. "X-ray photoelectron spectra of aluminium and oxidised aluminium." *Chem. Phys. Lett.* 1, 109–113.

16 Mannella, N., Gabetta, G., and Parmigiani, F. 2001. "Plasmon energy shift in porous silicon measured by X-ray photoelectron spectroscopy." *Appl. Phys. Lett.* 79, 4432–4434.

17 Pollak, R. A., Ley, L., McFeely, F. R. *et al.* 1974. "Characteristic energy loss structure of solids from X-ray photoemission spectra." *J. Electron. Spectrosc. Relat. Phenom.* 3, 381–398.

18 Auger, P. 1925. "Sur l'effet photoélectrique composé." *J. Phys. Radium.* 6, 205–208.

4

Instrumentation of XPS

4.1 Regular X-ray Source

X-ray source is the part producing X-ray for exciting electrons of subshells of atoms in the surface region of a sample in X-ray photoelectron spectroscopy (XPS) analysis. The principle of generation of X-rays from a metal anode was described in Chapter 3. This section focuses on the instrumentations of X-ray sources. In brief, X-rays can be generated from a specific metal or semiconductor (Table 3.3) while electrons with certain energy bombard surface of a specific metal or semiconductor. Energy, flux, brightness (flux density), spot size, and stability of the electron beam of X-ray gun are significant parameters of an X-ray source to determine the specifications of an X-ray beam. Table 3.3 lists the energies and line widths in terms of full width at half maximum (FWHM) of X-rays generated from different metals or semiconductors. To obtain high surface sensitivity, lower X-ray energy is preferred; however, a low energy of an X-ray beam limits the number of subshells and the number of elements to be analyzed with XPS (Figure 3.3). To analyze majority of elements of the periodic table, the X-ray energy must be higher than the binding energies of the most intensively emitted subshells of majority of elements of the periodical table (Figure 3.3). By considering these factors, Al Kα and Mg Kα are the commonly used X-rays on lab-based XPS systems since they allow XPS analyze surface with high surface sensitivity and cover the most intense subshells of majority of elements listed in Figure 3.3. As monochromation of X-ray radiations of an Al anode can be accomplished through a well-developed quartz crystal-based monochromator, the monochromated Al Kα is the mostly used lab-based X-ray.

An Al Kα X-ray source consists of three major parts: an electron source, an anode, and a circulated cooling system. Each of these significant components is controlled by electronics. These control electronics are not described here since they are beyond the focus of this book. The Al anode is an Al thin film with a thickness of about 10 μm deposited on the vacuum side of a Cu block; cooling water flows through the cavity of the Cu block during the function of the X-ray source to take away heat generated in the bombardment of electron beam on the Al thin film. It is noted that at least one-third of the energy of the electron beam is transferred to heat instead of fluorescence in terms of X-ray radiation. The heat generated from the electron bombardment could make the Al film melt if it is not taken away in time. Thus, flowing water before introducing the electron beam to the Al anode is requested. A typical practice is to keep water continuously cooling during electron bombardment and for additional half hour after turning off the electron bombardment, maximizing the life of Al anode. Al thin film must be melted and cracked at a microscopic scale if heat is accumulated due to the failure of cooling of the anode. A specific phenomenon resulting from breakage of the Cu block at microscopic scale is a leaking of water and air to a high vacuum environment of the chamber hosting the X-ray source. An early finding such a leaking is necessary. In general, such a leaking on an anode at its early stage does not obviously increase the base pressure of the chamber. Tracking composition of the background gas of the vacuum environment near to the X-ray source allows to find out whether the X-ray anode has been cracked or not. The diagnosis is typically done with a quadrupole mass spectrometer installed on the chamber near the X-ray source. If there is a micro leaking on the anode, a typical feature of the mass spectrum is that the partial pressures of H_2O and O_2 are higher than those in a routine high vacuum environment without any leaking of the X-ray anode.

The electrons to bombard the anode surface can be generated through a filament of thorium-containing tungsten as thorium-added tungsten has a lower work function in terms of higher emission efficiency of electrons than pure tungsten. Enough high electrical current and voltage applied to the filament are necessary to heat the filament to certain high temperature so that enough electrons can be emitted from the filament. When an X-ray source has not been used for over a week, the current and voltage should be slowly applied to the filament with enough patience while the base pressure of the vacuum environment near to the X-ray source is being watched closely. It is noted that electrons can be emitted only when the filament is heated to a relatively high temperature. As most X-ray source typically does not have a thermocouple to measure the temperature of the filament, one must watch the pressure gauge near the X-ray source while the electrical current or/and voltage is tuned up. Keeping a progressive increase of background pressure in XPS chamber but not a brust of base pressure near to X-ray source is the guidance for turning on a X-ray source. A high background pressure could make the filament burnt out. Although a filament kit is available from the manufacturer of the X-ray source, it is expensive, and its replacement needs much time and carefulness. Suggestion on keeping the Al Kα source for a longer life is to maintain a higher vacuum environment for the anode and efficient cooling of the anode.

Electrons emitted from the filament are dragged to strike the surface of the Al thin film by an electric field with an anode voltage of 15 kV for generating Al Kα. An appropriate voltage for a specific anode is determined by a threshold voltage of the electron beam requested for generating X-ray from this specific anode upon beam bombardment. The threshold voltage varies much, depending on the anode material. For generation of high-energy X-rays such as Cr Kα and Cu Kα, the energy of the electron beam needs to be >20 kV and >30 kV, respectively. The intensity of X-ray is strongly dependent on the electron beam energy. As an Al Kα source is typically operated at 100–500 W, the beam current is 10–100 mA. Notably, the beam current or called current of an electron beam is distinctly different from the electric current applied to the filament. The former is at the level of milli-Amperes (mA), but latter is at the level of Amperes (A). Along the increase of beam current and voltage, the base pressure around the anode typically increase obviously as the anode other than the filament is heated while more electrons are bombarding the surface of Al thin film. The higher the temperature is at an anode, the Al anode outgases more and then the base pressure around the X-ray source is higher.

Typically, a regular Al Kα source without a monochromator does not have a complicated lens to focus electron beam to a small spot. The spot size of X-ray beam of such a source could be at the level of 5–10 mm. The size of X-ray beam on a sample surface can be evaluated by mounting a piece of X-ray luminescent sample that emits visible fluorescence while it is irradiated with a X-ray beam. The size of the visible beam on the luminescent sample is approximately the size of the beam spot of X-ray irradiating on the sample surface.

Although a partially melted Al anode can still be used, Cu Lα X-ray with energy of about 930 eV can be generated since the electron beam can bombard the potentially exposed surface of the Cu block of a potentially melted Al anode. Unfortunately, Cu Lα could excite electrons of numerous subshells of elements, generating photoelectrons with kinetic energy (KE) of 0–930 eV. These photoelectrons appear as additional set of XPS peaks in the XPS spectrum of Al Kα. Sometimes they are called ghost peaks. These ghost peaks likely overlap with some of those normal XPS peaks generated by Al Kα, largely complicating the XPS analysis. In addition, the unwanted X-rays of Al anode including Al $K\alpha_3$, Al $K\alpha_4$, Al $K\alpha_5$, Al $K\alpha_6$, and Al Kβ generate extra peaks in XPS spectrum, making the analysis challenging as exemplified in Figure 3.4b. A monochromator can "filter out" these ghost peaks and only allow Al Kα diffracted, by which a pure Al Kα with a line width of 0.2–0.3 eV in terms of FWHM of 0.2–0.3 eV is generated.

4.2 X-ray Source with a Monochromator

Compared to a regular Al Kα source, the instrumentation of an X-ray source with a monochromator is more complicated. Its cost is more than twice of a regular Al Kα source without a monochromator. Figure 4.1c is a drawing of the external appearance of an Al Kα X-ray source with a monochromator prepared with a software called *solid work*. The

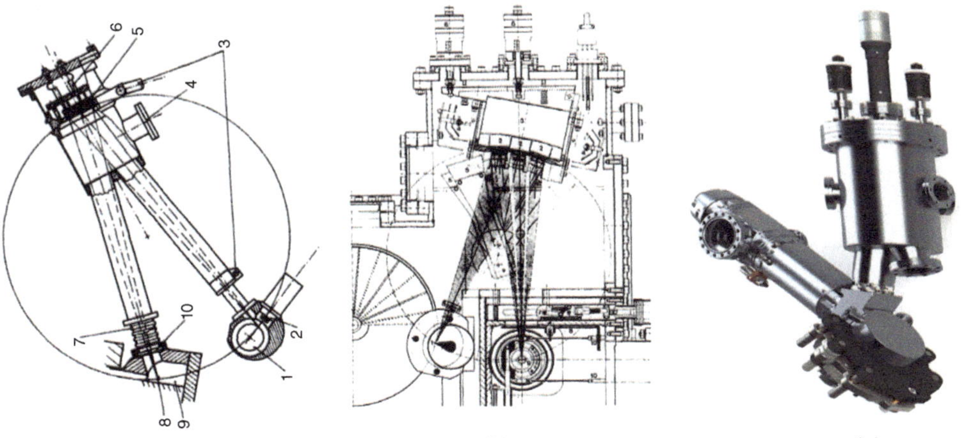

(a) (b) (c)

Figure 4.1 Published and prepared drawings of X-ray monochromator of lab-based X-ray source. (a) Structure of monochromator published in 1975. *Source:* Reproduced with permission from Baer et al.[1]/AIP Publishing. (b) Structure of monochromator published in 1984. *Source:* Reproduced with permission from Gelius et al.[2]/Elsevier. (c) Drawing for approximately representing external appearance of a regular monochromator.

monochromator is a critical component of an XPS system. To be compatible with a monochromator, the X-ray gun is quite different from a regular X-ray gun. Its X-ray beam spot is smaller, and thus its brightness is much higher. Notably, to collect enough photoelectrons from a sample, a higher flux of X-ray source is used in general. In other words, the decrease of spot size of X-ray beam in X-ray source with a monochromator is compensated by the increase of flux density. Flux density is calculated by dividing flux of an X-ray beam by its beam spot area. Thus, an X-ray with higher brightness and small beam spot offered by a monochromatic X-ray beam can typically generate more photoelectrons than that generated from an X-ray beam that has low flux density but a larger size of X-ray beam spot of an X-ray source without a monochromator. The advantage of using X-ray source with a monochromator is to gain Al $K\alpha_{1,2}$ without satellite X-rays, by which the XPS spectrum does not have ghost peaks and has better spectral resolution.

In terms of an X-ray gun compatible with a monochromator, additional electrostatic lenses including condenser lens and objective lens are typically installed between the filament and anode for focusing the electron beam on a smaller spot on the anode. To generate an X-ray beam with higher brightness, an electron beam with a higher current density is necessary since the number of photons produced from 1 mm^2 of Al thin film increases along the increase of the number of electrons bombarding 1 mm^2 of the Al thin film. For imaging spatial distribution of a constituting element on a sample surface, one of the methods is scanning a static sample surface with an X-ray beam. For this purpose, a smaller beam spot of X-ray (1–10 μm) is needed; In the study of the spatial distribution of an element or a specific chemical environment of an interested element, the X-ray beam scans the sample surface along X and Y directions in a raster scanning mode. To make X-ray beam scannable, the electron beam must perform a raster scan to the anode along X and Y directions. Thus, defection and scanning lenses must be included in the X-ray source. In addition, a LaB_6 filament with an ideal shape and size for best emission efficiency of electrons is typically used in an X-ray gun compatible with a monochromator.

A monochromator has been one of the parts for achieving X-ray with a higher resolution. Various monochromators have been available on the market in the past six decades. In terms of a monochromator of Al Kα, it consists of quartz crystal(s) and stainless steel scaffold to hold the quartz crystal and mini manipulators for tuning orientation of these crystals. These parts have been available at different manufacturers in Europe or Japan for several decades. How these parts were assembled into a monochromator is publicly available in literature such as the references published in 1975 and 1984.[1,2] Figure 4.1a and b are the representations of the assembly details of monochromated X-ray sources published in literatures in 1975 and 1984.[1,2] Figure 4.1c is a representation of the external appearance of an X-ray source having a monochromator.[4]

One function of a monochromator is removal of additional radiation lines except Al Kα, by which the resolution of XPS peak can be obviously improved since the FWHM of a monochromated Al Kα radiation can be as narrow as 0.2 eV instead of ≥1.0 eV of a regular Al Kα source without a monochromator. Several manufacturers have made different monochromators for Al Kα sources for decades. The principle of monochromations and configuration of the monochromators are very similar. As shown in Figure 4.2, the surface of X-ray anode, surface of a bent quartz crystal, and the sample surface are on the

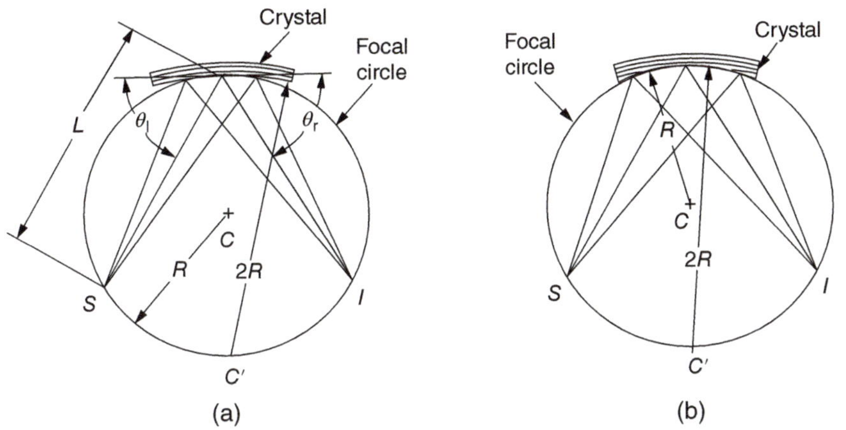

Figure 4.2 Geometry of focusing spectrometers: (a) Johann mode; (b) Johansson mode. Rays diverging from a source *S* are focused on image I. *R* and *C* are the radius and center of the focal circle, respectively. *C′* is the center of curvature of the crystal planes. θ_i and θ_r are the incidence and reflection angles, respectively. Reprinted from reference[3].

circumference of a Rowland sphere with a diameter of typically 500 mm.[1,2] The purpose of using bent quartz crystals is to focus the diffracted X-ray onto the sample.

It is well known the Bragg diffraction equation, $n\lambda = 2dsin\theta_B$, defines that only X-ray with a specific wavelength (λ) can be reflected constructively by a crystal with a specific interplanar distance (d), which meet the criteria of this equation. Based on the interplanar distance (*d*) of a quartz crystal, only Kα_1 was reflected and the rest of radiations from Al anode including Al Kα_2, Al Kα_3, Al Kα_4, Al Kβ, Bremsstrahlung radiation, Cu Lα, and Cu Kα were filtered. Through this diffraction, the X-ray irradiating the sample is only Al K$\alpha_{1,2}$ that has an FWHM of 0.2–0.3 eV. As λ of Ag Lα is nearly half of Al K$\alpha_{1,2}$, the monochromator of Al Kα can be used for monochromating radiations of an Ag anode as well. The FWHM of a monochromated Ag Lα X-ray source is about 1.2 eV. The advantage of using Ag Lα as the second X-ray source is to gain information from a deeper region of the sample surface requested in some studies. Notably, compared to Al K$\alpha_{1,2}$ the flux density of Ag Lα is weak in general.

For achieving constructive reflection and simultaneous focus of the reflected X-ray to a sample surface, lots of details must be considered as reported in literature. Figure 4.2 schematically presents the geometries of internal parts of two types of monochromators. They are mainly crystals assembled through a specific geometry developed by Johann or Johansson and Castaing.[5–7] By using the Bragg diffraction equation, the diffraction angle, θ_B, can be readily calculated for a specific X-ray with wavelength, λ, when the first order of diffraction ($n = 1$) is considered due to its least loss of intensity of incident X-ray. The incidence angle of X-ray to crystal plane, θ_i, is not necessary to be exactly as the Bragg diffraction angle. It can vary in the range of $\theta_B - \frac{w}{2} < \theta_i < \theta_B + \frac{w}{2}$, where w is the rocking curve width, a property of a crystal also an indicator of the quality in terms of crystallization and polishing of the crystal. As the reflection angle (θ_r) is equal to the incident angle, (θ_i), θ_r in Figure 4.2 falls into the range of $\theta_B - \frac{w}{2} < \theta_r < \theta_B + \frac{w}{2}$. A mosaic crystal typically has a larger w; thus, the reflected beam spreads wider if a mosaic or imperfect crystal is used; w of a quartz crystal is 5×10^{-5} radian.[9] To increase the effective reflection area for an incident X-ray, a cylindrically curved crystal is used. Certainly, focusing X-ray beam upon Bragg diffraction is another reason for using a curved crystal in the monochromator of the X-ray source. As shown in Figure 4.2b, the incident point of X-ray beam on a crystal, the source of X-ray in terms of an Al anode, and the focal point (I) decide a circle called the Rowland circle. Radius of the Rowland circle is R, which is marked on Figure 4.2b. In the case of only determining the radius of the Rowland circle, the crystal surface does not have to be curved. However, from optics point of view, to focus the X-ray beam, the crystal has to be curved and the radius of the crystal curvature has to be $2R$, which is marked on Figure 4.2b. In other words, the bent crystal surface is a part of a larger circle with a radius of about $2R$ as shown in Figure 4.2b.

The shape of a quartz crystal is critical for successful diffraction and focus. Either Johann (Figure 4.2a) or Johansson (Figure 4.2b) geometry can be adopted. In Johann geometry, L is defined as the distance from the anode surface to the central point of the curved surface of the crystal.[3] It is also the distance from this central point to the focal point of X-ray (I) after diffraction on the Rowland circle. In terms of a monochromator for XPS, this focal point is the center of a sample surface. L is $2R\sin\theta_B$.[3] To make an X-ray wave originating at anode hit to the crystal surface to be constructively diffracted and then to arrive at the focal point in terms of the center of the sample surface, the incident angle of X-ray, θ_i, must simultaneously fit $n\lambda = 2d\sin\theta_i$ and $\theta_B-\frac{w}{2}<\theta_i<\theta_B+\frac{w}{2}$. Then, only the center of the crystal surface can be the effective point (I) where it is tangent to the Rowland circle, by which the diffracted X-ray can be focused on the focal point on the Rowland circle (Figure 4.2a). With this geometry, only a quite small portion of X-rays emitted from X-source(s) can reach the focal point (I) in terms of the sample surface. This disadvantage was solved by Dumond and Kirkpatrick.[10] Dumond et al. proposed a solution to make the entire surface of the crystal in the plane of the Rowland circle.[10] Johansson is the first one who implemented such a geometry.[6] Later it was called Johansson geometry of monochromator for X-ray.

In the Johansson geometry (Figure 4.2b), the crystal is bent to a curvature with a radius, $2R$. Its inner surface in terms of the hitting point of incidence X-ray wave is on a curvature with a radius, R; here R is the radius of the Rowland circle. It is noted that the difference between the Johann and Johansson crystals is that the whole Johansson crystal is bent to overlap the Rowland circle, but Johann crystal is not. Here the quartz crystal is used as an example to demonstrate the difference of crystals between the two geometries. The (10–10) of quartz is chosen to diffract X-ray waves in the monochromator of Al Kα. The plane of (10–10) of the quartz crystal in Johansson geometry is not flat although it is atomically flat for the quartz crystal of Johann geometry. In practice, this geometry can be manufactured through first bending a crystal to a radius of $2R$, and then grinding the surface to a radius of R. As the diffraction of incident X-ray waves mainly performs in the surface region of a crystal, the crystal used in a monochromator of XPS is quite thin. Its thickness is typically <3 mm, favorable for easy bending the brittle quartz.

Figure 4.3a is a two-dimensional illustration of the Johann geometry. The surface of the Johann crystal is in fact three dimensional. By assuming the X-ray source is a point source, the diffraction region on the surface of a Johann crystal looks like an "X" as shown in Figure 4.3a.[3] The "X" shape results from the fact that these X-rays diverge from the plane of the Rowland circle and can meet the Bragg angle at points away from the

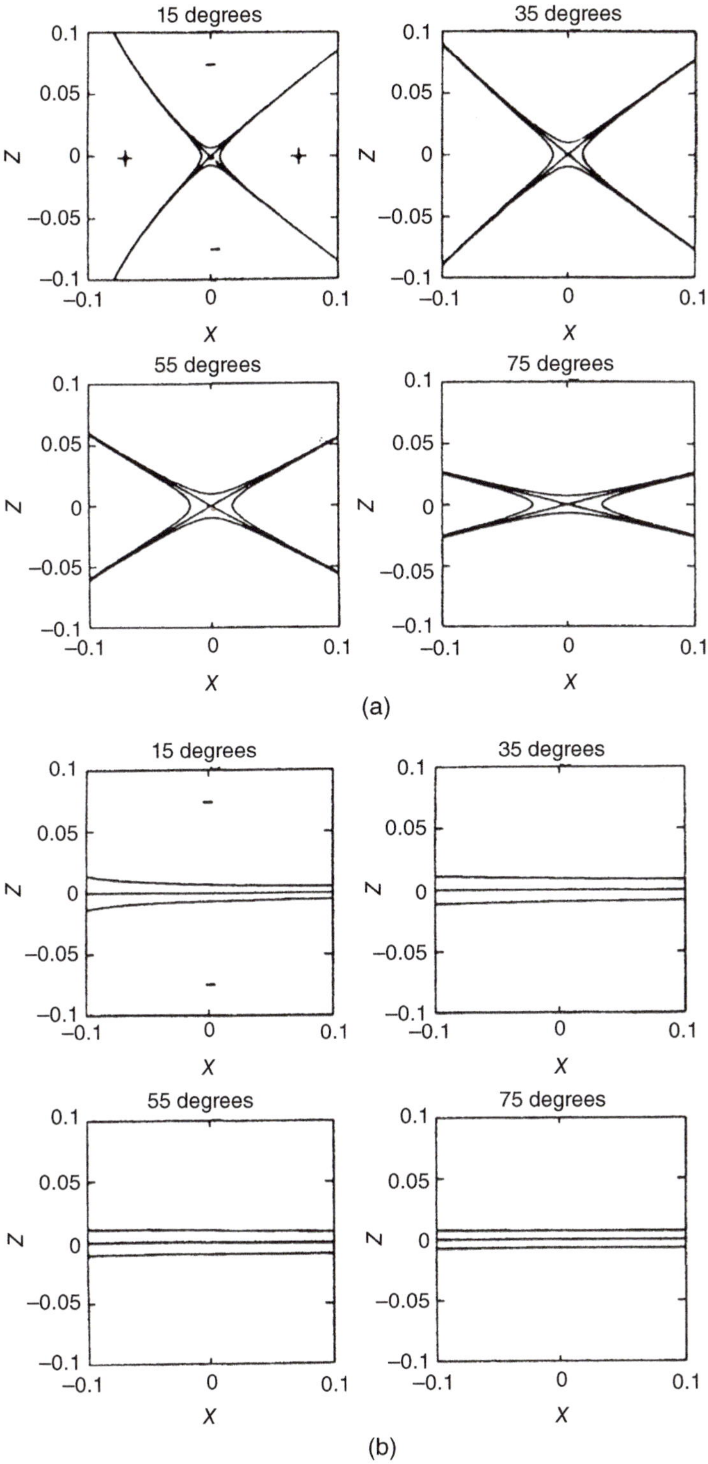

Figure 4.3 Represented effective area within which the incidence angle lies (θ_i) within the rocking curve for $w = 2 \times 10^{-4}$ radian, $\theta_i = \theta_B \pm \frac{w}{2}$ for Bragg angles of 15°, 35°, 55°, and 75°. (a) Johann crystal; (b) Johansson crystal. The X and Z values are dimensionless coordinates that are the actual values divided by the focal circle radius. *Source:* Reproduced with permission from Wittry and Sun[8]/AIP Publishing.

center of the curved surface of the crystal. In terms of the Johansson geometry, as shown in Figure 4.3b, the region diffracting X-rays appears as a long rectangular band parallel to the Rowland circle. The difference in the shapes of the diffracting regions of the two crystals was well documented in literature.[8] The size of the diffracting region is related to the rocking curve width, w. A larger w leads to a wider region. The different shape of the diffracting region on the crystal requires a different dimension of the crystal. As the diffracting region of a Johansson geometry is a long rectangle-like shape, the ratio of length to width of a Johansson crystal should be 3 or larger. For a Johann geometry, the ratio can be as low as 1.5 since the diffracting region of a Johann geometry is an "X" shape. Typically, the area of the diffracting region of a Johansson crystal is at least 3 or 4 times of that of a Johann crystal. As the area of the diffracting region on surface of a crystal basically determines the flux of X-rays being diffracted, the Johansson crystal gives a higher flux at focal point by 2–3 times at least than a Johann crystal (Figure 4.3a versus Figure 4.3b). Thus, obviously Johansson geometry is a better choice unless some crystal cannot be bent to this geometry.

Overall, there are several advantages of using a monochromator X-ray source for XPS studies. They offer X-ray with high energy resolution, a higher flux density, and a smaller beam spot convenient for raster scanning to generate a X-ray photoelectron image. In addition, a monochromator can effectively prevent the sample from being struck by electrons leaked from X-ray gun; without a monochromator, the impact of electrons of filament to a sample surface could interfere the collection of photoelectrons by an energy analyzer or may even negatively charge the sample surface.

4.3 Energy Analyzer

In instruments of surface analysis, at least three types of analyzers of electron energy are used. They are the retarding field analyzer (RFA), cylindrical mirror analyzer (CMA), and hemispherical sector analyzer (HSA) that is also called concentric hemispherical analyzer (CHA). RFA runs like an energy filter cutting off all electrons below certain KE. It is mainly used in an Auger Electron Spectroscopy (AES) of a Low Energy Electron Diffraction/Auger Electron Spectroscopy (LEED/AES) system; here the AES is different from an independent Auger electron spectrometer system.[11] CMA is mainly used in AES before 2000s although AES systems produced in this century have adopted a hemispherical analyzer. As most XPS systems use HSA, here only HSA is described. Figure 4.4 is a schematic of an HSA.

As photoionization events occur simultaneously in different subshells of all constituting elements of the surface region of a sample, numerous types of photoelectrons with different KEs are produced simultaneously when a sample is being irradiated. The original KEs of most photoelectrons are too large for HSA to separate them with high resolution of energy. Thus, majority of them are retarded to a low KE. The retardation is done by either a retard electrostatic lens before electrons enters the entrance slit of HSA or the HAS itself. If the retardation is done within the HSA itself, the HSA is run with a constant retard ratio (CRR) mode, by which all original KEs of all photoelectrons are decreased by the same ratio such as 10 to low KEs; at these low KEs, the HSA can separate them with a better energy resolution.

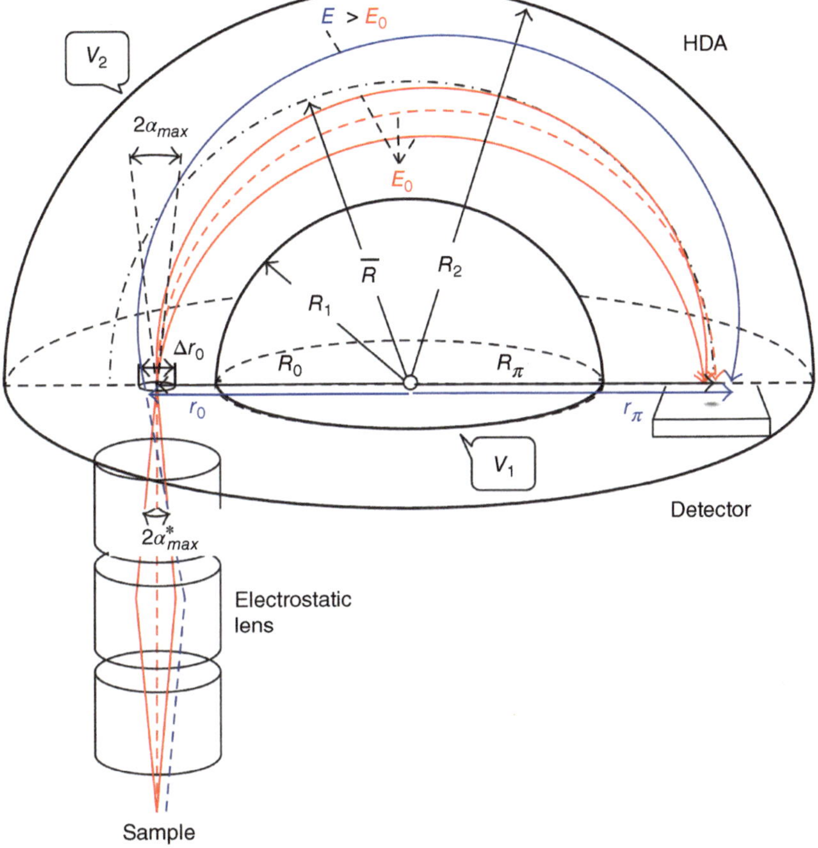

Figure 4.4 Schematic of HAS consisting of two concentric hemispheres with radii R_1 and R_2, an electrostatic lens at input, and a position-sensitive detector (PSD) at the exit. The HDA voltages are set to pass the reference trajectory (red dashed line) having nominal pass energy E_0 and launching angle $\alpha = 0$, with specified entry radius R_0 and exit radius R_π. Two additional trajectories (red lines) of the same energy E_0 are shown to originate from a point source, but with $\alpha = \pm\alpha_{max}$. Here, the entry is seen to be paracentric with R_0 displaced from R, the mean radius. The dash-dot line shows the conventional optical axis at R for central entry $R_0 = R$. A general trajectory is also shown (in blue) with pass energy E seen to start at entry radius r_0 and launching angle α, exiting the HDA at radius r_π. The PSD detector records the exit positions of the trajectories. *Source:* Reproduced with permission from Sise and Zouros[12]/Hindawi Publishing/CC BY-3.0.

As seen in Figure 4.4, an HSA consists of two concentric hemisphere shells made of μ metal with radius of R_{in} and R_{out}, respectively. It behaves as a narrow band *KE* filter. To make photoelectrons with a specific kinetic energy *KE* travel through the HSA along the hemisphere with radius of $\frac{R_{in} + R_{out}}{2}$, appropriate potentials, V_1 and V_2, should be applied to the inner and outer hemispheres, respectively. The difference in potential between the inner and out hemispheres, ΔV is $\frac{KE}{e}\frac{R_{out}^2 - R_{in}^2}{R_{out}R_{in}}$. The electrons allowed to be deflected through the HSA are these with KE of $KE = e\Delta V \frac{R_{out}R_{in}}{R_{out}^2 - R_{in}^2}$. At a moment $t = t_0$, $\Delta V = (\Delta V)_{t=t_0}$, only electrons having the specific kinetic energy, $KE = e(\Delta V)_{t=t_0} \times \frac{R_{out}R_{in}}{R^2_{out} - R^2_{in}}$ can pass through the HAS to arrive at the exit slit to enter the detector of electrons. By varying the potential difference, $(\Delta V)_t$ between the inner and outer shells as a function of time (t), photoelectrons with a different kinetic energy *KE* can travel through the HSA at a different moment; thus, photoelectrons with various kinetic energies can be readily separated. The intensity of photoelectrons measured by a detector is determined by the number of photoelectrons leaving the analyzer or entering a detector. At each moment, the intensity (I) of these photoelectrons with a specific kinetic energy (*KE*) is measured by the detector. By plotting intensity (I) as a function of its corresponding kinetic energy (*KE*) or binding energy ($BE = h\nu - KE - \Phi$), an XPS spectrum is produced.

In the CRR mode, KEs of photoelectrons are reduced with a constant ratio. All photoelectrons with different KEs simultaneously travel to the entrance slit of an HSA. At each moment, the potential difference (ΔV) is specific, but it is different from one moment to the next moment. Thus, at each moment only photoelectrons with KE equal to $e\Delta V \frac{R_{out}R_{in}}{R^2_{out} - R^2_{in}}$ can travel through the HSA to arrive at the exit of the HSA where a detector is installed. As photoelectrons whose energies are lower or higher than this specific KE takes a trajectory with a radius smaller or larger than the radius of $\frac{R_{in} + R_{out}}{2}$, they impinge on the inner and outer metal shells of the energy analyzer; these electrons do not arrive at the exit slit. If kinetic energies of some photoelectrons are close to the specific KE in terms of $e\Delta V \frac{R_{out}R_{in}}{R^2_{out} - R^2_{in}}$, these photoelectrons can travel to the exit of the HAS. These electrons with kinetic energies close to the specific KE can still escape the energy analyzer if the exit is a rectangular slit extending from the inner to the outer shell. If a series of detectors are arranged along the long direction of the rectangular exit slit, the numbers of all these electrons with kinetic energies close but some offset from the specific KE, $\Delta V \frac{R_{out}R_{in}}{R^2_{out} - R^2_{in}}$ can be found out by the array of detectors. By placing multiple arrays of detectors of electrons along another direction of the rectangular slit, numbers of electrons along two directions can be counted by multiple arrays of detectors and thus a 2-D image of surface information can be generated.

Other than the above-described CRR mode, an HSA can perform constant analyzer energy (CAE) mode. In CRR, the pass energy varies as a function of time; however, it is a constant in CAE. To allow electrons with different original KE to pass through the analyzer with the same pass energy, electrons are retarded typically or accelerated in the CAE mode to reach the same KE called pass energy, which is the energy of electrons traveling through the HSA. The acceleration or deceleration are carried out by a retard lens before photoelectrons enter HAS. As the energy of electrons traveling through the HSA is constant, the potentials on inner and outer hemispheres are fixed. In this mode, the separation of electrons with different KEs is done by varying acceleration/deceleration potential in the retard lens as a function of time. At each moment, only electrons with a specific KE are retarded to the pass energy for traveling through the HSA to be counted by a detector.

In the CAE mode, the energy resolution of HSA is independent of the original KEs of these photoelectrons which leave the sample surface with different KEs. In a CAE mode, the pass energy is the same for all photoelectrons once the pass energy is set for one experiment. Certainly, In another experiment, the pass energy can be set to a different value. Lower pass energy makes HSA have better energy resolution by scarifying the intensity in terms of the number of electrons detected at the exit of HSA. Thus, by using the CAE mode, the energy resolution of HSA for peaks in the whole energy window (0–1486.3) is the same since all photoelectrons entering HSA have the same KE.

Different from the CAE mode, the energy resolution contributed from the energy analyzer in the CRR mode depends on the original KEs of photoelectrons leaving the sample surface. In the CRR mode if HAS, along the increase of original KEs, the energy resolution of the HSA decreases. In other words, in an XPS spectrum collected with a CRR mode, an XPS peak with a higher binding energy has a relatively higher energy resolution. Thus, the energy analyzer working at CRR mode introduces unequal instrumental resolutions to different peaks in an XPS spectrum. This introduction makes us incapable of identifying the intrinsic peak resolution solely contributed by the nature of the sample surface. Compared to CRR mode, the energy resolution introduced by the energy analyzer working at CAE mode to XPS peaks with different binding energies is invariable. Since the CAE mode analysis introduces an equal instrumental resolution to resolutions of any peaks in an XPS spectrum, the XPS spectrum collected with the CAE mode better reflects the intrinsic information of the sample surface. Thus, CAE is the most commonly used mode in XPS studies.

Energy resolution of HSA is one of the major factors determining spectral resolution. The absolute uncertainty in energy measurement can be evaluated with the equation $\Delta E = \frac{E_{max} - E_{min}}{2}$, where E_{max} and E_{min} are the maximum and minimum energies transmitted and can be calculated through $E_{min} \cong E_0\left(1 - \frac{w}{2R_0}\right)$ and $E_{max} \cong E_0\left(1 + \frac{w}{2R_0} + \alpha_{max}^2\right)$. Thus, $\Delta E \cong E_0\left(\frac{w}{2R_0} + \frac{\alpha_{max}^2}{2}\right)$, where E_0 is the pass energy of electrons while these electrons are traveling through the analyzer, w is the slit width, R_0 is the average of the radii of inner and outer hemispheres, and α_{max} is the pencil angle. Obviously, in CRR mode the absolute energy resolution is lower at region of lower BE (higher KE). But it is the same for different peaks or different energy windows in CAE once a specific pass energy is chosen. In a CAE mode, the absolute energy resolution is higher or better when a low pass energy (1–5 eV) is used; it can be as low as 0.001 eV.

The relative energy resolution of an HSA, $\frac{\Delta E}{E} \cong \frac{w}{2R_0} + \frac{\alpha_{max}^2}{2}$, where $R_0 = (R_{in} + R_{out})/2$, is the specification to compare the energy resolution of two analyzers. It is mainly determined by the instrument parameters including the radii of the inner and outer hemispheres, slit width, and pencil angle. In general, a HAS having a large size (R_0) offers high energy resolution.

High resolution spectrum is the key for achieving a better understanding of the surface chemistry of a sample. If the energy distributions of X-ray source (ΔE_p), photoelectron emission (ΔE_n), and energy analyzer (ΔE_a) follow Gaussian distributions, energy resolution of an XPS peak can be determined by the equation $\Delta E = \sqrt{(\Delta E_p)^2 + (\Delta E_n)^2 + (\Delta E_a)^2}$; here the energy distributions of the X-ray source, photoelectron emission, and energy analyzer are ΔE_p, ΔE_n, and ΔE_a, respectively. If an Al Kα X-ray beam with a resolution of 0.25 eV (ΔE_p) is used, the spectral resolution in terms of energy resolution of an XPS peak is $\sqrt{(0.25)^2 + (\Delta E_n)^2 + (\Delta E_a)^2}$ under the assumption that all the home atoms of subshell electrons such as C 1s of all carbon atoms of a graphite sample have a homogeneous chemical environment. For XPS studies using synchrotron radiation, we can assume ΔE_p can go down to 0.01 eV; then, the spectral resolution, ΔE, energy resolution of an XPS peak can be evaluated by $\sqrt{(\Delta E_n)^2 + (\Delta E_a)^2}$. If both a monochromated Al source XPS and a synchrotron XPS use the same HSA analyzer having the same energy resolution of ΔE_a of 0.01 eV are used, the energy resolutions, ΔE is ΔE_n for synchrotron XPS and $\sqrt{(0.25)^2 + (\Delta E_n)^2}$ for Al Kα monochromated XPS. ΔE_n is determined by the lifetime of a core hole state formed from photoionization; it is the intrinsic property of a subshell; it can be calculated with Heisenberg's uncertainty principle, $\Delta E \geq h/(2\pi\Delta t)$ where h is the Plank constant and Δt is the lifetime of an interested core-hole; it is typically in the range of 0.3–0.9 eV. Obviously, the spectral resolution of synchrotron XPS is better than vacuum XPS. Since ΔE_p for synchrotron radiation can be taken as zero, the spectral resolution of XPS peak obtained by using both synchrotron radiation and high-resolution energy analyzer is approximately ΔE_n. Thus, difference in spectral resolution in terms of FWHM between synchrotron XPS and monochromated Al Kα XPS is no more than 0.25 eV. This is a comparison regardless of the heterogeneity of the sample surface.

In fact, a spectral resolution is also determined by the heterogeneity of chemical environment of atoms of an interested element such as rhodium. The heterogeneity of chemical environment could largely decrease the energy solution of the spectrum in terms of increase of ΔE. Here the heterogeneity is defined at two levels. One level is the coexistence of multiple chemical environments of an element at the same spot. For instance, the coexistence of rhodium oxide of surface layers and metal rhodium of the subsurface of a rhodium nanoparticle. Another level is the coexistence of multiple types of nanoparticles containing the same element at different chemical environments. For example, the sample is a mixture of rhodium particles, rhodium sulfide particles, and rhodium fluoride particles. In many cases, a sample has the two levels of heterogeneity. The comparisons in the above paragraph did not involve the issue of the heterogeneous chemistry of the sample. In fact, the better energy resolution of synchrotron XPS can only excel in studies of single crystal model catalysts such as Pt(111) since Pt(111) exhibits relatively high homogeneity. It does not necessarily give an obviously better resolution for an element in quite heterogeneous chemical environments. For instance, the FWHM of Rh $3d_{5/2}$ spectrum of a sample consisting of metal rhodium/rhodium oxide core/shell structured particles, rhodium sulfide particles, and rhodium sulfide particles obtained from a synchrotron XPS is similar to that obtained with a monochromated Al Kα XPS from the same sample.

A significant advantage of synchrotron XPS is the tunability of X-ray energy that allows for nondestructive depth-profile analysis through varying X-ray energy and for examination of the topmost atomic layers with low X-ray energy. Compared to Al Kα beam, the much smaller beam spot of synchrotron radiation can offer much better spatial resolution in X-ray photoelectron imaging analysis of elemental distribution on a sample surface. In addition, the much higher flux density of synchrotron radiation allows quick scan of the sample surface, which is significant for tracking the dynamics of a surface triggered by an external simulation such as heating, photoirradiation, pulse, charge transfer, type of gas, gas pressure, mechanic pressing, and electric or magnetic fields.

4.4 Detector

Detector of electrons used in XPS studies has been developed significantly in the past few decades. An energy analyzer separates electrons and measures their KEs. The number of electrons with a specific KE such as 1250.0 eV is measured by the detector of electrons. It is impossible at least by now to directly measure the number of electrons leaving exit of an energy analyzer without any magnification. In fact, the detector measures the number of electrons with a specific KE after a magnification since what matters is the relative intensities of peaks collected from the same sample. To make sure that the measured relative intensity of electrons of different elements at the detector can represent the relative intensity of these photoelectrons leaving the sample surface, the same magnification should be applied to all photoelectrons regardless of their kinetic energy when they entered the detector(s).

One type of detector is the electron multiplier (EM). It is operated by making electrons leaving the exit of an energy analyzer impinging onto the surface of an EM. Based on the shape or arrangement of EM, it can be dynode EM, channel EM, or microchannel EM. In terms of dynode EM, external electrons first strike the dynode, by which numerous secondary electrons are generated. Driving by the first acceleration electrode, these secondary electrons will be accelerated to strike the conversion dynode to produce more secondary electrons that will be accelerated by the second acceleration electrode to have third strike to generate many more secondary electrons and so forth. These multiple strikes in the specific geometry of this EM (Figure 4.5a) can gain a magnification of 10^6 or more for a coming electron per second. In *Y*-axis of an XPS spectrum, it is shown as 10^6 counts per second (cps). A channeltron EM (Figure 4.5b) uses the same principle to magnify the number of electrons. It looks like a horn. The larger opening faces to the exit of an analyzer. Electron emission materials with a low working function plated on the internal surface of the EM. When a coming electron strikes the region near the large opening, many electrons are generated. Due to cone-like shape of this EM, these electrons will strike other regions a little further from the large opening; such strikes occur multiple times until the generated electrons leave the channeltron EM. To drive these electrons to strike the internal surface, a voltage of a few kilovolts or higher is applied at the two openings of Figure 4.5b. A channeltron can offer a magnification up to 10^8 cps. The horn-like shape (Figure 4.5b) allows to place a series of channeltron EMs in parallel along certain orientation.

The dynode EM (Figure 4.5a) and channeltron EM (Figure 4.5b) can be considered as zero-dimension and one-dimension detector, respectively. Compared to dynode and channeltron EMs (Figure 4.5a and b), a 2D EM (Figure 4.5c) provides more functions and much higher magnification. A multichannel plate (MCP) is a 2D arrangement of glass capillaries;

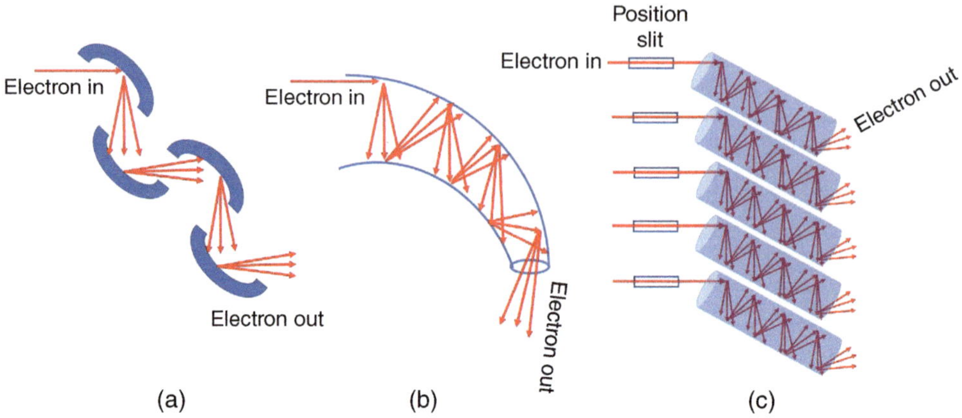

Figure 4.5 Schematic of three types of devices including (a) dynode, (b) channeltron, and (c) multichannel plate (MCP) for magnifying the number of electrons received at the exit of energy analyzer.

each of these capillaries is an EM such an EM (Figure 4.5c) is much smaller than a conventional channeltron (EM). It is at the dimension of a couple of tens of micrometers. Electron emissive material is deposited on the internal surface of the capillary for allowing electrons to strike under certain high voltage to achieve electron cascade. It is important to note that the magnification of an EM varies largely on the KE of the coming electrons received from the exit of energy analyzer. In order to make the measured relative intensity at different KEs represent the ratio of number of these electrons leaving the sample surface, it is necessary to accelerate or decelerate their original KEs to the same pass energy before these electrons strike the detector. In other words, for a quantitative analysis of surface concentration of an element or atomic ratio of two elements, it is necessary to use CAE mode instead of CRR mode to separate and collect electrons in HSA analyzer. Notably, the 2D array of MCP offers unique function in imaging the distribution of a specific element without scanning the sample or X-ray beam because MCP is already a type of position-sensitive detector (PSD).

Through these EMs, a small number of electrons leaving exit of HSA are magnified. In these EMs, the electron cascade results in electrical current pulse. The electrical current pulse can be collected with a phosphor screen, anode collector, or the latest developed delay line detector (DLD). In terms of phosphor screen, it measures the intensity of photons generated from impacts of secondary electrons produced from EMs; typically, a charge coupled device (CCD) camera system consisting of an optical lens and camera can image the signal in terms of the intensity of photons from a phosphor screen.

References

1 Baer, Y., Busch, G., and Cohn, P. 1975. "Monochromatized X-ray source for high resolution electron spectroscopy." *Rev. Sci. Instrum.* 46, 466.

2 Gelius, U., Asplund, L., Basilier, E. *et al.* 1984. "A high resolution multipurpose ESCA instrument with X-ray monochromator." *Nucl. Instrum. Methods Phys. Res., Sect. B* 1, 85–117.

3 Wittry, D. B. and Barbi, N. C. 2001. “X-ray crystal spectrometers and monochromators in microanalysis.” *Microsc. Microanal.* 7, 124–141.

4 Tao, F., Nguyen, L., and Tang, Y. 2022. “Report on photoelectron-induced excitation spectroscopy and its close correlation with fundamental studies of catalysis.” *J. Phys. Chem. C* 126, 13069–13087.

5 Johann, H. H. 1931. “Intense X-ray spectra obtained with concavecrystals.” *Z. Phys.* 69, 185–206.

6 Johansson, T. 1933. “New focusing X-ray spectrometer.” *Z. Phys.* 82, 507–528.

7 Castaing, R. 1952. “Application of electron probes to local chemical and crystallographic analysis.” PhD diss., University of Paris.

8 Wittry, D. B. and Sun, S. 1990. “X-ray optics of doubly curved diffractors.” *J. Appl. Phys.* 67, 1633.

9 Birks, L. S. 1971. *Electron Probe Microanalysis*. 2nd edn, Wiley-Interscience, 43.

10 Dumond, J. W. M. and Kirkpatrick, H. A. 1930. “The multiple crystal X-ray spectrograph.” *Rev. Sci. Instrum.* 1, 88.

11 Wolstenholme, J. 2015. *Auger Electron Spectroscopy: Practical Application to Materials Analysis and Characterization of Surfaces, Interfaces, and Thin Films*. Momentum Press.

12 Sise, O. and Zouros, T. J. 2015. “Position, energy, and transit time distributions in a hemispherical deflector analyzer with position sensitive detector.” *J. Spectrosc.* 2015, 153513.

5

Significance and Challenge of Studying Surface of a Catalyst in Gaseous Phase

5.1 Origin of Difference between Surface in UHV and Surface in Reactant Gas

Traditional vacuum surface science studied surfaces with adsorbed molecules in ultrahigh vacuum. Such a surface with adsorbed molecules is typically prepared by exposing a clean surface to certain amount of gas before a characterization to the surface in high vacuum or UHV is performed. The amount of the gas introduced is calculated with exposure. The exposure is defined by the product of multiplying the pressure of the gas exposed to the clean surface (in Torr) and the exposing time (in second). The unit of exposure is Torr s but it is typically presented in another unit, Langmuir (L), which is 1×10^{-6} Torr • s. Through dividing the exposure in Torr s by 1×10^{-6} Torr • s L^{-1}, the exposure is presented with the unit, Langmuir or L. Upon an exposure of a surface to gas the gas was subsequently purged to restore HV or UHV before a characterization is performed in HV and UHV. Such a surface with adsorbates in HV or UHV can be readily characterized with a great number of vacuum-based surface analytical techniques such as XPS and AES.

Compared with the above surface with adsorbed molecules in HV or UHV, a surface in gas phase is different in many ways.[1–3] One reason for the difference is the entropy contribution from gas phase to thermodynamics of the surface. This difference can be rationalized with the following analysis.[4,5] Chemical potential of a gas phase around a catalyst can be expressed as a function of gas pressure by the equation: $\mu_{gas}(T, p) = \mu^{o}_{gas}(T) + k_B T ln\frac{p}{p^{o}}$, where $\mu^{o}_{gas}(T)$ is the standard chemical potential of gas A. The chemical potential of a catalyst surface can be expressed with the equation: $\mu_{surf}(T,p) = N_M \mu_M + N_A \mu_A$, in which N_M is the number of exposed surface atoms of a catalyst and N_A the number of adsorbates (molecules or dissociated species); μ_M and μ_A are chemical potentials of a surface atom and adsorbate, respectively. Obviously, a change of pressure of gas A varies chemical potential of the gas phase around the catalyst surface. From thermodynamics point of view, a consequence of the variation of chemical potential of the gas phase surrounding the surface is to trigger a corresponding change of the chemical potential of the surface, $\mu_{surf}(T, p)$, since the gas phase is the source of adsorbed molecules and the gas phase is equilibrated with the

Application of Ambient Pressure X-ray Photoelectron Spectroscopy to Catalysis, First Edition. Franklin Tao.

surface adsorbate buried in the gas phase. A variation of surface chemical potential could be realized by change of one or more of the four variables including N_M, μ_M, N_A, and μ_A. For instance, along the increase of pressure of a reactant gas A, the surface coverage of molecules A or their derivative represented by N_A could increase, which was illustrated in the increase of CO coverage on a catalyst surface along increase of CO pressure; the number of exposed surface atoms of M denoted with N_M could increase, which was demonstrated in the reported restructuring of stepped Pt(111) surface along the increase of CO pressure at room temperature;[1] the chemical potential of an atom of the topmost layer on the surface, μ_M could change, which was exemplified in the change of coordination environment of atoms of the restructured Pt(557); the chemical potential of an adsorbed gas molecule or species derived from a gas molecule denoted with μ_A could change, which was demonstrated in the change of binding configuration of adsorbed CO molecules from the bridge to atop on the restructured Pt(557) along the increase of CO pressure.

5.2 Intrinsic Feature of Catalytic Sites on Surface: Environmental Sensitivity

Compared to a catalyst surface at room temperature in a reactant gas, the surface of a catalyst during catalysis could be different, driven by a different temperature of the catalyst or/and a different pressure or even composition of a gas.[6] Heterogeneous catalysis is typically performed in the mixture of reactants at a temperature >200 °C. Furthermore, in contrast to a catalyst at room temperature in UHV or at a catalysis temperature in UHV, the surface of a catalyst at a catalysis temperature in the mixture of reactants could be quite different. The difference can be minor for an inert surface such as a silica or alumina. However, it can be quite obvious for a metal surface.[1–3] In terms of heterogenous catalysis, a catalytic site consists of one, a few, or even several atoms of the surface of a particle or of the interface consisting of atoms from both a particle and its support. As a catalytic site directly participates in a reaction (A + B → Products), the catalytic site must form chemical binding with reactant molecules. Thus, these atoms of a catalytic site are typically undercoordinated to some extent during catalysis. These undercoordinated atoms are highly active. The chemical and coordination environments of these atoms can be retained only by some parameters such as the catalysis temperature or/and the gas mixture of the reactants at certain pressure.

Withdrawal of some or all the parameters of a catalytic condition (typically catalysis temperature, gas pressure, and gas composition) could make surface structure of the catalyst obviously different, called restructuring. Withdrawal of partial catalytic condition means removal of one or more but not all parameters of a catalytic condition. For instance, a semi-in-situ study was performed at room temperature in a mixture of reactants of the catalysis after cooling the catalyst from the catalysis temperature to room temperature. In fact, such a withdrawal of a part of a catalytic condition could have changed or created a structure different from the authentic catalyst surface during catalysis. In many cases, temperature is a key factor for remaining authentic surface of the catalyst during catalysis. Most catalytic reactions are performed at a relatively high temperature or even quite high temperature.[7] Thus, even if a catalyst is remained in the mixture of reactants, the surface structure of the catalyst at room temperature could be distinctly different from the authentic one during catalysis (high temperature in mixture of reactants).

It is true that surface of a catalyst at room temperature in UHV could never represent the surface of a catalyst during catalysis although it could still be some helpful in explaining the catalytic performance. A widely adopted approach in literature is the characterization of a used catalyst at room temperature in UHV (or in ambient environment). In this approach, a used catalyst is taken out from a reactor after the catalytic condition is completely withdrawn and then is (a) placed into a UHV environment that meets the requirement of many electron-based analytical techniques such as X-ray photoelectron spectroscopy (XPS) and transmission electron microscopy (TEM), or is (b) placed to ambient environment that fits many characterizations such as XRD or FTIR. In fact, this approach in terms of ex situ studies could introduce an additional change to the catalyst surface structure. In other words, exposure of the catalyst surface to air could make these undercoordinated metal atoms of a catalytic site be immediately oxidized by O_2, hydroxylated by H_2O, or carbonated by CO_2. Obviously, such an expected change just invalided the hypothesis that (1) surface of a catalyst before a catalysis or (2) surface of a catalyst at room temperature in UHV or air after catalysis represents the surface of the catalyst during catalysis. In brief, most authentic catalyst surfaces are highly sensitive to environment, which is an intrinsic feature of the active catalytic sites of a catalyst surface.

5.3 Ex Situ, Semi-in Situ, and In Situ/Operando Studies of Catalyst Surface at Ambient Pressure of Reactants

5.3.1 Difference among Ex Situ, Semi-In Situ, and In Situ/Operando Studies

Most materials function in ambient environment or a specific gaseous environment. Thus, the characterization of a material is assumed to be performed in working environment of the material such as the gaseous environment of reactants at high temperature. The feature of operando characterization is that the characterization of a material is performed when the material is functioning in its working environment. However, conventionally most electron-based techniques can only characterize a metal in UHV environment. Thus, a sample must be characterized in UHV environment although such a sample functions in gas phase or/and at high temperature. This compromised characterization is called ex situ characterization. As described in Section 5.2, surface of a catalyst in UHV or/and at room temperature can be very different from that under its working condition since the surface of a catalyst is sensitive to environment and it is even maintained by its working environment. Thus, without any doubt, in situ or operando characterization is necessary.

The differences among three types of characterizations which are ex situ, semi-in situ, and in situ/operando characterizations are schematically presented in Figure 5.1. Most catalytic reactions request a high temperature for achieving a high reaction rate from kinetics point of view. In addition, a high temperature is favorable for an endothermic reaction. Based on the gas pressure of reactants during catalysis, we can categorize them into ambient pressure reactions (Figure 5.1a) and high pressure reaction (Figure 5.1b). Here, "high pressure" is defined as a pressure higher than 1 atm.

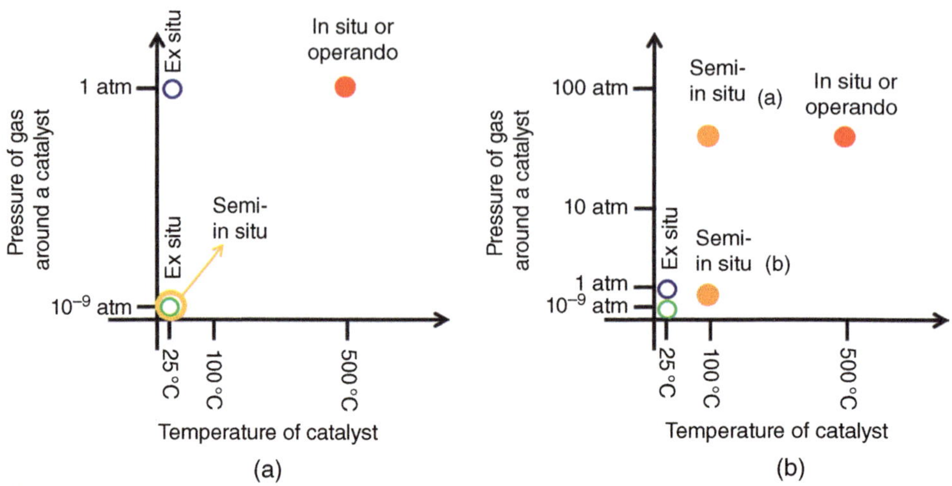

Figure 5.1 Schematic presentation of ex situ studies (empty green and blue circles), semi-in-situ studies (unfilled or filled yellow circles), and in situ/operando study (filled red circle) as a function of temperature of catalyst and pressure of reactant gases during a characterization of a catalyst. (a) Ambient pressure reaction. (b) High pressure reaction. *Source:* Reproduced with permission from Tang et al.[8]/AIP Publishing.

An in situ characterization or an in situ study is defined as a characterization of a catalyst during catalysis without simultaneous measurement of its kinetics. In addition, an operando characterization or an operando study is referred to a characterization of a catalyst during catalysis while catalytic kinetics is being studied or reaction rate is being measured. The difference between in situ and operando is whether there is a simultaneous measurement of the kinetics. Under an assumption that the catalysis represented by in situ/operando studies is performed under a kinetics-controlled regime the in situ and operando characterizations are the same in nature. In other words, the structural information obtained from in situ characterization is the same as operando characterization when the catalysis is being performed under a kinetics-controlled regime although its kinetics is not being measured in in-situ studies. Thus, we do not distinguish in situ and operando characterizations here. The red spot in Figure 5.1 exemplifies the condition of in situ or operando characterization of a catalyst working at high temperature in the mixture of reactant gases of a reaction at approximately 1 atm (Figure 5.1a) or high pressure (Figure 5.1b).

In terms of ambient pressure catalytic reaction, historically ex situ study is the main approach to characterize catalysts. It is typically done at room temperature in ambient environment for photon-in–photon-out-based analytical techniques such as XRD or/and in UHV environment for electron-based analytical techniques such as XPS and TEM. Ex situ study is the most convenient characterization since it is readily done by placing a fresh catalyst (before catalysis) or a used catalyst (after catalysis) to the characterization instrument without introduction of gas of reactants. Compared to ex situ study, a catalyst can be characterized by transferring it to a characterization instrument in UHV or still remaining in gas of reactants while the catalyst is at room temperature. The feature of such a characterization is to prevent the catalyst from being exposed to ambient environment; this type of characterization is called semi-in situ study as marked with unfilled yellow circle in Figure 5.1a; in practice, it can be done through connecting a reaction cell to the characterization instrument (such as a UHV-XPS) while a gate value is placed between the cell and characterization instrument; before catalysis, the gas environment in the reaction cell is isolated from high vacuum of instrument through the gate valve. After catalysis, the catalyst is cooled down to room temperature in reactant gas and then the reactant gas is purged, and the gate valve is opened to allow the sample to be transferred to the characterization instrument (such as UHV-XPS) which is in HV or UHV. Although this semi-in situ approach can prevent a catalyst surface from being oxidized by O_2 in air, without an in situ or operando study it is still hard to know whether the information provided by such a semi-in situ study could represent the authentic surface of the catalyst during catalysis.

5.3.2 Example of Surface Structures Only Formed and Maintained by Reactant at a Relatively High Pressure

It is true that a semi-in situ study cannot provide information on surface structures of some catalysts in reactant gases in many cases. Particularly, it can not provide information on the authentic surface of a catalyst whose structure is pressure or/and temperature dependent. Our previous studies uncovered that the surface structure of a model catalyst Pt(557) significantly restructures along increase of pressure of CO gases.[1] More importantly, its surface structure changes back to its original structure after CO gas is purged. As shown in Figure 5.2, along the increase of CO pressure from 5×10^{-8} to 1 Torr, the catalyst surface is massively broken from curl step edges (Figure 5.2b) to nanocluster surface (Figure 5.2d). Surprisingly, once 1 Torr CO gas is purged to 1×10^{-8} Torr, the nanocluster surface (Figure 5.2e) reverts to a step surface with curl edges (Figure 5.2f) that is highly similar to the surface early formed in the CO gas at 5×10^{-8} Torr (Figure 5.2b). It suggests the surface structure of Pt(557) is pressure dependent in CO even at room temperature and the restructuring from a step surface to nanocluster surface is reversable. This pressure-dependent reversible restructuring clearly shows the nanocluster surface is maintained by the gas pressure at a pressure of 1 Torr or higher. Clearly, the nanocluster surface cannot be observed through a semi-in situ study (Figure 5.1a). The reason is that the surface formed at 1 Torr (Figure 5.2e) reverts to the high vacuum structure (Figure 5.2f) once CO is purged before a semi-in situ study. The surface information achieved through a semi-in situ study (Figure 5.2f) is definitely different from the authentic surface information of Pt(557) in 1 Torr CO (Figure 5.2d or e).

Ambient pressure X-ray photoelectron spectroscopy (AP-XPS) studies confirmed the above conclusion that the semi-in situ studies cannot represent the surface chemistry of Pt(557) in 1 Torr CO after the CO gas in the reaction cell is purged.[1] As shown in Figure 5.3, the photoemission feature of Pt 4f evolved significantly along the increase of pressure of CO gas around Pt(557) (Figure 5.3a–d). At 0.5 Torr, the pressure-dependent peak C is over 1/3 of the intensity of the overall Pt $4f_{7/2}$ spectrum (Figure 5.3d1). Interestingly, the Pt 4f photoemission feature changed largely after the CO gas was purged to 2×10^{-8} Torr (Figure 5.3e). In fact, the photoemission feature of Pt 4f at 2×10^{-8} Torr (Figure 5.3e) is quite similar to Figure 5.3b which was collected in CO at 5×10^{-9} Torr. Figure 5.3e is the data of a semi-in situ study. Definitely, the data collected from the semi-in situ study in 1×10^{-8} Torr CO (Figure 5.3e) is distinctly different from that in 0.5 Torr (Figure 5.3d). Interestingly, when the gas pressure of the CO environment was filled up to 0.5 Torr, the

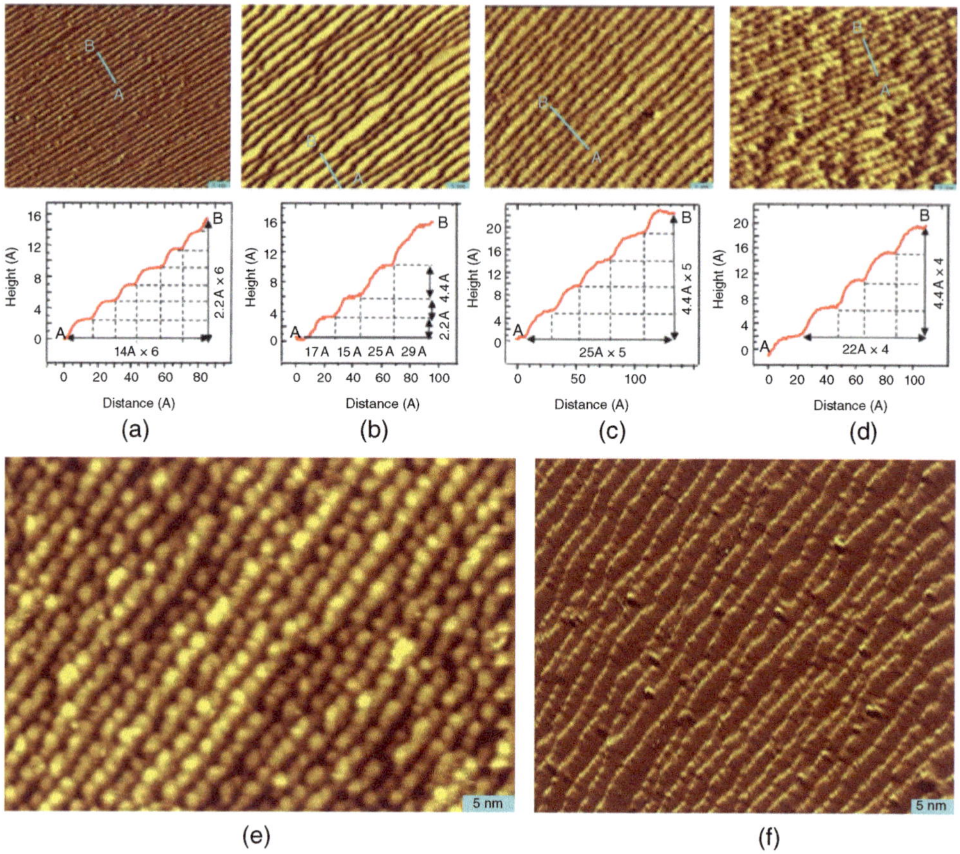

Figure 5.2 Surface structure of Pt(557) catalyst in UHV (a), 5×10^{-8} Torr CO (b), 5×10^{-7} Torr CO (c), 1 Torr CO (d), 1 Torr CO (e), and 1×10^{-8} Torr CO (f) collected with high pressure scanning tunneling microscope (HP-STM); (d) and (e) were collected at the same pressure 1 Torr CO but different area of Pt(557). The plots below (a), (b), (c), and (d) are the line profiles from point A to B marked in the image (a), (b), (c), and (d), respectively. *Source:* Tao et al.[1] Reprinted from American Association for the Advancement of Science – AAAS.

portion of component C increased immediately (Figure 5.3f1). Thus, clearly, the surface structure of Pt(557) is definitely pressure dependent. The nanocluster surface of the Pt(557) formed at 0.5–1 Torr cannot be identified through a semi-in situ study since it is only observable in 0.5 Torr instead of 10^{-9}–10^{-8} Torr CO. This is the earliest example suggesting that the structural information uncovered by a semi-in situ study cannot represent that from an in situ study.

5.3.3 Example of Catalyst Structure Only Observable during Catalysis

Other than the reactant pressure-driven restructuring of catalyst surface, the recent work shows that pressure of a product can be the key factor for maintaining an active catalytic phase, based on both computational and experimental studies of the catalyst structure at atomic scale during catalysis of steam reforming of methane, $CH_4 + H_2O \rightarrow CO + 3H_2$.[7]

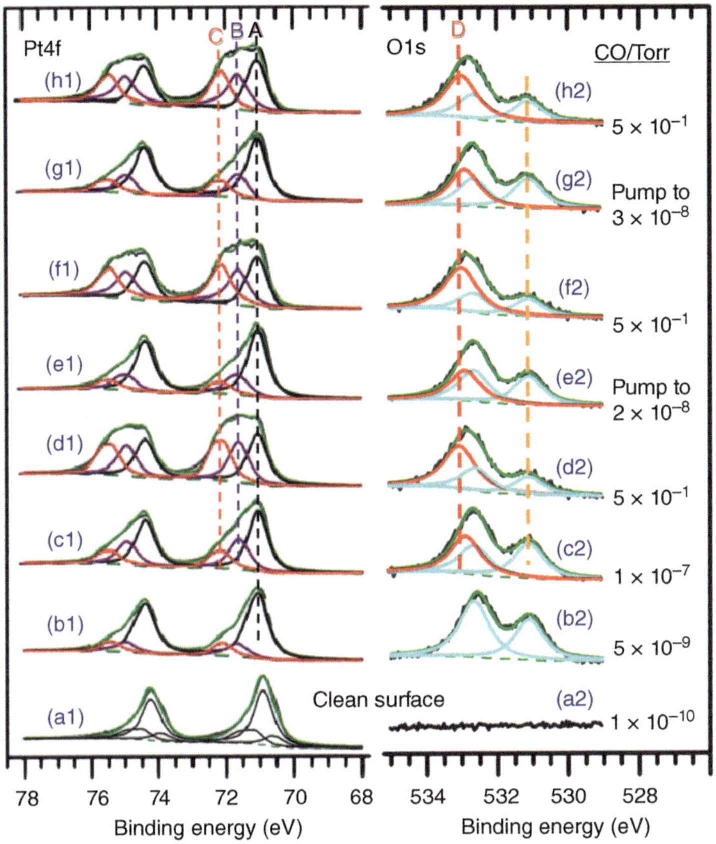

Figure 5.3 AP-XPS peaks of Pt 4f and O 1s of Pt(557) catalyst at 25 °C in UHV or CO at different pressure. (a) UHV, (b) 5×10^{-9} Torr CO, (c) 1×10^{-7} Torr CO, (d) 5×10^{-1} Torr CO (in-situ study), (e) purged to 2×10^{-8} Torr CO (semi in-situ study), (f) refilling to 0.5 Torr CO, (g) purged to 2×10^{-8} Torr CO, (h) refilling to 0.5 Torr CO. Each Pt 4f7/2 was deconvoluted to three components A, B, and C. Component C is contributed from under-coordinated Pt atoms that are Pt atm at step edge and Pt atoms at edge of nan[1] oclusters formed at 0.5 Torr. *Source:* Reproduced with permission from Tao et al.[1]/American Association for the Advancement of Science.

The thermodynamics computational studies show that the size and coordination environment of Rh atoms in $Rh_m(CO)_n$ nanoclusters depend on temperature of catalyst and pressure of product CO in gas environment where the catalyst is placed. As shown in Figure 5.4a, in the right-bottom region the pressure of product CO is relatively high and the temperature of catalysis is relatively low; in this phase region, the thermodynamically stable phase is single-atom site stabilized by two coordinated CO molecules in $Rh_1(CO)_2$ as shown in Figure 5.4g. Along the decrease of CO pressure or/and increase in catalysis temperature, the $Rh_1(CO)_2$ restructures to $Rh_3(CO)_4$, $Rh_3(CO)_3$, and $Rh_3(CO)_2$, which follow an ordering of decreasing molar ratio of CO to Rh in $Rh_m(CO)_n$. It suggests the catalyst structures at atomic scale vary along the decrease of pressure of CO; here, CO is one of the products of steam reforming of CH_4. In other words, the catalytic site, $Rh_m(CO)_n$ has to be stabilized under a specific pressure and temperature. Based on Figure 5.4a, CO-free Rh nanoclusters are expected to form when CO is purged. The in situ/operando spectroscopy

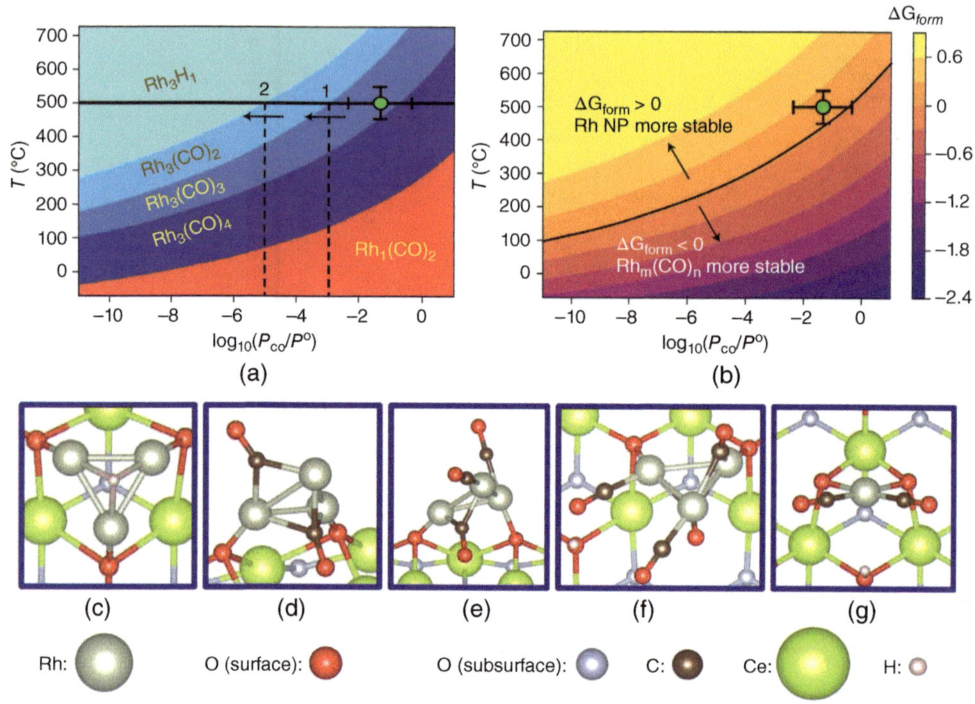

Figure 5.4 Structure and stability of Rh active sites. (a, b) Theoretical modeling of the evolution of the Rh active site under catalytic conditions as a function of catalysis temperature (*T*) and CO pressure (P_{CO}) by holding the chemical potentials of gas phase H_2 and H_2O constant as those during steam reforming of methane (SRM) at 500 °C. (a) Theoretically proposed stable Rh species; the SRM condition at 500 °C (P_{CO} = 0.045 bar) is denoted by a green dot, with bars representing potential variation in *T* and P_{CO}. As P_{CO} decreases, $Rh_3(CO)_3$ is transformed to $Rh_3(CO)_2$ and then Rh_3H_1. (b) Gibbs free energy of formation (ΔG_{form}) of the most stable Rh-based sites as a function of catalysis *T* and P_{CO}; solid black line denotes transition between supported Rh nanoparticles and highly dispersed $Rh_m(CO)_n$ clusters, below which $Rh_m(CO)_n$ species are more stable than Rh NPs. Here ΔG_{form} denotes the formation energy of these active sites. (c–g) Atomic-scale geometries of Rh sites present in (a): Rh_3H_1 (c), $Rh_3(CO)_2$ (d), $Rh_3(CO)_3$ (e), $Rh_3(CO)_4$ (f), and $Rh_1(CO)_2$ (g). *Source:* Reproduced with permission from Yan et al.[7]/Springer Nature.

in Figure 5.5 confirmed that, once product CO is formed and reach a pressure at 0.01 atm or so at 500 °C, $Rh_3(CO)_3$ was identified with diffuse reflectance infrared Fourier transform spectroscopy (DRIFTS) (Figure 5.5b1). This is consistent with the results of density functional theory (DFT) calculation (Figure 5.4a and e). More importantly, the in situ/operando spectroscopy confirmed that the $Rh_3(CO)_3$ did restructure along the decrease of CO pressure driven by replacing CH_4 with flowing Ar. The vibrational frequency of CO decreases along the decrease of CO pressure and eventually all CO vibrational features disappeared (Figure 5.5b8 and b9). This is consistent with computational results that $Rh_3(CO)_3$ at 500 °C is only stable in CO gas with a pressure higher than 0.001 atm and transferred to CO-free Rh nanocluster when CO pressure is decreased. *Both Figures 5.4 and 5.5 show the authentic catalyst structure during catalysis cannot be observed in a semi-in situ study or a post-catalysis study.*

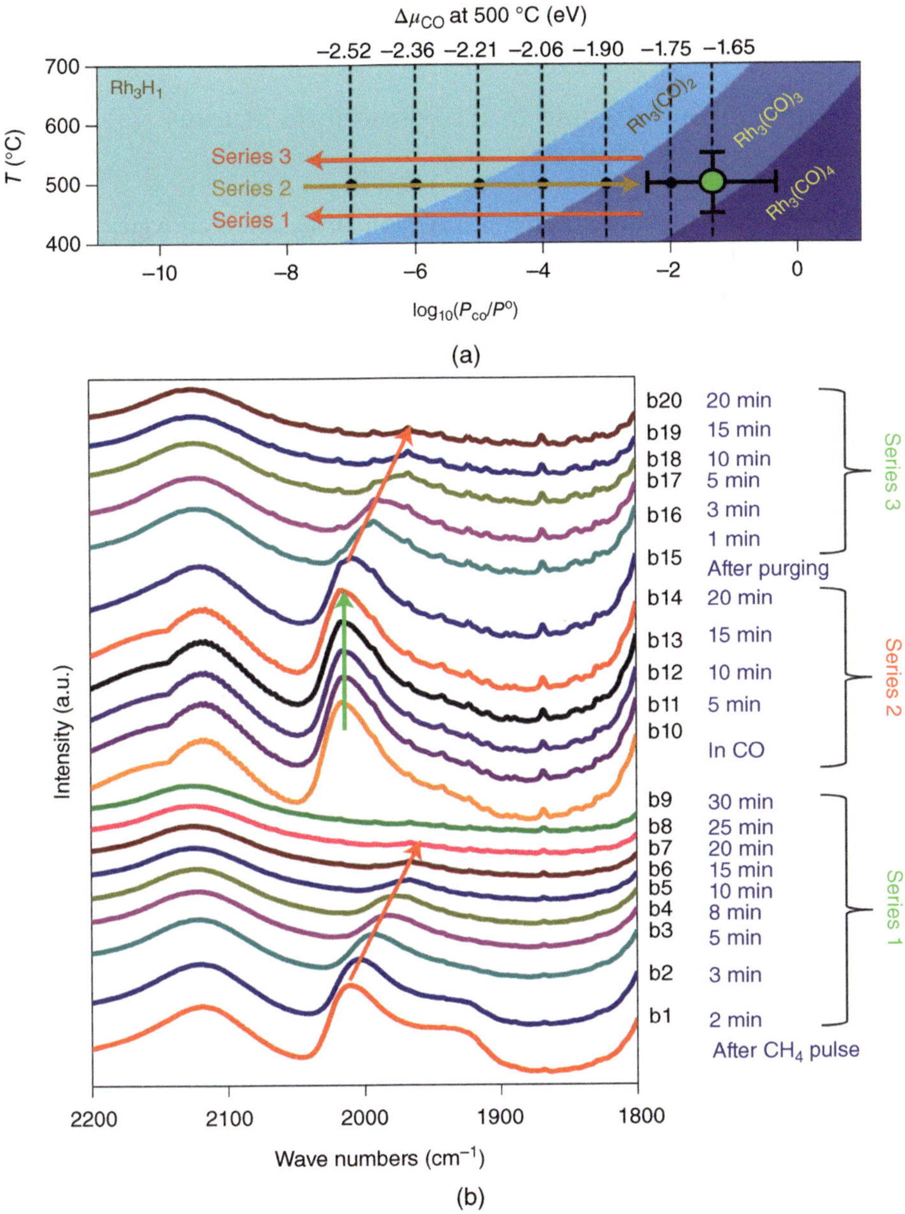

Figure 5.5 Reversible restructuring of Rh-based active sites for steam reforming CH_4. (a) Three series of variations of CO pressure at 500 °C marked on the theoretical P_{CO}-T phase diagram. (b) DRIFTS spectra at 500 °C under SRM catalytic condition (b1–9) and under other conditions (b10–20). Series 1: b1–9, a transient pulse of CH_4 reactant (equivalent to 0.1 ml CH_4, at 25 °C, 1 atm) composed of 1% CH_4 and 99% Ar, was introduced to the catalyst by mixing the pulse with flowing water vapor formed by flowing Ar (99.999%) through liquid water at 25 °C. No external CO source was introduced before and during series 1. Series 2: b10–14, spectra in a flow of CO (1%) balanced with Ar. Series 3: spectra (b15–20) being purged with Ar as a function of time. The background spectrum for the DRIFTS study was collected on the catalyst surface under Ar and water vapor flow at 500 °C. In series 1, 20 ml/min water vapor containing 3% vapor was continuously flowed through the reaction cell; CH_4 gas (1% CH_4 in Ar) was pulsed to mix with water vapor. DRIFTS spectra were taken as a function of time without CH_4 in the flow. At t = 2 min (b1), the vibrational features in the 2050–1850 cm^{-1} region, assigned to $Rh_3(CO)_3$, were clearly observed because CO was generated from SRM at 500 °C. These experiments show the formation of $Rh_3(CO)_3$ to be reversible, and it is maintained by high temperature and CO pressure (a.u., arbitrary units). *Source:* Reproduced with permission from Yan et al.[7]/Springer Nature.

5.4 Ex Situ, Semi-in Situ, and In Situ/Operando Studies of Catalyst Structure at High Pressure

The above is for the catalysis reaction performed at 1 atm. In fact, there are a great number of reactions performed at a pressure higher than 1 atm even up to hundreds of atms. The red spot in Figure 5.1b exemplifies the condition of in situ or operando characterization of a catalyst working at high temperature in mixture of reactants at high pressure (HTHP).

One of the core tasks in the field of catalysis performed at an HTHP condition is the fundamental understanding of catalytic mechanism at a molecular level because such a profound understanding would inspire a rational design of a catalyst with a better catalytic performance under HTHP condition. However, this task has been always challenging since in many cases there has been a lack of in situ or operando characterizations of a catalyst during catalysis under HTHP conditions.

Fischer–Tropsch synthesis (FTS) and ammonia synthesis are two well-known examples of catalysis at HTHP. An in situ/operando study of high temperature high pressure (HTHP) catalysis is more challenging than catalysis at 1 atm. The semi-in situ characterization is a compromised approach for characterization of catalysts working at HTHP conditions. There are at least two types of semi-in situ characterizations (the filled yellow circles in Figure 5.1b) frequently used in literature. To easily understand the difference between a semi-in situ characterization and an in situ characterization, an extensively studied reaction, FTS, is taken as an example to suggest the difference between semi-in situ study and in situ/operando study of HTHP catalysis. One type of semi-in situ characterization is a characterization of an FTS catalyst in H_2 with a pressure of 1 bar H_2 or a mixture of H_2 and CO at a temperature near room temperature (25–100 °C) immediately upon cooling the catalyst from pretreatment temperature. Another type of semi-in situ study is the characterization of an FTS catalyst at a temperature near room temperature (25–100 °C) in the high pressure of H_2 or high pressure mixture of CO and H_2 after an FTS is performed at HTHP. One obvious difference between the two types of semi-in situ studies is the pressure (ambient pressure versus high pressure) between the two semi-in situ characterizations. Notably, neither of the two semi-in situ characterizations of FTS is a characterization of a catalyst during catalysis since the catalysis is performed at high temperature of catalyst in a mixture of reactants at high pressure. The two types of semi-in situ studies do not necessarily give the same information as the structure of a catalyst under an HTHP condition since high temperature or high pressure of reactants could restructure surface or even bulk structure of catalyst nanoparticles. Thus, without an in situ or operando characterization of an FTS catalyst during catalysis at HTHP conditions, it is hard to exclude any potential difference between structure of a catalyst offered from a semi-in situ characterization and the authentic structure of a catalyst during catalysis under an HTHP condition. In fact, our recent studies using XANES and EXAFS show that the catalyst structure under a semi-in situ study at 300 °C in flowing H_2 (Figure 5.6b2) and the authentic catalyst structure during FTS catalysis at 300 °C in the mixture of 12 bar H_2 and 6 bar CO (Figure 5.6b3) are different. The *r*-space spectrum of Ru K-edge in Figure 5.6b suggests a new phase formed on surface of the supported Ru nanoparticle during catalysis in a flowing mixture 12 bar H_2 and 6 bar CO at 250 °C. The formation of a potential new phase was suggested by the observation of a new peak at 1.96 Å marked with a red dashed line in Figure 5.6b3, which was not observed in the semi-in situ study (Figure 5.6b2).

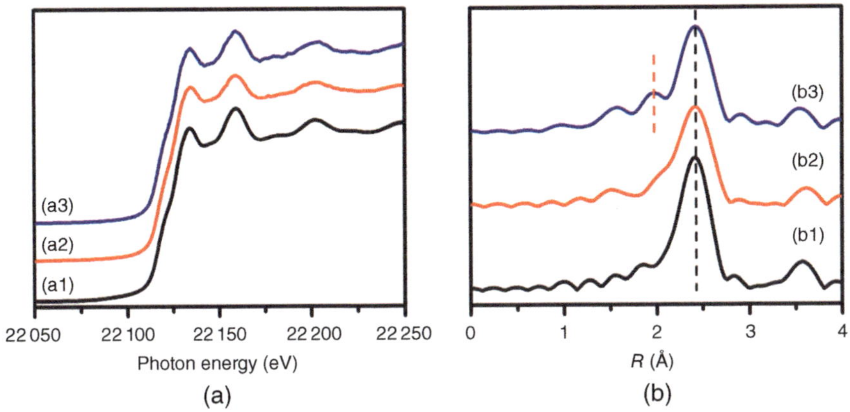

Figure 5.6 Operando studies of Ru catalyst for Fischer–Tropsch synthesis using XANES (a) and EXAFS (b). (a) XANES spectra and (b) Fourier transformed EXAFS spectra of Ru K-edge of Ru catalyst; data were collected at the conditions as following: (a1) and (b1) Ru foil at ambient environment (at 25 °C in air); (a2) and (b2) Ru catalyst at 300 °C in flowing H_2 (20 ml/min at 1 atm); (a3) and (b3) Ru catalyst at 250 °C in flow mixture of 18 bar (12 bar H_2, 6 bar CO with a total flow rate of 30 ml/min). *Source:* Reproduced with permission from Tang et al.[8]/AIP Publishing.

5.5 Technical Challenges in Studying Surface of a Catalyst in Gas Phase

Most characterization techniques can be roughly categorized into one of the four types of surface analytic techniques based on the incident beam of photon or charged particles and the detected photon or charged particles. Here the charged particles are typically electrons or ions in some cases. They are (i) analytical technique of photon-in and photon-out, (ii) analytical technique of photon-in and charged particle-out, and (iii) analytical technique of charged particle-in and photon-out, and (iv) analytical technique of charged particle-in and charged particle-out. For instance, DRIFTS and Raman spectroscopy belong to the type of photon-in and photon-out techniques; XPS is a technique of photon-in and charged particle-out; TEM is a technique of a charged particle-in and charged particle-out. To realize in situ or operando study of catalysts, techniques of incident charged particle-in or/and charged particle-out such as XPS and TEM face enormous challenges as catalysis typically performs on catalyst at a high temperature in a mixture of reactants at 1 atm or above. Obviously, a characterization using techniques of incident charged particle-in or/and charged particle-out at high temperature and high pressure can be extremely challenging. A main issue is the inelastic scattering of charged particles by gas molecules; such inelastic scattering largely decays the intensity of charged particles when they arrive at an analyzer of these charged particles. For instance, photoelectrons generated from a catalyst surface are largely inelastically scattered by gas molecules around the catalyst surface before these photoelectrons could be collected and analyzed by an energy analyzer of AP-XPS system. After traveling through a gaseous region at certain pressure and with certain thickness, only a portion of photoelectrons retain their original kinetic energy (Figure 5.7b). KE_0 is the original kinetic energy of photoelectrons generated from photoionization event of a catalyst atom before it loses any kinetic energy. I_0 is the intensity of photoelectrons

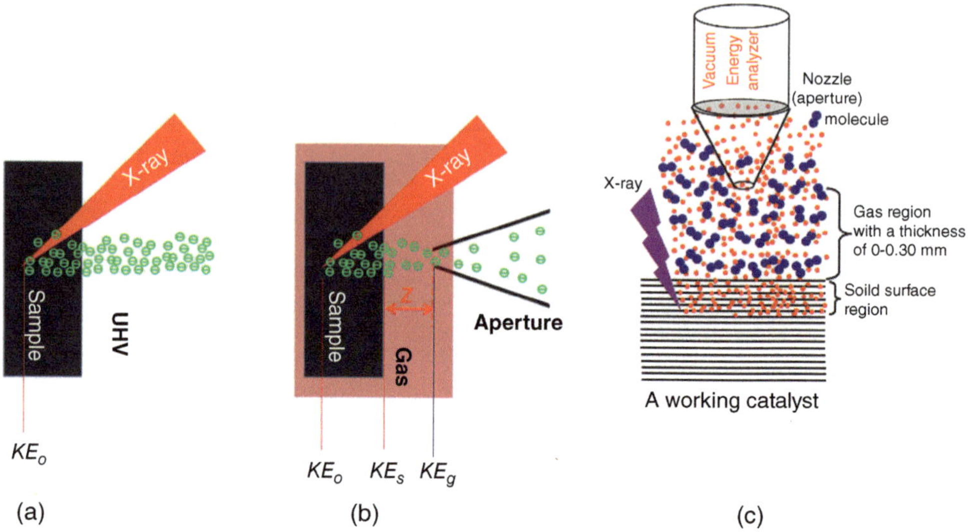

Figure 5.7 Schematic showing the difference in using a technique of photon-in charged particle-out in ultrahigh vacuum (a) and in gas phase (b and c). In (b and c), the photoelectrons generated from subsurface of a sample must scatter these atoms between their home atoms and the sample surface and then scatter gas phase molecules if the technique of photon-in charged particle-out is used for studying a solid surface in a gas phase. (c) A detailed version of (b); in (c), the scatterings of photoelectrons with atoms of the catalyst surface region and with molecules in gas phase are schematically shown. *Sources:* (a, b) Reproduced with permission from Tao and Nguyen[9]/Royal Society of Chemistry. (c) Reproduced with permission from Tao et al.[10]/American Chemical Society.

with KE_0 when they just leave the topmost atomic layer of a surface. KE_g is the kinetic energy of photoelectrons when they leave the gas phase region to enter the vacuum environment of an energy analyzer after traveling through the gas region. Notably, only photoelectrons whose KE_g equals to KE_0 can contribute to the signal of an XPS peak. The intensity of these photoelectrons leaving the gas phase remaining their kinetic energy (KE_0) is designated as I_p. Then, the portion of photoelectrons that elastically scatters gas molecules can be defined and evaluated with the following equation:

$$F_p = \frac{I_p}{I_0} = e^{-\frac{z\sigma_{KE} p}{kT}}$$

If F_p is larger, the intensity of a collected XPS peak intensity is higher. F_p is determined by the travel distance (z), gas pressure (p), gas temperature (T), and kinetic energy-dependent cross section (σ_{KE}). The thickness and pressure of the gas region are the key factors determining F_p. In UHV, F_p is nearly 100% as p is nearly zero. Notably, p is a factor we have to keep for catalysis since a catalyst must work in a gas at certain pressure. Thus, only z is a variable. It needs to be minimized for maximizing F_p. Figure 5.8 presents the exponential decay of F_p along the increase of thickness in terms of the travel distance in gas region and pressure of the gas region. Such an *exponential* dependence of F_p versus travel distance of electrons in gas phase, z or thickness of gas region requests a very accurate

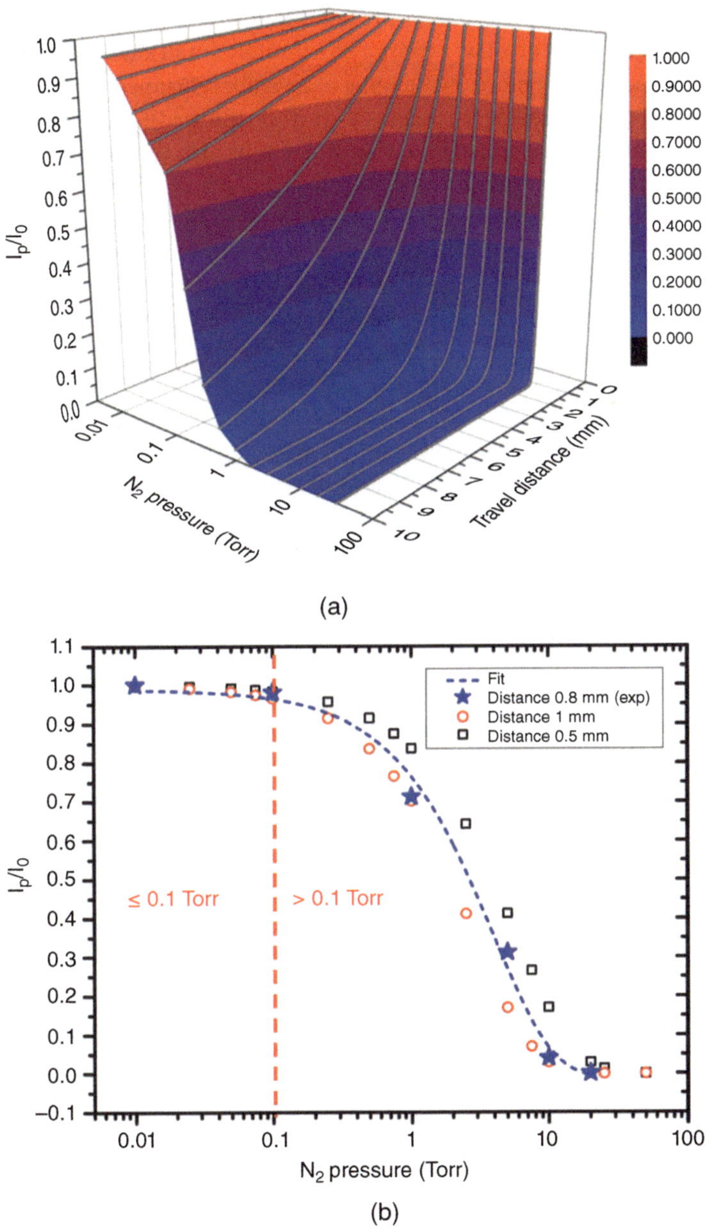

Figure 5.8 Attenuation of photoelectrons of 1000 eV traveling in the gas phase with different distances (0–10 mm) between sample surface and aperture in gas phase at different pressures (0.01–100 Torr). (a) Map of I_p/I_0 of 1000 eV photoelectrons as a function of pressure (0.01–100 Torr) and sample–aperture distance (0–10 mm) at 300 K. (b) Comparison of the experimentally measured attenuation (blue stars and the fitted blue dashed line) to the calculated attenuations of photoelectron intensity at a distance of 1.0 mm (red dots) and 0.5 mm (black squares) as a function of N_2 pressure in 0.01–100 Torr. An obvious feature is that there is no attenuation of photoelectron intensity when the pressure of gas phase is 0.1 Torr or lower. *Source:* Reproduced with permission from Nguyen et al.[9]/Royal Society of Chemistry.

control of the thickness of gas region in terms of the sample aperture distance, z, in Figure 5.7b. Thus, from technique point of view, z cannot be infinitely small since an accurate control of the distance is technically more challenging at a smaller distance.

Based on the above equation, the gas region thickness (z) must be thinner if the pressure in the gas region (p) is higher since F_p is proportional to e^{-zp}. To minimize the loss of photoelectrons when they travel through a gas region, the best travel distance of photoelectrons should be equal to or shorter than the inelastic scattering mean path of the photoelectrons in the gas phase, designated as λ_{gas}. λ_{gas} is determined by gas pressure, kinetic energy of the photoelectrons, and mass and structure of gas molecules. At sub-Torr pressure, λ_{gas} is a few millimeters (mm) by assuming the kinetic energy of photoelectrons is 1000 eV. At 1 Torr of N_2, it is approximately 1 mm. At 1 atm of N_2, it is only a few micrometers (μm). Technically, it is extremely challenging in remaining a gas region as thin as a few micrometers while the pressure of the gas phase is 1 atm. By using Al Kα or soft X-ray, it is extremely challenging to perform XPS analysis of a sample surface in gas at 1 atm. Simply speaking, this is because it is extremely challenging, if not impossible, to create a well-controlled gas region with a thickness of only a few micrometers; the challenge also results from shading effect of the front cone (or called aperture) near the sample surface and the lack of repeated, precise control of the distance between sample surface and the aperture.

References

1 Tao, F., Dag, S., Wang, L. W. *et al.* 2010. "Break-up of stepped platinum catalyst surfaces by high CO coverage." *Science* 327, 850–853.

2 Tao, F., Grass, M. E., Zhang, Y. *et al.* 2008. "Reaction-driven restructuring of Rh-Pd and Pt-Pd core-shell nanoparticles." *Science* 322, 932–934.

3 Tao, F. and Salmeron, M. 2011. "In situ studies of chemistry and structure of materials in reactive environments." *Science* 331, 171–174.

4 Nguyen, L., Tao, F. F., Tang, Y. *et al.* 2019. "Understanding catalyst surfaces during catalysis through near ambient pressure X-ray photoelectron spectroscopy." *Chem. Rev.* 119, 6822–6905. doi:10.1021/acs.chemrev.8b00114.

5 Tao, F. F. and Crozier, P. A. 2016. "Atomic-scale observations of catalyst structures under reaction conditions and during catalysis." *Chem. Rev.* 116, 3487–3539. doi:10.1021/cr5002657.

6 Zhang, S., Nguyen, L., Zhu, Y. *et al.* 2013. "In-situ studies of nanocatalysis." *Acc. Chem. Res.* 46, 1731–1739. doi:10.1021/ar300245g.

7 Yan, G., Tang, Y., Li, Y. *et al.* 2022. "Reaction product-driven restructuring and assisted stabilization of a highly dispersed Rh-on-ceria catalyst." *Nat. Catal.* 5, 119–127.

8 Tang, Y., Nguyen, Y., Li, T. *et al.* 2022. "In situ and operando study of catalysts during high-temperature high-pressure catalysis in a fixed-bed plug flow reactor with X-ray absorption spectroscopy." *Rev. Sci. Instrum.* 95, 054102.

9 Tao, F. and Nguyen, Y. 2018. "Interactions of gaseous molecules with X-rayphotons and photoelectrons in AP-XPS study of solid surface in gas phase." *Phys. Chem. Chem. Phys.* 20, 9812–9823.

10 Tao, F., Nguyen, L., and Tang, Y. 2022. "Report on photoelectron-induced excitation spectroscopy and its close correlation with fundamental studies of catalysis." *J. Phys. Chem. C* 126, 13069–13087. doi:10.1021/acs.jpcc.2c00209.

6

Instrumentation of Ambient Pressure X-ray Photoelectron Spectrometer

6.1 X-ray Source for AP-XPS Studies

6.1.1 Brief of X-ray Sources

In general, based on energy of X-ray there are three types of X-ray sources. They are low-energy X-ray source that is also called soft X-ray, intermediate energy X-ray, and high-energy X-ray source. The latter is often called hard X-ray. For the same subshell of an element, the kinetic energy of their electrons upon excitation increases along increase of X-ray energy based on the equation, $KE = h\nu - BE - \varphi$. These photoelectrons with different kinetic energy (KE) have different inelastic scattering mean free path (λ) in a solid. λ is a parameter indicating how far electrons with their original energy KE_0 can travel in solid before they lose their energies. It is defined by the distance these electrons can travel before their intensity decreases to $\frac{1}{e}$ in terms of 36.8%; mathematically, the base of this definition is $I(d)= I_0e^{-d/\lambda(KE)}$. λ is mainly determined by kinetic energy (KE), the density of the solid they travel through, and the masses of the constituting atoms of the solid.

As mentioned above, the inelastic mean free path of electrons, λ, depends on their *KE*. As shown in the universal cure (Figure 6.1), the inelastic scattering mean free path of electrons in a solid first decrease along the increase of kinetic energy in the range of 1–100 eV and then increases along the increase of kinetic energy. Obviously, λ is minimum when the kinetic energy of electrons is in the range of 10 –100 eV. For instance, by using soft X-ray (<600 eV) to excite electrons of subshells of atoms with binding energy (BE) of 500 eV, the kinetic energy of these photoelectrons is <100 eV. Based on the universal curve, such photoelectrons have minimum λ, about 5 Å that is the thicnkness of the top 2–3 atomic layers of a metal such as Pt, Rh, Pd, or Cu. Thus, soft X-ray XPS offers highest surface sensitivity since the intensity of an XPS peak is contributed from these photoelectrons without any loss of their initial energy. The soft X-ray has extensively used in AP-XPS studies at torr or sub-torr pressure range.

The mostly commonly used X-ray with intermediate energy is Al Kα ($h\nu$=1486.3eV). In principle, an Al Kα source used for high vacuum X-ray photoelectron spectroscopy (XPS) can be used for AP-XPS studies upon isolating the gaseous environment of reaction cell/chamber from the high vacuum environment of Al Kα source. The isolation can be done by

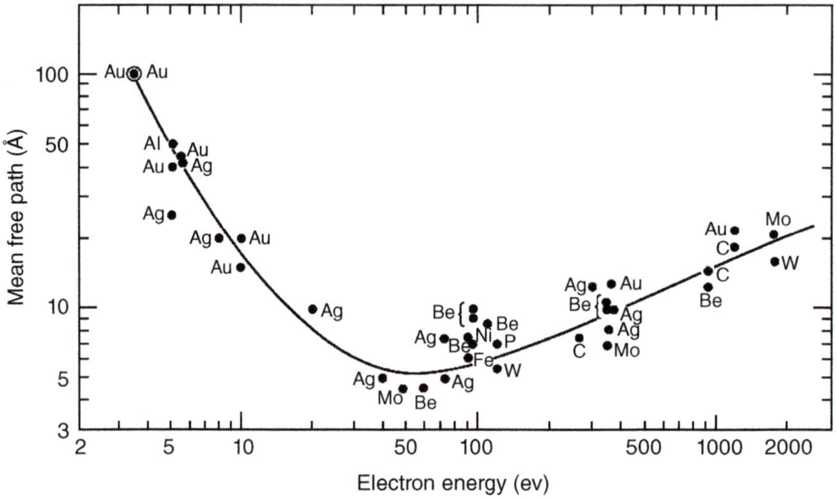

Figure 6.1 Universal curve of metal elements and carbon. *Source:* Reproduced with permission from Somorjai[1]/Cornell University Press.

installation of an aluminum membrane on the port between Al Kα X-ray source and reaction cell/chamber or an ultrathin silicon nitride window on the reaction cell.

Hard X-rays are available at synchrotron centers. In addition, they can be generated through lab-based Cu Kα, Cr Kα, or Ga Kα which are X-ray sources having been available on market decades ago. Upon excitation with a hard X-ray, photoelectrons have much higher kinetic energy. Based on the universal curve (Figure 6.1), their inelastic mean free paths are obviously longer than those generated from Al Kα and soft X-ray.

6.1.2 Soft X-ray for AP-XPS and Its Limitation in High Pressure Studies

Soft X-ray cannot be used for high pressure purpose. As shown in Figure 6.2a and b, X-ray must go through a gas phase before it irradiates the catalyst surface since the catalyst has to be placed in a gaseous environment of a reaction. The absorption (A) of X-ray beam by gas phase results in attenuation of X-ray beam intensity. The absorption of X-ray by gas follows the Lambert–Beer law, $A = lg\frac{I_0^{X-ray}}{I^{X-ray}} = \varepsilon \times b \times C$. Here I_0^{X-ray} is the intensity of X-ray before irradiating a gas phase. I^{X-ray} is the intensity of X-ray after irradiating a gas phase. Here b is the thickness of the gas phase that absorbs X-ray. In all the calculations in Figure 6.2c–e, the thickness, b is taken as 20 mm as marked in Figure 6.2a for simplicity. Alternatively, the X-ray absorption can be expressed with the equation[2], $I^{X-ray} = I_0^{X-ray}e^{-\mu\rho v}$; In this equation, I_0^{X-ray} is the intensity of the X-ray before entering the gas phase. μ is the mass absorption coefficient of gas to X-ray, cm^3/g; ρ is the density of the gas; v is the irradiation volume of the gas which can be evaluated with the product of the thickness of gas phase in terms of travel distance in the gas phase (b) and beam cross-section area (A). Based on ideal gas law, the gas density, ρ, is proportional to pressure of gas, p. Thus, I_0^{X-ray} exponentially decays along the increase of gas pressure in the reaction cell.

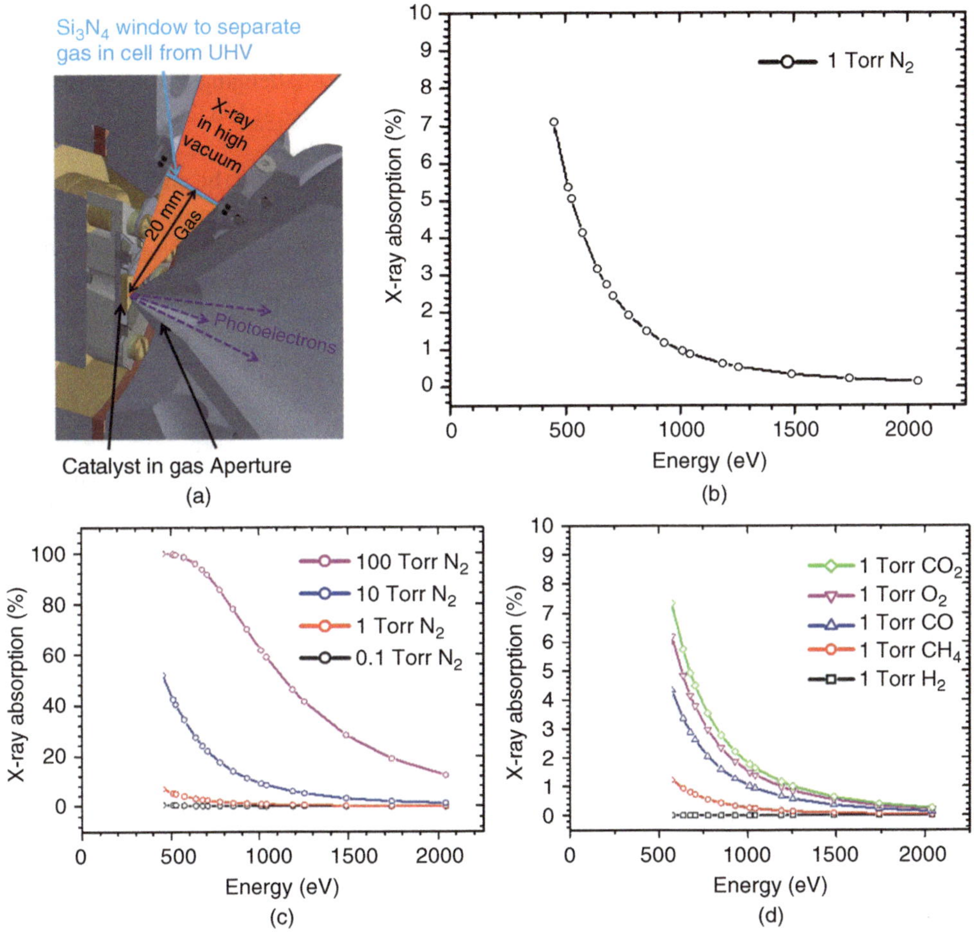

Figure 6.2 Schematic of gas region around a catalyst and X-ray absorption by gas molecules of N_2, CO_2, O_2, CO, CH_4, and H_2 in the pressure range of 0.1–100 Torr as a function of X-ray energy. X-ray absorption in b–d are generated using equation $I^{X-ray} = I_0e^{-\mu\rho x}$, described in the text with mass absorption coefficients from Tao and Nguyen[2] (a) Drawing showing the volume of gas molecules irradiated by X-rays, transmitted through the Si_3N_4 window; the Si_3N_4 window separates the gas phase in the cell from the high vacuum environment of the analysis chamber where the reaction cell is located. (b) Calculated absorption of X-rays with different energies by 1 Torr N_2 at a travel distance of 20 mm and irradiation volume of 10 mm^3. (c) calculated absorption of X-rays with different energies by N_2 at different pressures (0.1–100 Torr) at a travel distance of 20 mm and irradiation volume of 10 mm^3. (d) Calculated absorption of X-rays with different energies by N_2, CO_2, O_2, CO, CH_4, and H_2 at a pressure of 1 Torr at a travel distance of 20 mm and irradiation volume of 10 mm^3. *Source:* Reproduced with permission from Tao and Nguyen[2] and Tao and Crozier[3]/Royal Society of Chemistry.

As shown in Figure 6.2b, the absorption rate exponentially increases along the decrease of X-ray energy. For X-ray of 500 eV in 1 Torr N_2, about 7% of intensity of the X-ray was absorbed by the gas phase. Figure 6.2d lists similar exponential increase of absorption rate as a function of X-ray energy in several different gases at 1 Torr. One piece of valuable information we can see from these plots in Figure 6.2c is that the absorption rate of X-ray beam of 500 eV by N_2 sharply increase along the increase of the gas pressure of N_2. For instance, over 50% of

X-ray of 500 eV is absorbed by 10 Torr N_2 with a thickness of 20 mm (Figure 6.2c). Surprisingly, the absorption rate of 100 Torr N_2 with a thickness of 20 mm for X-ray of 500 eV is as high as 99% (Figure 6.2c). It suggests that it is extremely challenging, if not possible, to use soft X-ray AP-XPS to study surface of a catalyst in a gas phase at 100 Torr.

Based on the calculation, over 75% of intensity of Al Kα ($h\nu = 1486.3$ eV) still remain after absorption by N_2 gas at 100 Torr with a thickness of 20 mm (see the curve of 100 Torr N_2 in Figure 6.2c). In fact, certain portion of intensity of Al Kα can be retained upon absorption by 1 atm N_2 with a thickness of 20 mm. By decreasing the thickness of N_2 of 1 atm from 20 mm to 5 mm, it is expected that a reasonable portion of Al Kα can arrive at sample surface for AP-XPS studies at 1 atm. Thus, in principle Al Kα X-ray source can be used for performing AP-XPS studies at 1 atm.

6.1.3 Al Kα for AP-XPS and Its Challenge in Working at Higher Pressure

Al Kα is the most commonly used X-ray source on lab-based AP-XPS systems. One notable difference in X-ray source between high vacuum XPS and AP-XPS is the much higher flux density of Al Kα used on AP-XPS. This is driven by the purpose of gaining enough photoelectrons for analysis in AP-XPS. Compared to high vacuum XPS (Figure 6.3c), the area of a sample contributing photoelectrons collectable by energy analyzer of AP-XPS (Figure 6.3d) is much smaller in AP-XPS since it is limited by the size of aperture installed at the front of the energy analyzer of AP-XPS. In addition, only a portion of photoelectrons generated from surface of a catalyst can travel through the gas region as schematically shown in Figure 6.3d. To counter these factors decreasing the number of photoelectrons remaining their original kinetic energies in AP-XPS, the flux density of the X-ray source used in AP-XPS must be higher than that of high vacuum XPS. Thus, a monochromator X-ray source is typically used in AP-XPS since it provides X-ray beam with high flux density in terms of a larger number of X-ray photons arriving to 1 mm^2 per second than a nonmonochromatic X-ray source. Numerous commercial manufacturers have designed different types of monochromator X-ray sources providing X-ray of high flux density.

Another difference related to the X-ray source is that the X-ray source of AP-XPS must be separated from the gas environment of a sample. How they are separated depends on the mode of an AP-XPS study. One is called a backfilling mode of AP-XPS studies as schematically shown in Figure 6.3a. In the backfilling mode, a mixture of reactants of a reaction is introduced to the whole chamber. The separation of gas phase from vacuum environment of X-ray source is done by installing a thin aluminum foil to cover the port of the flange installed between a monochromator X-ray source and the chamber filled with gas and placed a sample in the gas. The thin aluminum foil can persist the gas pressure up to tens of Torr. Another AP-XPS study mode is the reaction cell mode (Figure 6.3b). In this mode, AP-XPS is performed by assembling a sample to a reaction cell installed in a UHV chamber (Figure 6.3b). A momochromator Al Kα source is installed on the UHV chamber. To isolate the gaseous environment in the reaction cell from the UHV chamber, a Si_3N_4 window is installed on the reaction cell. X-ray beam can transit the Si_3N_4 window on the reaction cell and gas in the cell and then irradiate surface of the sample placed in the reaction cell (Figure 6.3b). The thickness of a Si_3N_4 window used for this purpose is several to a couple of tens of nm. It is available on market. Construction of a reaction cell for AP-XPS will be described in the flowing section.

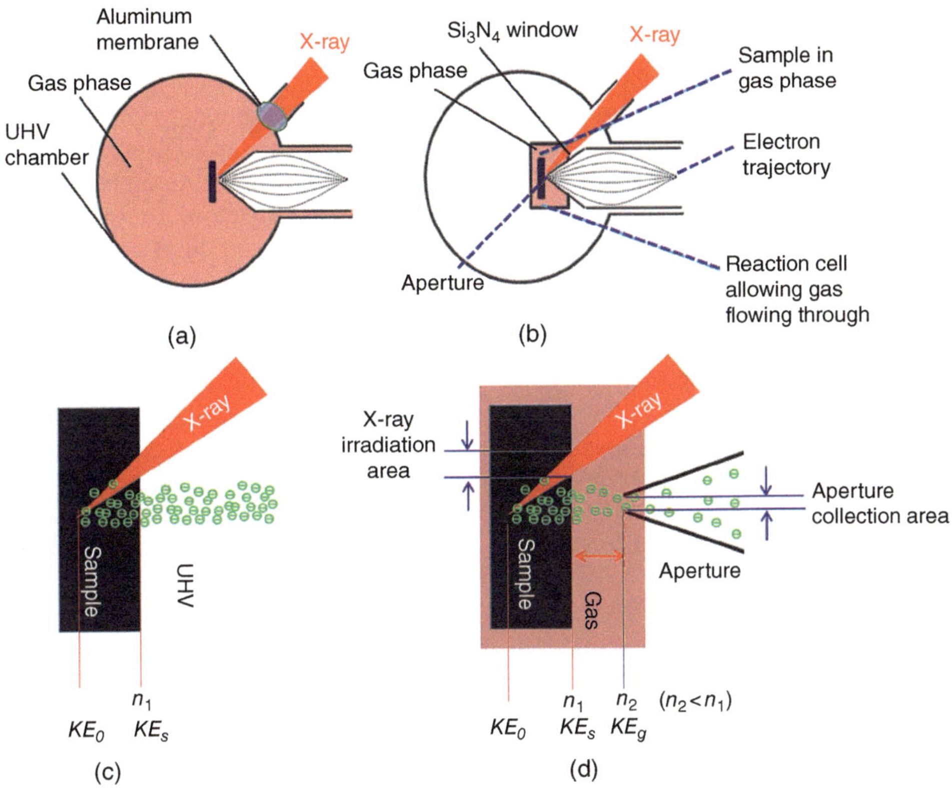

Figure 6.3 Schematics (a and b) showing two modes of AP-XPS studies and (c and d) evolution of kinetic energies of photoelectrons and decrease of number of photoelectrons remaining original kinetic energy KE_0. (a) Backfilling mode of AP-XPS studies; X-ray source is separated from backfilling chamber through aluminum membrane. (b) Reaction cell mode; the gas phase in cell is separated from X-ray source by ultrathin Si_3N_4 window. *Source:* Reproduced with permission from Tang et al.[4]/Elsevier. (c) Schematic showing collection area of photoelectrons while a sample is in UHV. (d) Schematic showing collection area of a sample in terms of X-ray irradiation area, sample-aperture distance (z), aperture size in terms of aperture collection area. Reproduced with permission from Tao et al.[2]/Royal Society of Chemistry.

As discussed in Chapter 5, F_p is the transmission rate of photoelectrons after they go through a gas phase at pressure p with a thickness of z (Figure 6.3d). It can be calculated with the equation, $F_p = \frac{I_p}{I_0} = e^{-\frac{z\sigma_{KE}p}{kT}}$;[2] here T is the gas temperature and σ_{KE} is the cross section depending on kinetic energy of photoelectrons. I_0 is the intensity of photoelectrons with a specific kinetic energy before entering the gas phase; I_p is the intensity of photoelectrons of the photoemission feature in terms of XPS peak collected after scattering with the gas phase at a pressure p with a thickness of z. To maximize the F_p, it is necessary to minimize z. It is well known that the transmission could reach the maximum when the thickness of a gas, z, is less than or nearly equal to the inelastic scattering mean free path of photoelectrons in the gas phase. In gas of 1 atm, λ_{gas} of photoelectrons with kinetic energy of 1000 eV is in the range of 10 μm. To perform AP-XPS studies of a sample in a gas at 1 atm of N_2, it is necessary to keep the thickness of 1 atm N_2 as small as 10 μm. In other words, the sample-aperture distance, *z*, in Figure 6.3d should be as short as 0.01 mm. Control of a distance of 0.01 mm with high accuracy is possible though it is definitely challenging.

6.1.4 Hard X-ray for AP-XPS and its Application to High Pressure Studies

It is understood that surface of a catalyst in gas phase at a pressure of 1 atm can be studied with XPS with a very specific graphene-based gas cell without using a hard X-ray source.[5,6] In other words, from the point view of proving a concept, AP-XPS working at a pressure of 1 atm has been reported.[5,6] These demonstrations request unique designs and dedicated fabrication of gas cells. To perform AP-XPS experiments routinely at a pressure obviously higher than 25 Torr or even up to 1 atm, an alternative method is the use of high-energy X-ray. It is called hard X-ray AP-XPS. Hard X-ray XPS was developed several decades ago and has been commercialized for a few decades.[7–9] Hard X-ray AP-XPS has been available and been used extensively in various fields of solid materials and devices including catalysts in gas phase at a pressure of up to 1 atm in the last couple of decades. At least one concise review article[10] and one book chapter[11] summarizing the progress of applications of hard X-ray XPS and hard X-ray AP-XPS to physical and chemical sciences were published in the last decade.

The difference between Al Kα AP-XPS and hard X-ray AP-XPS is the energy of X-ray source. A hard X-ray source having a much higher energy can generate photoelectrons with a much higher kinetic energy; these photoelectrons with a higher kinetic energy have larger inelastic scattering mean free path and thus a larger portion of these high-energy photoelectrons can travel through a gas phase at a high pressure while remaining their original kinetic energies. Hard X-ray AP-XPS systems have been available at synchrotron centers over a couple of decades.[10] For instance, Spring-8 synchrotron center at Japan has at least one hard X-ray AP-XPS system publicly accessible since the last decade. Lab-based hard X-ray AP-XPS systems are available from manufacturers in European countries and Japan. Very similar to an Al Kα AP-XPS, a hard X-ray AP-XPS consists of X-ray source, stainless steel chambers, energy analyzer, reaction cell, and many other parts.

Retaining a sample-aperture distance at 0.01 mm requested for using Al Kα AP-XPS to study a sample is definitely challenging since an accurate control of the sample-aperture distance is a difficult task. To perform AP-XPS studies at 1 atm routintely, a reliable solution is the use of a hard X-ray instead of Al Kα to generate high-energy photoelectrons that have a larger inelastic scattering mean free path in a gas phase. This larger λ allows for performing AP-XPS at 1 atm while the thickness of gas phase in terms of sample-aperture distance can be larger and thus be controlled easiler. For instance, a study of Rh catalyst in gas of 1 atm can be realized by using a hard X-ray to excite electrons in Rh 3d subshell of a catalyst surface, generating high-energy photoelectrons whose inelastic mean free path in gas at 1 atm allows for a relatively large sample-aperture distance which is practical.

Cu Kα and Cr Kα are hard X-ray sources readily available from vendors in Europe countries and Japan. Their energies are 8048 and 5415 eV, respectively. Both Cu Kα or Cr Kα X-ray sources are available from numerous manufacturers in Europe and Japan such as Chemitronics of Japan. To generate hard X-ray, a source of high-energy electron beam is needed. For example, a high-energy electron gun of Kimball Physics can be used as a source of electron beam. The function of this electron source is to excite atoms of a metal target to generate hard X-ray.

Notably, an disadvantage using a hard X-ray AP-XPS is a relatively low surface sensitivity. This is because the high kinetic energy of photoelectrons excited by hard X-ray unfortunately makes their inelastic scattering mean path λ_{solid} obviously larger based on the equations,[12] $\lambda_{element} = \frac{538}{KE^2} + 0.41\alpha^{1.5}KE^{0.5}$ for elements or $\lambda_{inorganic} = \frac{2170}{KE^2} + 0.72\alpha^{1.5}KE^{0.5}$ for inorganic compounds, where α is a particular thickness of a monolayer of a solid material and calculated with the equation: $\alpha = \left(\frac{A}{\rho n N_A}\right)^{1/3} \times 10^8$ nm. The increase of $\lambda_{element}$ and $\lambda_{inorganic}$ in a solid along the increase of kinetic energy of photoelectrons (KE>100 eV) is clearly suggested from Figure 6.1. In summary, hard X-ray AP-XPS is capable of studying a sample in gas at 1 atm while it has a lower surface sensitivity than Al Kα AP-XPS.

6.2 Reaction Cell with Capability of Flowing Gas

6.2.1 Necessity of Having a Reaction Cell for Performing In Situ/Operando Studies of Catalysis

In backfilling mode of AP-XPS studies, the volume of gas filled in the UHV chamber of an AP-XPS system is typically as large as tens of liters (Figure 6.3a). The volume of reactant gas of a reaction cell of an AP-XPS system is in the range of tens of milliliters (Figure 6.3b). Thus, the dead volume around the catalyst in the backfilling AP-XPS is tens or even hundreds of times higher. Such a large dead volume of the filled gas in the chamber of backfilling mode makes gas flow around a catalyst quite slowly or nearly impossible. An immediate consequence of the slow flow of gas around the catalyst is the limited diffusion of product molecules. Due to the limited diffusion, the formed product molecules could readsorb on surface of the catalyst, blocking the access of reactant molecules to catalytic sites. Thus, it is important to minimize the dead volume around the catalyst, by which gas around the catalyst can flow much faster under the same pumping capacitance of the AP-XPS system. Installation of a reaction cell with a volume of 10–100 ml, by which the mixture of all reactants is introduced and flows through the catalyst surface and then exits the cell, is an approach to reduce the dead volume largely. Thus, running AP-XPS studies in a reaction cell can avoid the issue of large dead volume encountered in catalysis studies. However, the large dead volume should not be a concern if uncovering surface chemistry of a catalyst ***during catalysis*** is not the task. In other words, the large dead volume is not an issue for studying phase-dependent adsorption on a surface in the gas phase.

Another advantage of running AP-XPS studies in a reaction cell with small dead volume is the capability of performing AP-XPS studies of the sample in gas phase at a higher pressure in contrast to the backfilling mode. The use of a reaction cell can largely reduce the travel distance of X-ray beam in gas (Figure 6.3b versus a), minimizing absorption rate of X-ray beam by gas phase. As shown in Figure 6.3d, the absorption of X-ray beam by gas phase results in attenuation of X-ray beam (Figure 6.3a). The attenuation can be expressed with the equation,[2] $I^{X-ray} = I_0^{X-ray} e^{-\mu\rho v}$. I_0^{X-ray} is the intensity of the X-ray before entering the reaction cell. μ is the mass absorption coefficient of gas to X-ray, cm^3/g; ρ is the density of the material; v is the volume of the gas. Here, based on ideal gas law, the density, ρ, is proportional to the pressure of gas, p. Thus, $I^{X\text{-}ray}$ exponentially decays along the increase of mass of gas X-ray travels through in the reaction cell; the mass, m is the product of gas density ρ and volume v; the volume is the space the X-ray beam occupies in the cell; it can be evaluated by the product of the beam cross-section (A) and travel distance of the beam in gas phase (d). The travel distance of X-ray beam, d, in the chamber of the backfilling mode is about a couple of tens to a few tens or even several tens of centimeters which is a couple of tens of times of the travel distance (20 mm) in the reaction cell (Figure 6.3b). For the same mass of gas ($m = pVM/RT = pAdM/RT$), the pressure of gas in the reaction cell can be much higher than the pressure in the chamber of backfilling mode. Thus, a reaction cell mode allows performing AP-XPS studies at a much higher pressure than backfilling mode under the assumption of the same X-ray source and energy analyzer without consideration of the increased loss of photoelectron.

6.2.2 Structure of Reaction Cell

Different types of reaction cells were reported in literature.[5,13,14] The common features of these reaction cells are presented here. First, its volume is typically a couple of tens of milliliters or up to one hundred milliliters. Second, the internal wall is coated with inert materials, typically Au thin film to prevent surface of stainless steel wall from participating into the catalysis; this is driven by the catalytic function of Ni element of stainless steel in many reactions since the steel typically contains up to 35% Ni. Third, the cell has one port for inserting a sample before AP-XPS studies and removing it after AP-XPS studies. Fourth, it has a window of ultrathin Si_3N_4 (mostly) or Al thin membrane (occasionally) to allow X-ray to transit into the cell and gas phase and eventually irradiate surface of a catalyst to excite photoelectrons from the catalyst surface; thickness of this Si_3N_4 window or Al foil is chosen by considering the pressure of gas in the reaction cell and its transmission to X-ray. A thicker window or membrane can remain gas at a higher pressure but a much lower transmissivity to X-ray. Definitely, its thickness is also determined by the energy of X-ray; based on Figure 6.2c, X-ray with higher energy has a high transmissivity and thus a relatively thicker window or membrane can be used. Typically, for the X-ray Al Kα, Si_3N_4 of 50 nm is used as window of a reaction cell to allow X-ray to transmit. In addition, compared to backfilling AP-XPS, the absorption of X-ray by gas phase filled in the reaction cell is largely reduced. In addition, an aperture is either installed as a part of a reaction cell or engaged to a reaction cell.

An aperture plays an important role in AP-XPS. It has dual functions. It allows gas to flow through the reaction cell and photoelectrons escape the reaction cell. Its size is crucial. A larger opening allows more photoelectrons to escape the gas phase to enter the energy analyzer, by which the number of photoelectrons being detected is higher. However, a larger aperture size results in inefficient pumping to the gas leaked from the aperture, thus limiting the flow rate of gas in the reaction cell. More seriously, from fluidics point of view a larger aperture in front of the sample surface makes gas near sample surface be preferentially pumped, leading to the actual pressure of gas near the catalyst surface lower than the pressure read at the entrance or exit tubing of the reaction cell.[15] Based on the calculation of fluidics, to make the pressure of gas near the catalyst surface 95% of the pressure of gas

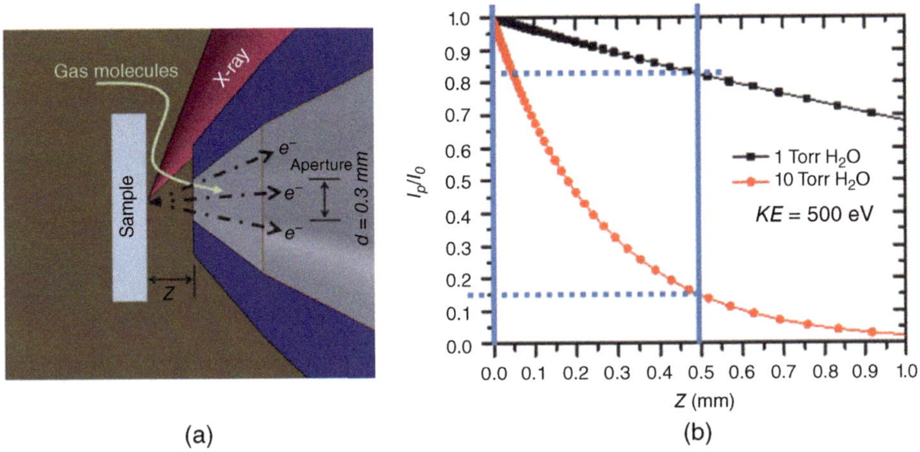

Figure 6.4 Schematics showing the configuration of sample and aperture in a gas phase and the ratio I_p/I_0 as a function of sample-aperture distance. (a) Schematic showing the X-ray beam, sample surface, and aperture. (b) Plot of collection rate (I_p/I_0) as a function of sample-aperture distance (mm), z, after photoelectrons of 500 eV travel through a H_2O vapor with a pressure at 1 and 10 Torr. *Source:* Reproduced with permission from Tang et al.[4]/Elsevier.

measured at the wall, entrance, or exit of the tubing connected to the reaction cell, it is necessary to retain the aperture size smaller than half of the distance between sample surface and the aperture.[15] The relation between the aperture diameter (d) and sample-aperture distance (z) are schematically shown in Figure 6.4a. In other words, to make the pressure reading from gauges installed at the entrance or exit tubing of the reaction cell represent the pressure of the gas near the sample surface, the distance between the sample surface and aperture is at least two times the aperture size. However, the transmission of the gas phase to the photoelectrons generated from the sample surface decays exponentially as the increase of sample surface-aperture distance due to the inelastic scattering of photoelectrons to gas molecules. Figure 6.4b presents the decay of transition factor in terms of F_p calculated with equation $F_p = \frac{I_p}{I_0} = e^{-\frac{z\sigma_{KE} p}{kT}}$ as a function of sample-aperture distance while the aperture diameter is fixed. In 1 Torr water vapor, 83% of photoelectrons with KE of 500 eV can travel through a gas region with a thickness of 0.5 mm ($z = 0.5$ mm) based on Figure 6.4b. Based on the relationship between the aperture size and sample-aperture distance, the aperture size should be half of the distance, 0.25 or smaller. However, if the vapor pressure of water in the cell is increased to 10 Torr, with the setup of the aperture of 0.25 mm and sample surface-aperture distance of 0.5 mm, 15% of photoelectrons leaving the sample surface can travel through the 10 Torr gas without loss of kinetic energy (Figure 6.4b). Clearly, the inelastic scattering between photoelectrons and gas molecules sharply depends on the molecular density in the gas phase in terms of gas pressure in the aperture. Thus, in general, the sample surface-aperture, the gas phase pressure decides the sample surface-aperture distance, z, for collecting enough photoelectrons without losing kinetic energies for AP-XPS analysis; Then, the sample surface-aperture distance, z, decides the size of aperture since the aperture size is typically a half of z.

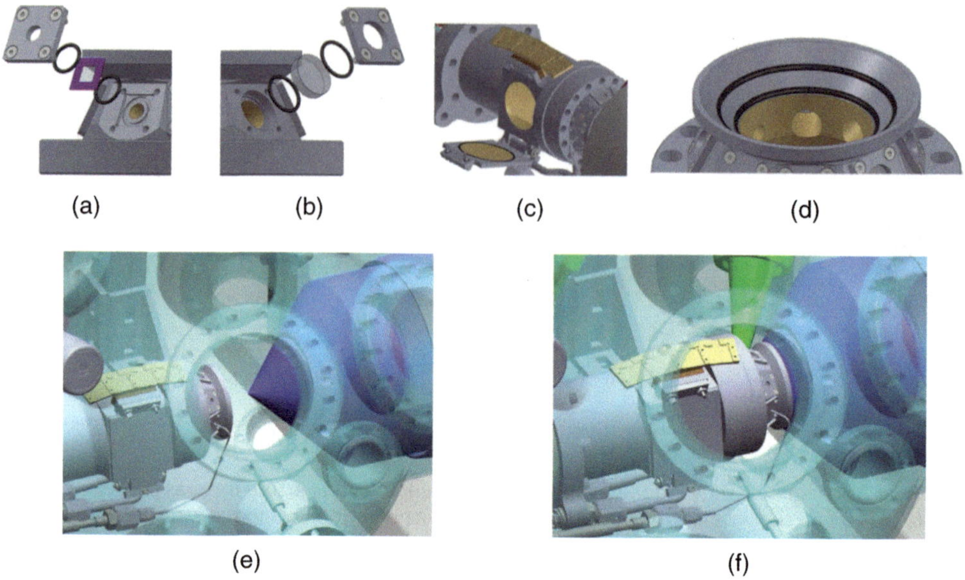

Figure 6.5 Sealing mechanism for different interfaces with the UHV environment of chamber. (a) Sealing mechanism for X-ray Si_3N_4 window; black circles are two identical O-ring anchored on two sides of a Si_3N_4 window (the pink part). (b) Sealing mechanism of fused silica windows; the transparent disk is the window. (c) Sealing of the sample door of the reaction cell; the internal surface of the door was coated with Au for passivation of activity of the stainless steel. (d) Sealing of the cell neck through a perfect alignment between the female cell neck and the male front cone of the ambient pressure energy analyzer; the alignment does not need any locker; the alignment can be automatically optimized by applying a force from a linear manipulator. Reaction cell retracted away from the front cone of the energy analyzer, respectively. (f) Reaction cell is well aligned and engaged to the front cone of the energy analyzer. *Source:* Reproduced with permission from Nguyen and Tao[14]/AIP Publishing LLC.

6.2.3 Sealing of a Reaction Cell and its Engaging Mechanism

A reaction cell, accurately speaking, the center of a sample surface, must be positioned at the cross point between the X-ray beam and the central line of a differently pumped energy analyzer. Figure 6.5e and f is the reaction cell positioned at the cross point of X-ray beam direction and the central line of a differently pumped energy analyzer. Figure 6.5d is the head of the reaction cell. The cell was engaged to the energy analyzer front cone through two O-rings shown in Figure 6.5d. This O-ring sealing prevents the gas of the reaction cell from leaking to the chamber hosting the reaction cell since the environment in the chamber should be UHV or HV. Figure 6.5e illustrates the status of the reaction cell before it engages the front cone of the energy analyzer for sealing. The front cone is shown in blue in Figure 6.5e and f. Driven by a precisely controlled motor, the reaction cell engage controllably moves to the front cone to fully engage to the front cone of the energy analyzer.

Figure 6.6 illustrates how the reaction cell is sealed to the energy analyzer based on the autocorrection mechanism of an engaging scanning tunneling microscope (STM) tip holder to fit a magnetic bowl of STM scanner Tao et al. reported in literature.[16] As the wall of the reaction cell is welded to a bellow (Figure 6.7d) and the bellow is flexible to some extent, the orientation of the head of the cell can be readily autocorrected to fit the front cone of the energy analyzer, making sure the cell is aligned to the central line of the differentially

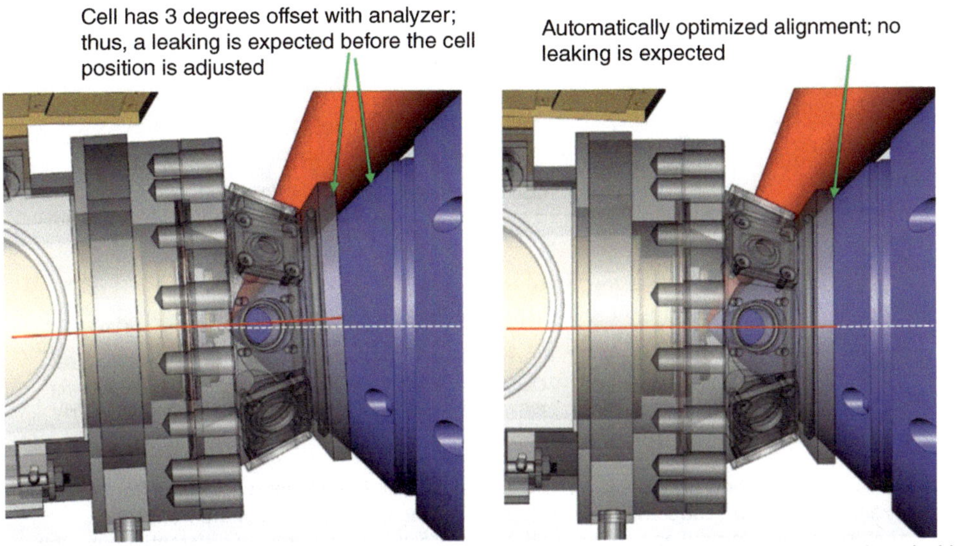

Figure 6.6 Important alignment of X-ray beam and trajectory of photoelectron and the mechanism of sealing of reaction cell at the interface of cell neck and aperture/front cone of ambient pressure energy analyzer. (a) Misaligned cell and front cone of energy analyzer. (b) Perfectly aligned cell and energy analyzer. *Source:* Reproduced with permission from Nguyen and Tao[14]/AIP Publishing LLC.

pumped energy analyzer. For instance, the head of the reaction cell (Figure 6.6a) is offset from the central line (white dashed line in Figure 6.6a and b) by 3°. This offset resulted in severe leaking since the whole cell was placed in the high vacuum environment of the XPS chamber. The bowl-like engaging mechanism makes the head of the reaction cell self-adjust to gain an ideal alignment between the head of the reaction cell and the front cone of the energy analyzer (Figure 6.6b).[16]

6.2.4 Function of a Reaction Cell for AP-XPS Studies of Catalyst

Quality of a reaction cell can be evaluated on the following three aspects. The first aspect is how good the sealing at the door on the cell and that at the head connecting to the front cone of the energy analyzer are. The base pressure in the main chamber is typically lower than 1×10^{-9} Torr such as 3×10^{-10} Torr. In a routine operation of the cell, a used sample is taken out from the cell and then a new sample placed on a sample holder is transferred to the cell through a door on the cell. Then, the door is closed with the wobble stick and subsequently reactant gas(es) is introduced. Figures 6.7e and 6.7g show the cell door at the statuses of the door closing and door opening, respectively. Quality of this sealing of the door determines the base pressure of the main chamber since the reaction cell is placed in high vacuum environment of the main chamber. Typically, a cell is placed into a main chamber through a movable manipulator. The base pressure in the main chamber is typically lower than 1×10^{-9} Torr such as 3×10^{-10}.

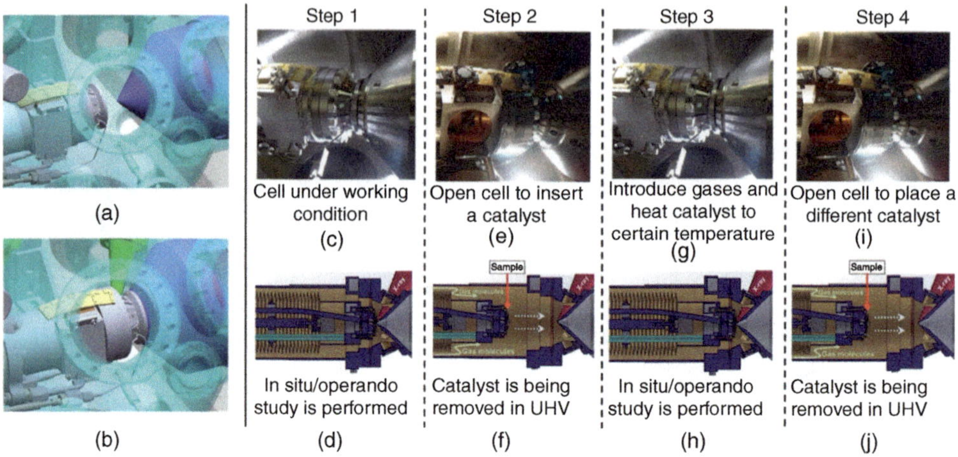

Figure 6.7 Drawing to show how a reaction cell installed in the AP-XPS approaches the front cone of an energy analyzer and how the sample holder is approaching to an aperture of the front cone. (a) State before the reaction cell approaches to the aperture of the front cone of differential pumping stage. (b) State after the cell has engaged to the front cone. (c), (e), (g), and (i) are photos of the cell. (d), (f), (h), and (j) are drawings of the internal structure of the reaction cell. (c) Photos of the cell at a working condition. (d) Internal structure of the cell at working conditions of panel c. (e) Photo of cell in vacuum after in situ study is done and gas is purged and then door is opened; the cell is ready to take one sample. (f) Internal structure of the cell in the open state; the sample stage has been retracted to the position to take sample. (g) Photo of cell ready for in situ studies after a new sample is inserted, gas is introduced, and sample is heated to an ideal temperature. (h) Internal structure of the cell in a ready state. (i) Photo of the cell in a waiting state for next sample after in situ studies. (j) Internal structure of the cell. *Source:* Reproduced with permission from Nguyen et al.[13]/American Chemical Society.

When the door of the cell is open, the vacuum in the cell is the same as the main chamber of AP-XPS. At this status, the base pressure can be as low as 3×10^{-10} Torr. After a new sample is assembled to the reaction cell and the door is closed, gas starts to flow to the cell and the base pressure of the main chamber increases. As shown in Figure 6.8d, the base pressure basically remained at 7.5×10^{-10} Torr even though the gas pressure in the reaction cell was increased from 10^{-10} to 10 Torr. Maintaining a high vacuum environment in the main chamber is significant for a long-time function of the monochromated X-ray source since the X-ray source was installed on the main chamber. Notably, even if the pressure of gas in the reaction cell is increased to 1 atm (760 Torr), the base pressure of the main chamber remained at 3.5×10^{-7} Torr (Figure 6.8c), which is still a safe pressure for running X-ray source. Overall, the quality of the sealing of this cell demonstrated in Figure 6.8 is quite high. The main reason is the bowl-based self-correction of the position of the reaction cell (Figure 6.6) and the double-ring sealing at the interface between the reaction cell and the front cone of the energy analyzer (Figure 6.5d).[14]

The second aspect determining the quality of a reaction cell is how heating a catalyst sample in the reaction cell is managed since this heating could largely influences the temperature of reaction cell. In terms of the function of sample heating, an ideal heating is to heat only the catalyst and reactants around the catalyst to a desired temperature but retain all parts of the cell at a temperature as close to room temperature as possible. As shown in Figure 6.9a, the cell can be quickly heated up to 600 °C within just two to three minutes as an infrared laser beam was introduced to the back of sample through a focusing lens which is irradiated

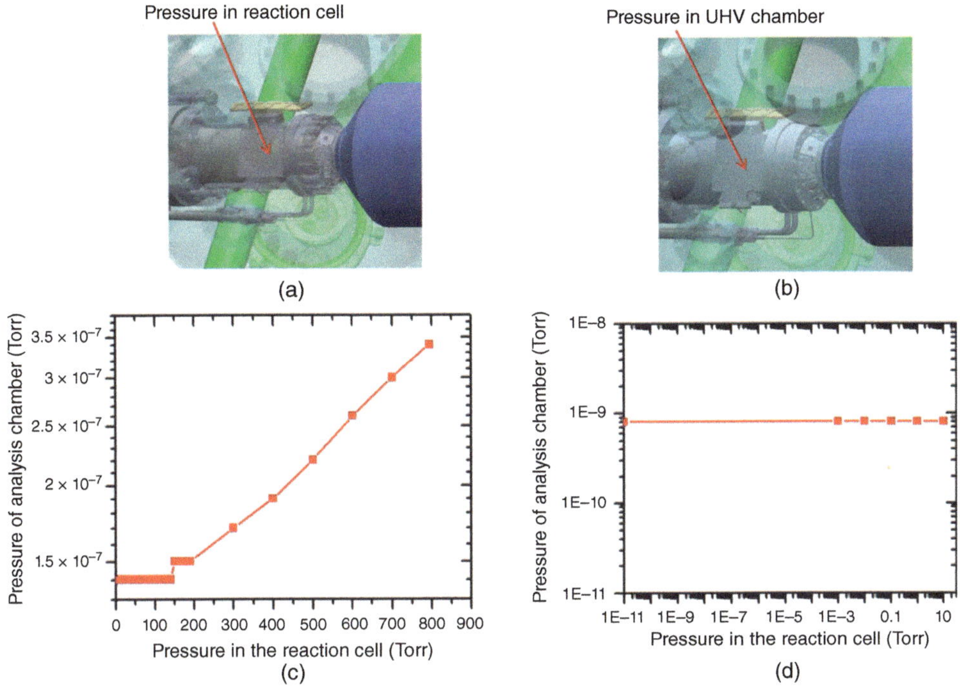

Figure 6.8 Sealing performance of the in situ reaction cell. (a) Location of the pressures measured in the reaction cell. (b) Location of the pressure measured in the main chamber. (c) Plot of the measured pressures in UHV as a function of pressures of the introduced N_2 up to 800 Torr in the cell; the base pressure of the main chamber is 1.4×10^{-7} Torr before any N_2 was introduced to the reaction cell. (d) Plot of the measured pressures in the main chamber as a function of pressures of N_2 in the reaction cell up to 10 Torr; the base pressure is 8×10^{-10} Torr before any N_2 was introduced to the reaction cell. *Source:* Reproduced with permission from Nguyen and Tao[14]/AIP Publishing LLC.

by the laser beam from one end of a fiber glass connecting to a laser source. Through such a focused laser beam heating, all parts of the cell except the back of the sample holder loading catalyst powder or a single crystal model catalyst are not heated at all. The sample holder was assembled on a high-density alumina sample stage sitting on a flange of an IR transparent glass. The temperatures of the sample holder and the region around the sample holder were recorded simultaneously along the sample holder was being heated. Figure 6.9c shows the temperature of the area around the sample holder increased by only 20 °C when the sample holder was heated up to about 620 °C. The lack of obvious heating to the area close to the sample holder shows the sample holder is well insulated from its surrounding area in the reaction cell. The excellent insulation also benefits from the direct heating of sample holder by the IR laser beam. The sample holder was installed on a nonremovable sample stage. The temperatures of both the flange holding the sample stage and the reaction cell body were nearly at room temperature when the sample holder was even heated up to 700 or 800 °C. In addition, the high-quality insulation of the sample holder from its surrounding area is contributed from the high-dense alumina-based heating stage that largely blocks potential heat transfer from the sample holder to its adjacent area. Figure 6.9e plots both temperatures of sample and reaction cell as a function of time while the sample was heated

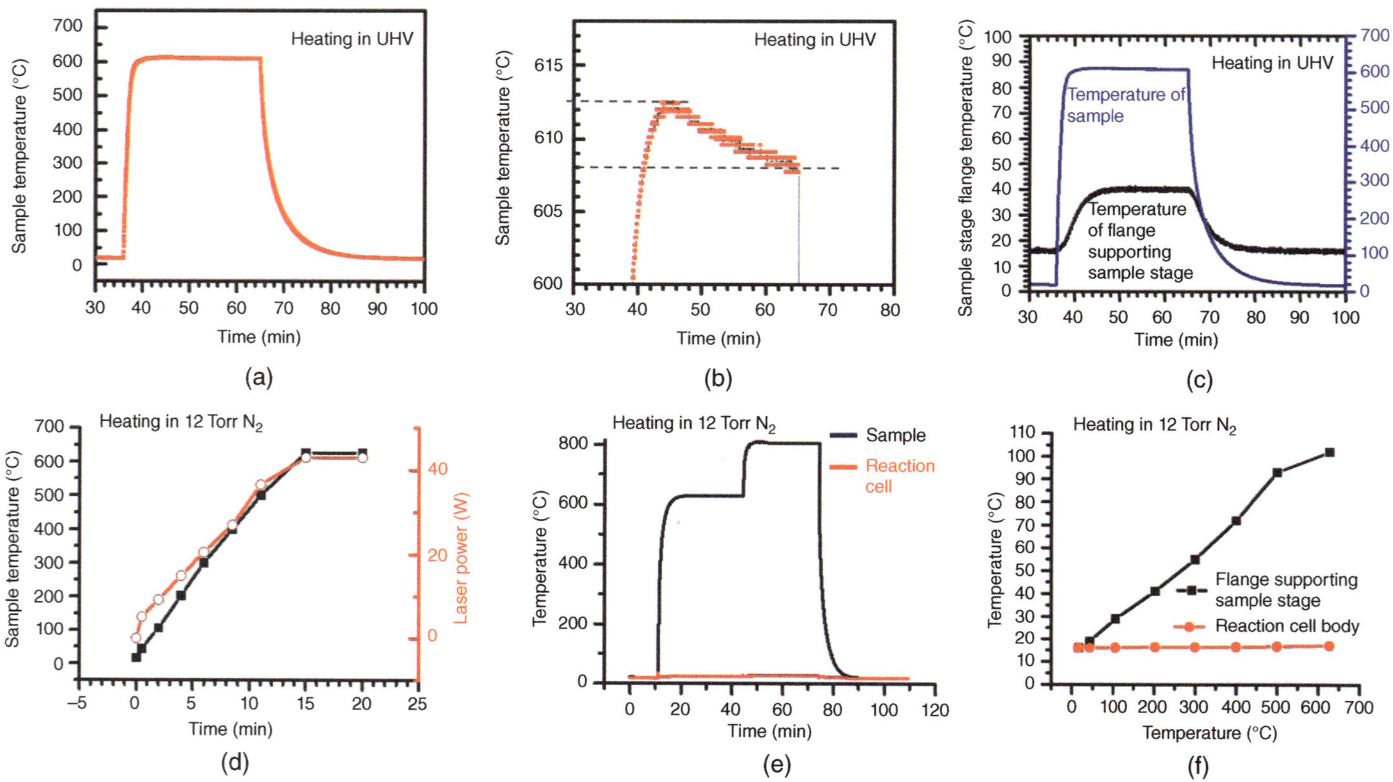

Figure 6.9 Sample heating characteristic of the reaction cell. (a) Plot of the sample temperature in UHV as a function of time when 20W of laser was turned on and remained at 20W for half hour and then turned off; the sample can reach the desired temperature very quickly (in less than 5 minutes) and cooled to room temperature quickly. (b) Plot the temperature of sample when the power supply of the laser beam stayed at 20W without any tuning up or down; within about 20 minutes, the temperature of the sample only decayed by 5 °C. (c) Plot of both temperatures of the sample in UHV (blue curve) and the flange supporting the sample stage (which hold the sample or sample holder) (black) as a function of time when power supply of laser beam was turned to 20W, remained at 20W, and turned off. (d) Sample temperature (black) and the applied laser power (red) as a function of time when the sample was heated in flowing N_2 (4 ml/min, 12 Torr of N_2). Basically, the sample temperature increased linearly with heating laser power. (e) Plot of temperature of both the flange supporting the sample stage (black) and the reaction cell (red) as a function of sample temperature when the sample was heated in flowing N_2 (4 ml/min, 12 Torr of N_2). Laser power used was 42 W. (f) Evolutions of temperatures of the sample (black line) and the reaction cell (red line) when the sample was heated from room temperature to 613 °C within 5 minutes and then remained at a temperature >608 °C for about 20 minutes and then heated up to 800 °C within 2 minutes and then remained at a temperature >795 °C for 30 minutes and then cooled down to room temperature. *Source:* Reproduced with permission from Nguyen and Tao[14]/AIP Publishing.

in 12 Torr N_2 to 608 °C and then to 795 °C, confirming the excellent thermal insulation of the reaction cell from the high-temperature sample in a flowing gas.[14]

The third aspect is whether the metal of the reaction cell could bring "fake" catalytic activity to an unknown sample. Most parts of a reaction cell of AP-XPS were made of stainless steel consisting of up to 35% Ni and other metals potentially active for catalyzing some reactions such as CO oxidation and activations of C—H and C—C bonds. To passivate the internal wall of a reaction cell, Au thin film is typically deposited on the internal surface of this reaction cell.[14] By introducing CO and O_2 to a reference reaction cell without passivated Au thin film, CO_2 peak was clearly observed when a sample holder was heated to 100 °C (Figure 6.10a). It suggests Ni or other metals of some parts of the cell must have catalyzed the CO oxidation.[14] Different from the reference reaction cell, no CO_2 was formed at 100 °C from the reaction cell where Au thin film was deposited on the internal wall of the reaction cell (Figure 6.10b). It confirms the passivation of surface of the stainless steel by the deposition of Au thin film is necessary and effective.

These tests confirmed the excellent sealing of gas, reliable heating of sample holder while retaining temperatures of other parts at low temperature, and inertness of reaction cell for catalysis. This cell had served for the research projects of Tao group for several years. One example is the in situ studies of oxidative coupling of methane (OCM) in the mixture of CH_4 and O_2, where AP-XPS study was performed at 800 °C in the reaction cell for several hours in the mixture of CH_4 and O_2.[17] More information on the instrumentation of this reaction cell can be found in the literature.[14]

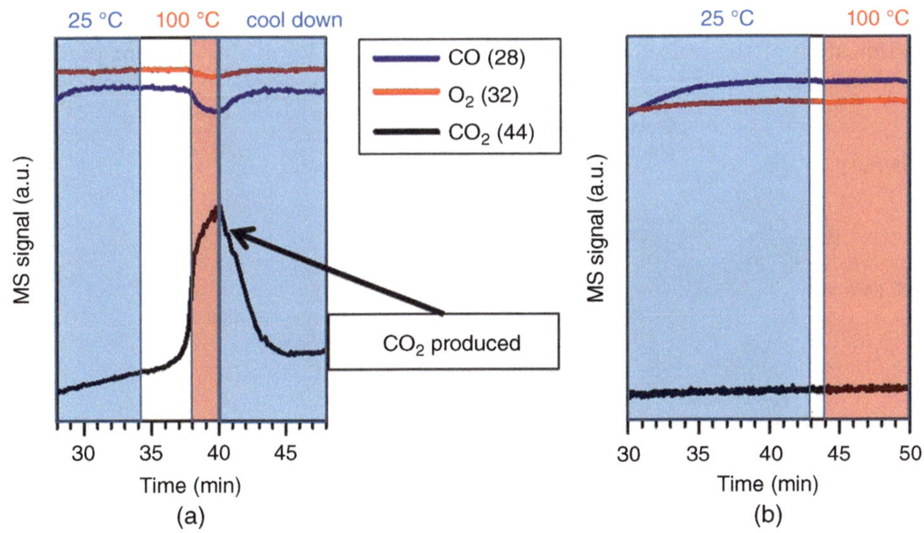

Figure 6.10 Mass spectrometer data from two types of reaction cells with different sample heating mechanisms. Mixture of 0.1 Torr CO and O_2 was introduced into the cells. The sample holder was a blank tantalum sheet without any catalyst. In (a), sample holder was heated indirectly through the sample stage, which was conductively heated through the external surface of the reaction cell bombarded by electron beams in UHV. CO and O_2 were being flown through the reaction cell. Production of CO_2 was observed at 100 °C even though there was no catalyst present. In (b), the sample was heated directly through infrared laser beam with the reaction cell of reference[14] here. No CO_2 was observed at 100 °C while both CO and O_2 were being flown through the reaction cell. *Source:* Reproduced with permission from Nguyen and Tao[14]/AIP Publishing.

6.3 Differential Pumping Energy Analyzer with High Transmission

Differential pumping stage of the energy analyzer was requested for effectively collecting photoelectrons in AP-XPS. One of its main functions is to pump gas leaking from the aperture of the reaction cell to the energy analyzer. As schematically shown in Figure 6.11, photoelectrons leaving the gas phase to enter the front cone of the energy analyzer through the aperture spread out in the front cone region. The front cone region is the gray area in Figure 6.4a. The wide spread of photoelectrons in the front cone region makes collection of photoelectrons by a regular energy analyzer ineffective. To enhance the collection of photoelectrons, focusing lenses are needed.[18] Thus, additional electrostatic focusing lenses are typically a part of an ambient pressure energy analyzer although they are not necessary for a high vacuum energy analyzer. To increase the overall collection rate, there are more than one focusing lenses that are the modulated electric/magnetic fields to make photoelectrons travel along certain trajectory to focus on focal points. Red lines in the middle of Figure 6.11 schematically present the trajectories of photoelectrons in focusing lenses in the ambient pressure energy analyzer. The earliest example of using focusing lens in Torr pressure on an AP-XPS instrument was reported in 2002.[18] Figure 6.11 schematically shows the trajectory of photoelectrons in each electrostatic lens that is modulated through tunable voltages applied to +X, −X, +Y, and −Y deflection plates at each stage. Between two adjacent differential stages, there is a focal point, mentioned above, where photoelectrons are focused and then subsequently enter the next differentially pumped region where photoelectrons are focused again, by which the spreading photoelectrons entering the vacuum through the aperture can be focused and travel in the energy analyzer without a major loss. The optimization of the trajectory is done with the electronics of the ambient pressure

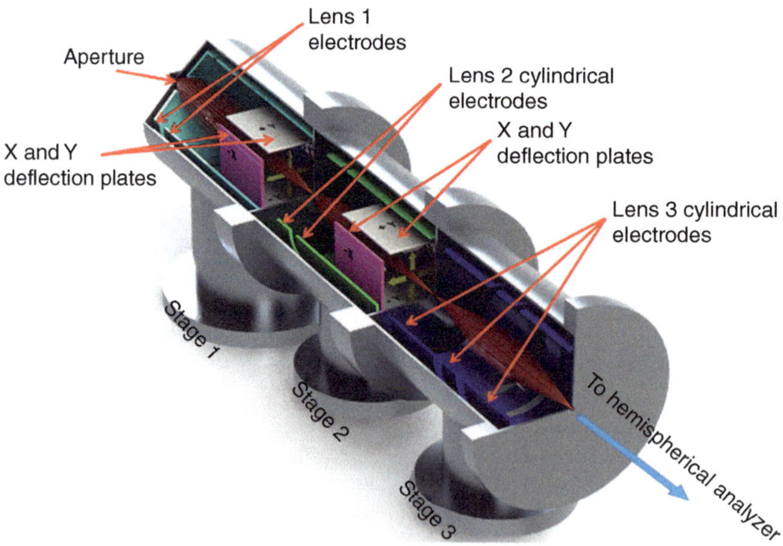

Figure 6.11 Schematic showing the working principle of focusing lenses in differential pumping stages of a commercial energy analyzer of AP-XPS produced by a manufacturer in Europe. *Source:* Reproduced with permission from Nguyen et al.[13]/American Chemical Society.

energy analyzer available on the market a decade ago. Notably, Figure 6.11 is an quite simplified schematic for the trajectory of photoelectrons in the energy analyzer.

6.4 Mass Spectrometer with Capability of Measurement of Catalytic Performance

A mass spectrometer can be installed on a pumping stage where a base pressure is 10^{-7}–10^{-8} Torr. As the gas around the sample in the reaction cell is pumped to all pumping stages, the mass spectrometer installed on a differential stage, typically the second stage, can provide information on the identities and composition of gases in this stage which originate at the reaction cell. It could be arguable that the catalytic activity and selectivity measured with a quadruple mass spectrometer installed on the second pumping stage of the ambient pressure energy analyzer are representative for the identity and composition of the gaseous species on the on-site gas proximal to the surface of a catalyst during catalysis (Figure 6.12)[19]. Compared to the evaluation of catalytic activity and selectivity with a mass spectrometer, the identity and composition of gaseous species proximal to the catalyst surface can be measured with a new spectroscopy recently reported by Tao et al., termed photoelectron-induced excitation (PEIE) spectroscopy.[19] The spectrum of PEIE is automatically generated by photoelectrons inelastically scattered by molecules or even radicals in the gas phase while the catalyst surface during catalysis is being characterized with AP-XPS. Since the parent peaks of extended PEIE spectrum represent surface chemistry of the catalyst and the PEIE spectrum reflects molecular identity and partial pressure of constituting gases proximal to a working catalyst surface, an integration of PEIE spectroscopy to ambient pressure XPS offers unique, dual function, characterization of a catalyst surface during catalysis with AP-XPS along with simultaneous measurement of on-site catalytic performance with PEIE spectroscopy; this integration of AP-XPS and PEIE is significant for fundamental understanding of a catalytic reaction in a real-time manner of spatial proximity and temporal simultaneity.[19]

Mass spectrometry is a technique that can qualitatively identify the constituting gases and quantitatively measure the gas composition through analyzing the gas in a differentially pumped energy analyzer that is away from the catalytic event by a couple of meters. The travel distance for gas above the surface of the catalyst installed in the reaction cell to the mass spectrometer is about 1–2 m. Compared to PEIE, a significant difference is the temperature of the gas to be analyzed by mass spectrometer far away from the high-temperature catalyst and the temperature of gas to be analyzed by PEIE.[19] As different constituting gases such as H_2 and CH_3OH in a gas mixture exhibit quite different diffusion rates in gas phase and sticking coefficient on stainless steel surface, these gases with high boiling points such as C_2H_5OH are partially lost in the travel from a high-temperature catalyst surface to a mass spectrometer, the composition in terms of partial pressure of each constituting gas measured by a mass spectrometer at the location of the ionizer of the mass spectrometer installed on the second pumping stage of a differentially pumped energy analyzer could be different, even very different from that of the on-site gas phase proximal to the catalyst surface in the reaction cell. Figure 6.12 schematically shows the difference between (a) the analysis of product in an off-site gas phase with gas chromatography (GC)

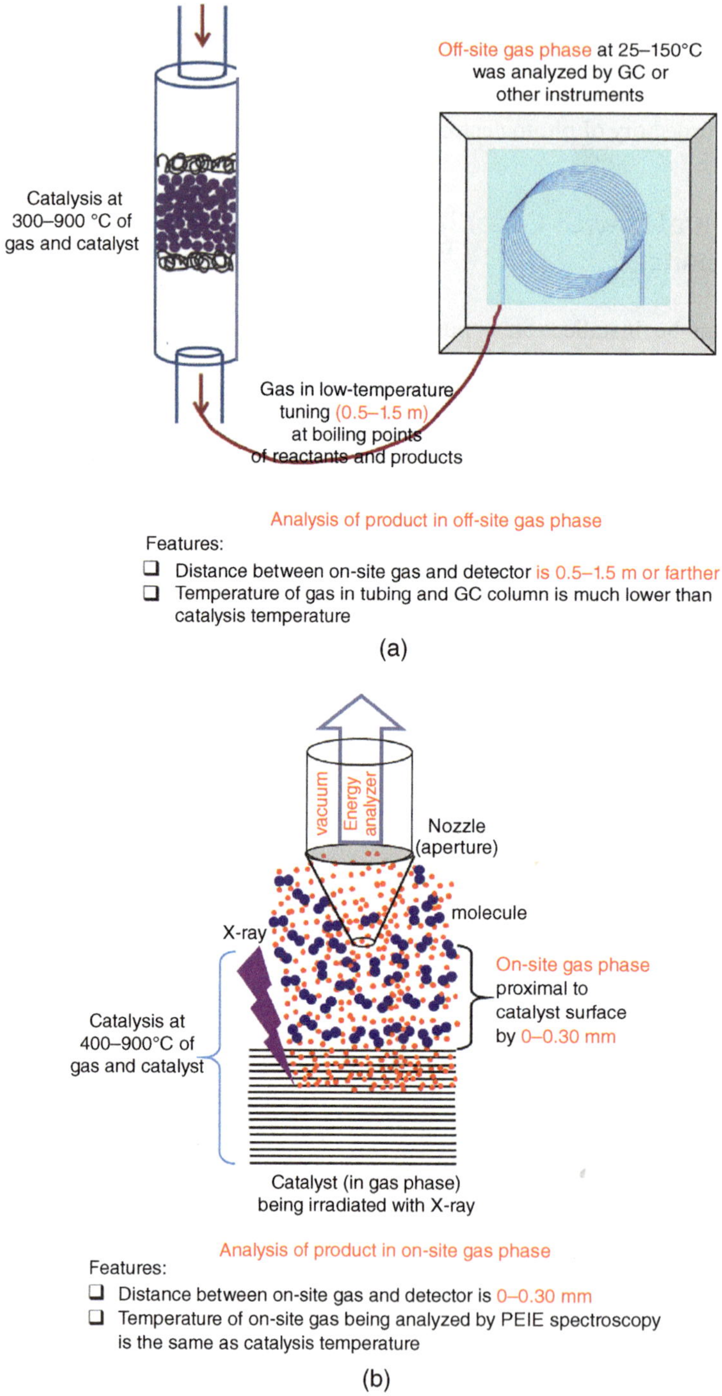

Figure 6.12 Schematics showing the difference between (a) the analysis of product in an *off-site gas phase* with gas chromatography (GC) or other analytic techniques and (b) the analysis of products in an *on-site gas phase* using photoelectron-induced excitation (PEIE) spectroscopy. In the analysis of off-site gas phase, the distance between the off-site gas phase of products and the catalyst is in the range of 0.5–1.5 m or farther. In the PEIE analysis of on-site gas phase, the distance between on-site gas phase and catalyst surface is only 0–0.30 mm. *Source:* Reproduced with permission from Tao et al.[19]/American Chemical Society.

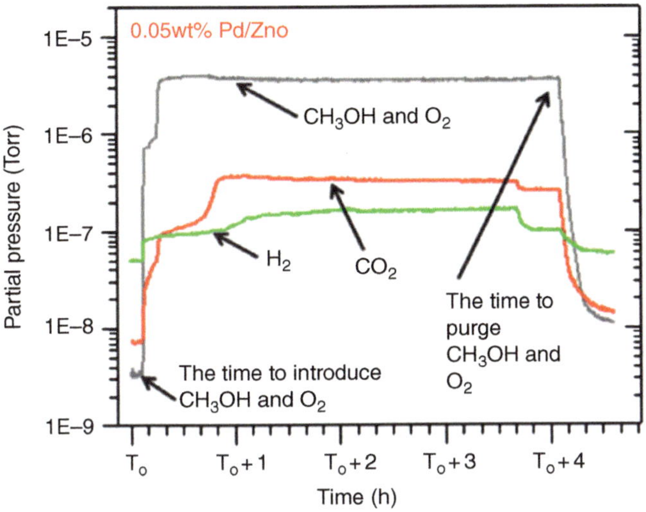

Figure 6.13 Analysis of composition of gas in the reaction cell during catalysis at 290 °C with mass spectrometer on online mode while AP-XPS was analyzing the catalyst surface during methanol partial oxidation. *Source:* Reproduced with permission from Tang et al.[20]/American Chemical Society.

or other analytic techniques such as a mass spectrometer and (b) the analysis of products in an on-site gas phase using PEIE spectroscopy reported recently.[19]

Although mass spectrometry could not perform an on-site analysis of the gas proximal to catalyst surface, it is still highly valuable to track how partial pressure of a reactant or a product changes as a function of reaction parameters such as catalytic temperature or the composition of the feed-in gas. For instance, Figure 6.13 demonstrates the capability of mass spectrometer in tracking the evolution of the partial pressures of CH_3OH (or O_2) (gray), H_2 (green), and CO_2 (red) as a function of a catalytic temperature.[20] Such a correlation between the change of partial pressure of a specific product such as H_2 with the perturbation such as variation of temperature, gas composition, or partial pressure of a reactant (temperature represented with time on the horizontal axis) is significant for fundamental understanding of catalytic reactions.

A unique function of mass spectrometry is the identification of isotope substitutes. This unique capability offers the opportunity of identifying products, stable intermediates, and even catalytic sites and even how they form. Typically, replacement of ${}^{1}_{1}H$ with ${}^{2}_{1}H$, ${}^{12}_{6}C$ with ${}^{13}_{6}C$, and ${}^{16}_{8}O$ with ${}^{18}_{8}O$ of a specific function group of a reactant or even catalytic sites of a catalyst can allow for identifying whether these replaced atoms or the function groups can participate in the reaction or not. For instance, whether the surface lattice oxygen atoms of $NiCo_2O_4$ can directly participate in complete oxidation of CH_4 was elucidated with the isotope-replacement function of the mass spectrometry. ${}^{18}_{8}O$ isotope-labeled catalyst, $NiCo_2{}^{16}_{8}O_{4-x}{}^{18}_{8}O_x$, was used for elucidating whether surface lattice oxygen atoms of $NiCo_2O_4$ are a part of the catalytic sites to participate into the catalytic reaction (Figure 6.14).[21] The isotope-labelled catalyst was prepared by annealing $NiCo_2O_4$ in ${}^{18}_{8}O_2$. While flowing CH_4 and ${}^{16}_{8}O_2$ through this isotope-labeled catalyst, both $C{}^{16}_{8}O{}^{18}_{8}O$ ($m/z = 46$)

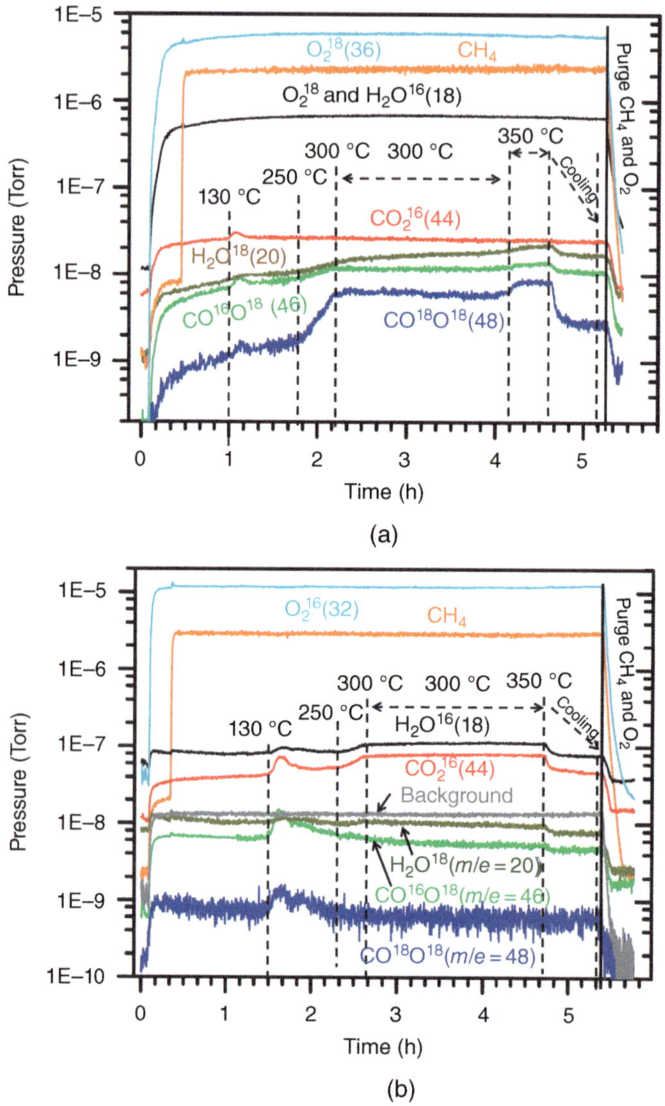

Figure 6.14 Evolution of products during (a) CH_4 complete oxidation in O_2^{18} on $NiCo_2O_4^{16}$, and (b) CH_4 complete oxidation in O_2^{16} on isotope-labeled catalyst $NiCo_2O_{4-x}^{16}O_x^{18}$ along the increase of catalysis temperature. *Source:* Reproduced with permission from Tao et al.[21]/Springer Nature.

and $C^{18}_{8}O_2$ ($m/z = 48$) were observed by mass spectrometry (Figure 6.14a). It suggests that the surface lattice oxygen atoms of the catalyst $NiCo_2O_4$ directly participated in this complete oxidation of CH_4. Clearly, a mass spectrometer installed on the differential pumping stage can provide important information for understanding catalytic mechanism. Particularly, its function can be enhanced once isotope-labeling experiments are incorporated to the studies using the online mass spectrometer of AP-the XPS system.

References

1 Somorjai, G. A. 1981. *Chemistry in Two Dimensions: Surfaces*. Cornell University Press.

2 Tao, F. F. and Nguyen, L. 2018. "Interactions of gaseous molecules with X-ray photons and photoelectrons in AP-XPS study of solid surface in gas phase." *Phys. Chem. Chem. Phys.* 9812–9823.

3 Tao, F. and Crozier, P. A. 2016. "Atomic-scale observations of catalyst structures under reaction conditions and during catalysis." *Chem. Rev.* 20, 3487–3539.

4 Tang, Y., Nguyen, L., Li, Y. *et al.* 2016. "Surface of a catalyst in a gas phase." *Curr. Opinion. Chem. Eng.* 12, 52–61.

5 Velasco-Vélez, J., Pfeifer, V., Havecker, M. *et al.* 2016. "Atmospheric pressure X-ray photoelectron spectroscopy apparatus: bridging the pressure gap." *Rev. Sci. Instrum.* 87, 053121.

6 Nguyen, L., Tao, P., Liu, H. *et al.* 2018. "Studies of surface of metal nanoparticles in a flowing liquid with XPS." *Chem. Commun. (Camb.)* 54, 9981–9984.

7 ViolBarbosa, C. E., Ouardi, S., Fecher, G. H. *et al.* 2013. "Magnetic dichroism in angular resolved hard X-ray photoelectron spectroscopy from buried magnetic layers." *J. Electron Spectrosc.* 189, 146–151.

8 Weiland, C., Rumaiz, A. K., Pianetta, P. *et al.* 2016. "Recent applications of hard X-ray photoelectron spectroscopy." *J. Vac. Sci. Technol. A* 34, 030801.

9 Cancellieri, C. and Strocov, V. N. 2018. *Spectroscopy of Complex Oxide Interfaces: Photoemission and Related Spectroscopies*. Vol. 266. Springer.

10 Takagi, Y., Uruga, T., Tada, M. *et al.* 2018. "Ambient pressure hard X-ray photoelectron spectroscopy for functional material systems as fuel cells under working conditions." *Acc. Chem. Res.* 51, 719–727.

11 Woicik, J. C. and Woicik, J. 2016. Kobayashi, K. "Chapter 18 HAXPES Applications to Advanced Materials", in *Hard X-ray Photoelectron Spectroscopy (HAXPES)*. Springer.

12 Seah, M. P. and Dench, W. A. 1979. "Quantitative electron spectroscopy of surfaces: a standard data base for electron inelastic mean free paths in solids." *Surf. Interface Anal.* 1, 2–11.

13 Nguyen, L., Tao, F. F., Tang, Y. *et al.* 2019. "Understanding catalyst surfaces during catalysis through near ambient pressure X-ray photoelectron spectroscopy." *Chem. Rev.* 119, 6822–6905.

14 Nguyen, L. and Tao, F. F. 2016. "Development of a reaction cell for in-situ/operando studies of surface of a catalyst under a reaction condition and during catalysis." *Rev. Sci. Instrum.* 87, 064101.

15 Salmeron, M. and Schlogl, R. 2008. "Ambient pressure photoelectron spectroscopy: a new tool for surface science and nanotechnology." *Surf. Sci. Rep.* 63, 169–199.

16 Tao, F., Tang, D., Salmeron, M. *et al.* 2008. "A new scanning tunneling microscope reactor used for high-pressure and high-temperature catalysis studies." *Rev. Sci. Instrum.* 79, 084101.

17 Takanabe, K., Khan, A. M., Tang, Y. *et al.* 2018. "Integrated in situ characterization of a molten salt catalyst surface: evidence of sodium peroxide and hydroxyl radical formation." *Angew. Chem. Int. Ed. Engl.* 56, 10403–10407.

18 Ogletree, D. F., Bluhm, H., Lebedev, G. *et al.* 2002. "A differentially pumped electrostatic lens system for photoemission studies in the millibar range." *Rev. Sci. Instrum.* 73, 3872.

19 Tao, F., Nguyen, L., and Tang, Y. 2022. "Report on photoelectron-induced excitation spectroscopy and its close correlation with fundamental studies of catalysis." *J. Phys. Chem. C* 126, 13069–13087.

20 Tang, Y., Zhang, S., Rawal, T. B. *et al.* 2020. "Atomic-scale structure and catalysis on positively charged bimetallic sites for generation of H2." *Nano Lett.* 20, 6255–6262.

21 Tao, F., Shan, J., Nguyen, L. *et al.* 2015. "Understanding complete oxidation of methane on spinel oxides at a molecular level." *Nat. Commun.* 6, 7798.

7

Experimental Methods of AP-XPS Studies

7.1 Leak Test of Reaction Cell

Leak test is a regular practice in terms of maintenance of a vacuum environment in surface science and catalysis communities. Regarding AP-XPS, a type of leak test is to check the sealing of a reaction cell. For the reaction cell described in Section 6.2, there could be two potential leaking places including the door on the reaction cell for exchanging catalyst sample and the interface between the cell head and the front cone of a differentially pumped energy analyzer (Figure 6.5–6.7 in Chapter 6). The reaction cell is placed in high vacuum environment of the main chamber of an AP-XPS system. This main chamber is connected to X-ray source, energy analyzer, and other parts. A leak at the reaction cell could result in a serious problem to the X-ray source.

Whether the rection cell leaks or not can be found out by monitoring the pressure, $p_{UHV\ chamber}$, in the main chamber where the reaction cell is placed in, as a function of time while gas at a pressure, p_{cell}, such as 10 Torr is flowing through the reaction cell. The specific leak test is typically done through a mass spectrometer installed on the UHV chamber. Figure 6.8 is such a leak test. It presents the pressure in a UHV chamber as a function of time while the pressure of flowing gas in the cell remains at 10 Torr. Clearly, when gas pressure in the cell remains at 10 Torr, the pressure in the UHV chamber is about 3×10^{-9} Torr that is slightly higher than the base pressure of an UHV XPS chamber (Figure 6.8d). This leak test suggests the sealing of the reaction cell is excellent[1].

7.2 Exclusion of Catalysis by Reaction Cell

As the cell body is made of stainless steel, Ni or other transition metal of a stainless steel could be an unwanted catalyst. This is because these transition metals are active for catalyzing reactions such as CO oxidation, water-gas shift, hydrogenation, methane steam reforming, and methane dry reforming. These unwanted catalysts on the cell body could give misleading information on catalytic activity or selectivity of a reaction being studied because the observed activity and selectivity could partially or entirely contributed from

Application of Ambient Pressure X-ray Photoelectron Spectroscopy to Catalysis, First Edition. Franklin Tao.

these unwanted catalysts. Thus, it is necessary to check whether internal surface of the reaction cell has been passivated fully for studying the interested reactions. For testing whether this passivation is effective or not, CO oxidation is taking as a probe reaction. A mixture of CO and O_2 was flowing into the cell when a sample holder without a catalyst was heated from 25 to 100 °C. Clearly, the one without passivation of the internal surface of the reaction cell can catalyze CO oxidation since CO_2 was clearly observed by the mass spectrometer (Figure 6.10a). Obviously, it is important to passivate the inner surface of a reaction cell. Another method of minimizing the activity of the internal surface for an interested reaction is to keep the reaction cell at a temperature as low as possible; For this purpose, the reaction temperature should be lower than 100 °C; However, this is not feasible since most reactions on heterogeneous catalysis are performed at a temperature higher than 300 °C. A practical solution is to keep the reaction cell wall at a temperature as close to room temperature as possible through a good thermal insulation.

A good suggestion is always to run a blank experiment while a mixture of reactants of the reaction was flowing through the cell and the sample holder without a catalyst is heated to a catalysis temperature. If products of this reaction are detected by a mass spectrometer, two solutions can help to correct. One short-term solution is to make an appropriate calibration; for instance, the amount of products catalyzed by the "false catalyst" in terms of the reaction cell at a catalysis temperature should be subtracted from the total amount of the product generated from both the cell (the false catalyst) and the placed catalyst at this temperature. A long-term solution is to fully passivate the internal surface of the reaction cell with Au thin film.

7.3 Tunning and Control of Sample-Aperture Distance

Control of sample-aperture distance is the key for a successful AP-XPS experiment. It is the base for quantitative analysis and for comparing XPS peak intensity performed at different temperatures, different pressure, or different molar ratios of reactant gases. AP-XPS studies can be categorized into (i) sample-aperture distance unrelated analysis and (ii) sample-aperture distance related analysis. For instance, the measurement of atomic ratio of Rh to Pd of $Rh_{1-x}Pd_x$ nanoparticles in gas phase is a sample-aperture distance ***unrelated*** analysis because the Rh 3d and Pd 3d have very similar binding energies and are always collected under the same condition in terms of the same temperature and pressure.[2] In this type of studies, the sample-aperture distance does not change the measured Rh/Pd atomic ratio. However, exploration of how the absolute intensity of Rh $3d_{5/2}$ peak in surface region in the mixture of NO and CO changes as a function of catalysis temperature is a sample-aperture distance related analysis since the sample-aperture distance varies the targeted information, absolute intensity of Rh $3d_{5/2}$ XPS peak significantly. In addition, as to be discussed in Chapter 8, the sample-aperture distance is also significant for measurement of the atomic ratio of elements X to Z in a sample if the chosen subshell levels of elements X and Z have quite different binding energies. Thus, the measurements of atomic ratio X/Z as a function of temperature are sample-aperture distance related analyses.

As a matter of fact, the quantitative analysis of a surface composition of surface in gas phase should be based on the intensities of photoelectrons leaving surface before scattering with gas phase molecules such as $I_{A(surf)}$ and $I_{X(surf)}$; it should not be calculated with the intensities of photoelectrons collected at aperture, $I_{p(A)(aperture)}$ and $I_{p(X)(aperture)}$. However, only the $I_{p(A)(aperture)}$ and $I_{p(X)(aperture)}$ are measurable. Thus, the accuracy of measuring atomic ratio of X to Z of sample surface by AP-XPS depends on whether the decay factor of element A, $F_{p(A)}$ is the same as $F_{p(X)}$ of element X. Then, we need to understand what determines the decay factor F_p when photoelectrons pass through a gas phase.

As described in Sections 5.5 and 6.1.3, the decay factor of photoelectron intensity, F_p, is determined by the equation, $F_p = \frac{I_p}{I_0} = e^{-\frac{z\sigma_{KE}p}{kT}}$. Here z is the thickness of the gas region; it is in fact the distance between the catalyst surface and aperture. Under the assumption that σ_{KE} is the same for the photoelectrons of A and X, z is the main factor in determining F_p. To have the same F_p for XPS peak of A and X, z must be the same while XPS peaks of elements A and X are collected. As F_p exponentially increases or decreases as a function of z, a precise control of z is significant for accurately measuring the atomic ratio of A to X of the catalyst surface. Inequal distances (z) while collecting XPS peaks of A and X could give a large accidental error bar for the quantitative analysis of the atomic ratio A to X.

How significant a change of sample-aperture distance could impact the absolute intensity of the XPS peak can be seen from Figure 7.1. Figure 7.1 is a plot of the Ag $3d_{5/2}$ peak

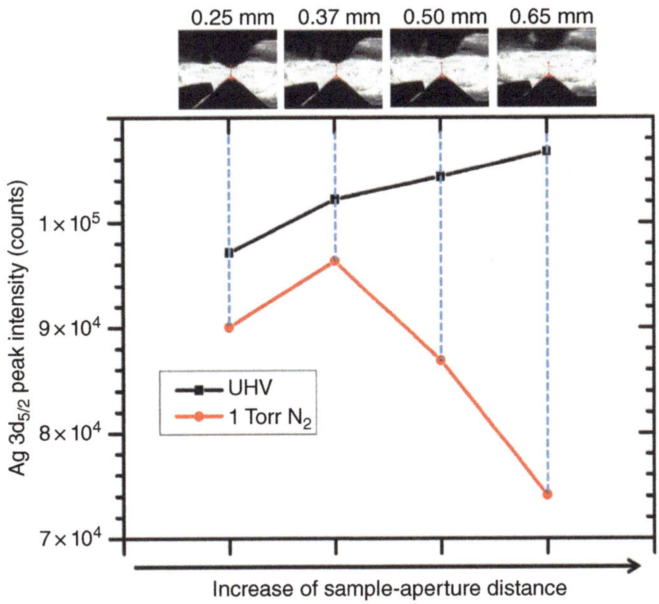

Figure 7.1 Test results of Ag $3d_{5/2}$ peak intensity as a function of sample-aperture distance in UHV and in 1 Torr N_2. Slight change in sample-aperture distance can cause great fluctuation in XPS signal intensity. Analyzer slit size was 4 mm × 30 mm. Pass energy used was 100 eV with standard lens mode. *Source:* Reproduced with permission from Tao et al.,[1]/AIP Publishing LLC.

intensity as a function of the sample-aperture distance when the Ag foil is at 25 °C in 1 Torr N_2 (red line). For comparison, the data collected in UHV was also plotted as a function of the sample-aperture distance and this plot is shown in a black line. The black line shows the Ag $3d_{5/2}$ intensity in UHV increases along the increase of the sample-aperture distance, suggesting a shadow effect of the aperture on X-ray irradiation to the sample surface. The red line in Figure 7.1 is the evolution of Ag $3d_{5/2}$ as a function of sample-aperture distance while the silver foil is in 1 Torr N_2. At 0.37 mm, the Ag $3d_{5/2}$ intensity is maximum. Along the increase of sample-aperture distance, the intensity dropped quickly. For instance, an increase of a distance by 0.28 mm (from 0.37 to 0.65 mm) results in a decrease of the intensity of the collected photoelectrons by 20% although the authentic intensity of photoelectrons leaving the catalyst surface is the same. Again, the red line in Figure 7.1 clearly shows the sample-aperture distance dramatically impacts the intensity of XPS peak collected at the aperture. It suggests the significance of an accurate control to the sample-aperture distance during data acquisition.

If the distance for collecting XPS peak of element A is different from that for collecting XPS peak of element X, a large error of the atomic ratio of A to X that is calculated through the ratio $\frac{I_A}{I_Z}$ is expected. In addition, in some cases it is necessary to investigate how the absolute intensity of element A, $\boldsymbol{I}_{\mathbf{A}}$ changes as a function of temperature, gas pressure, or gas composition. If experiment 1 is performed at a sample-aperture distance (z_1) different from that at a sample-aperture distance in experiment 2 (z_2), the additional variable, z is unfortunately introduced to $I_{A(T_1)}$. Under this condition, there are in fact two variables, temperature and sample-aperture distance varying $I_{A(T_2)}$. Then, the found dependence of intensity, $\boldsymbol{I}_A$, on temperature, T could be incorrect.

An accurate measurement of the distance between sample surface and aperture is not a trivial task. Figure 7.2 and 7.3 shows the region of sample and aperture can be monitored with a video camera combined with an optical lens. The lens was positioned on the air side of a window of the main chamber facing to the sample. A high-resolution digital video camera was installed above the lens. The real-time image was transmitted to a computer simultaneously. Then, the region of sample and aperture and its movement (if any) was watched in real time. Notably, the sample-aperture distance can be measured with the digital video camera with a combined lens. With this setup, the frame of the region consisting of the sample surface and aperture with a physical size of 1 mm × 1 mm level can be enlarged to show on the computer screen with a size of 100 mm × 100 mm on the screen. This magnification-based visualization makes the measurement of the distance between the sample surface and aperture accurate. Figure 7.2 shows the object and its image. The separation of sample from aperture at the physical distance of 1 mm can be imaged on the computer screen. The size of an image on the screen is much larger than the size of an object (the sample and aperture). The enlargement factor, M, can be defined by $M = \frac{d_{sample-aperture\ distacne\ on\ screen}}{d_{sample-aperture\ distance\ (real,object)}}$. Then, the real sample-aperture distance in the reaction cell in another specific experiment can be readily calculated by the equation $\frac{d_{sample-aperture\ distance\ on\ screen}}{M}$ with the distance between sample and aperture on the computer screen ($d_{sample-aperture\ distance\ on\ screen}$).

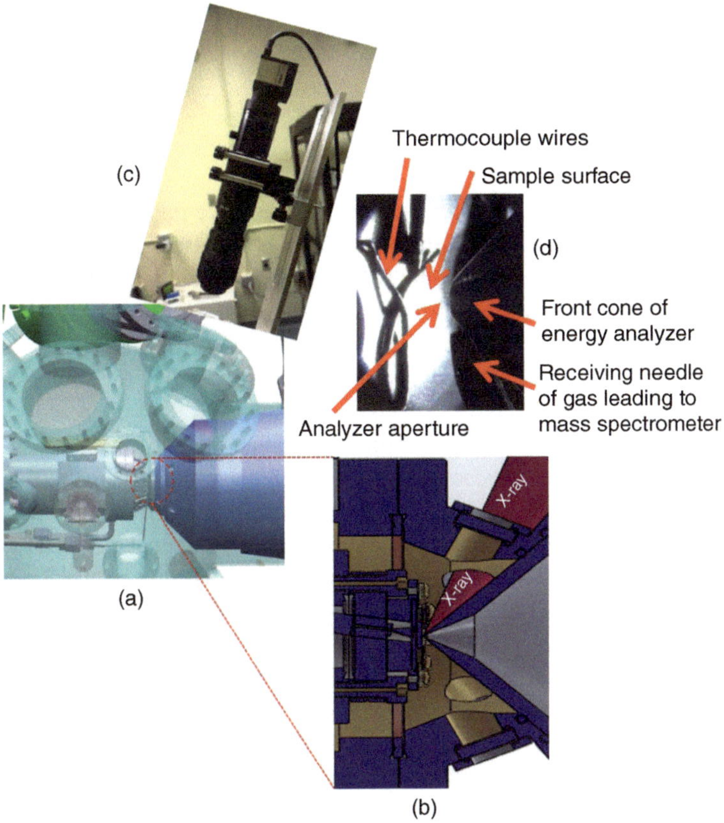

Figure 7.2 Schematic and photo of distance between the sample and the front cone aperture. (a and b) Status of sample and aperture when the neck of the reaction cell matches the font cone. (c) Video camera to visualize the sample and aperture through a 4.5″ flange window of UHV chamber and a glass window of the reaction cell. (d) Photo of the screen on a computer showing the magnified region of sample and aperture and their surroundings. *Source:* Reproduced with permission from Tao et al.,[1]/AIP Publishing LLC.

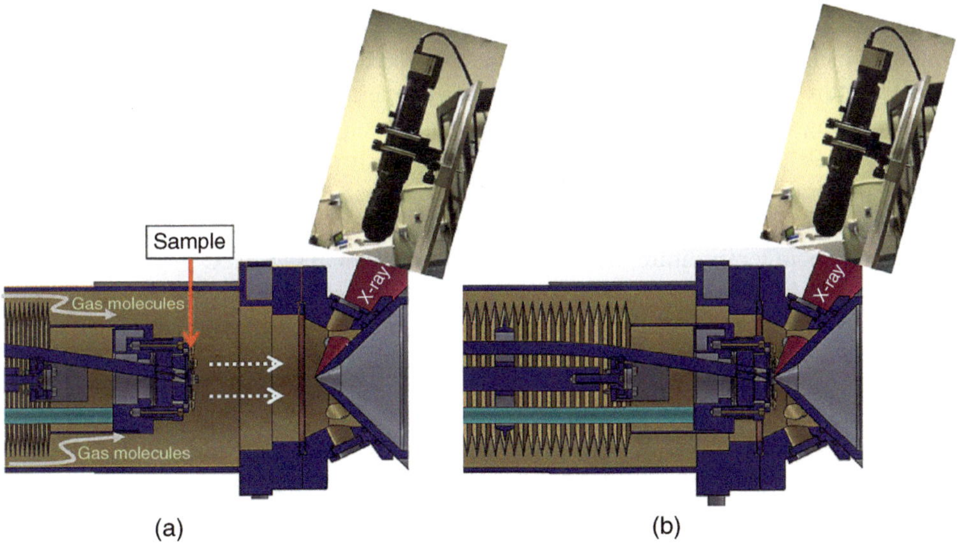

Figure 7.3 Schematic showing the process of approaching the sample to the front cone aperture of an energy analyzer. A high-resolution camera is used to monitor the distance between the sample and the aperture to make sure a reproducible sample-aperture distance can be achieved. (a) Sample is away from the aperture. (b) Sample is close to the aperture with a distance of 0.5 mm. *Source:* Reproduced with permission from Nguyen et al.,[1]/AIP Publishing LLC.

7.4 Sample Heating and Temperature Control

Temperature of most reactions performed on heterogeneous catalysts are performed in the temperature range of 200–800 °C. Controllable heating of a catalyst is important for a successful AP-XPS study of surface of a catalyst during catalysis. A well-designed heating system should be able to heat a sample and the gas surrounding the sample to an ideal catalysis temperature, but the around the catalyst should remain at temperature as close to room temperature as possible. A catalyst in the cell can be heated by heating the external wall of the stainless steel cell to conductively heat the sample that is assembled on the internal wall of the reaction cell. In such an indirect heating to the catalyst, the heating source is a filament installed (at UHV side) near to the external surface of the reaction cell. By applying a high voltage to the filament, electrons emitted from the filament, typically Th-doped tungsten wire are accelerated to bombard the external surface of the reaction cell, generating heat. The heat received by the reaction cell is conductively transferred to the sample stage, then sample holder, and finally the sample installed on the sample holder. The disadvantage of this indirect heating is the sloppy control of temperature of the reaction cell since the sink to be heated in this method is quite large and the thermal conduction from the sink to the sample is slow. Here the sink is the mass of the specific region of the reaction cell being heated. This filament heating method is still used on some AP-XPS systems since it does not need expensive experimental setup.

Compared to the above filament heating, a much better method is an infrared (IR) laser-based heating of sample in a reaction cell developed in Tao group.[1] It is a direct, non-contact heating to the back side of a sample holder by infrared laser beam (Figure 6.9). The IR laser is generated by a power supply and then introduced to the IR-transparent window of the reaction cell through optical fiber. Then, the infrared laser beam is transmitted through this window to heat the back side of the sample holder holding a catalyst such as powder of nanoparticles or a single crystal model catalyst. This method exhibits several advantages. First, laser heating is much faster than filament heating. As shown in Figure 6.9a, b, and c, a sample in a reaction cell can be heated from room temperature to 600 °C within just a few minutes, which is much faster than the filament-based heating. Further tests showed that the temperature of the reaction cell can remain at a temperature as low as 30 °C while 12 Torr N_2 was flowing through the reaction cell when the sample remains at a high temperature (Figure 6.9e). More importantly, laser heating is highly localized, which minimized any potential thermal diffusion to the surrounding area of the sample.[1] The good thermal insulation of the laser-heated sample is evidenced by the low temperature of reaction cell at 40 °C, very close to room temperature, while the sample was heated to 600 °C (Figure 6.9e). The highly localized heating was realized by the excellent thermal insulation between the sample holder and the wall of the reaction cell.[1] Particularly, the excellent thermal insulation benefits from the use of fused silica for designing a sample stage since the fused silica has a quite low thermal conductivity.

7.5 Online Measurement of Reactants and Products

The flowing gas surrounding the catalyst surface in the reaction cell can be analyzed spontaneously during catalysis through an online mass spectrometer while the surface of the catalyst is being characterized with AP-XPS during catalysis. The online mass spectrometry is performed through analyzing the gas typically in the second differential stage of the energy analyzer where a small portion of the gas in the reaction cell is detected. The gas is obviously originated at the region of sample in the reaction cell. Alternatively, the gas can be sampled from the reaction cell during catalysis through a nozzle to an additional HV chamber wherein a mass spectrometer is installed. Figure 7.4a schematically shows this extraction of gas. For both methods, in principle to measure product(s) or reactant(s) that

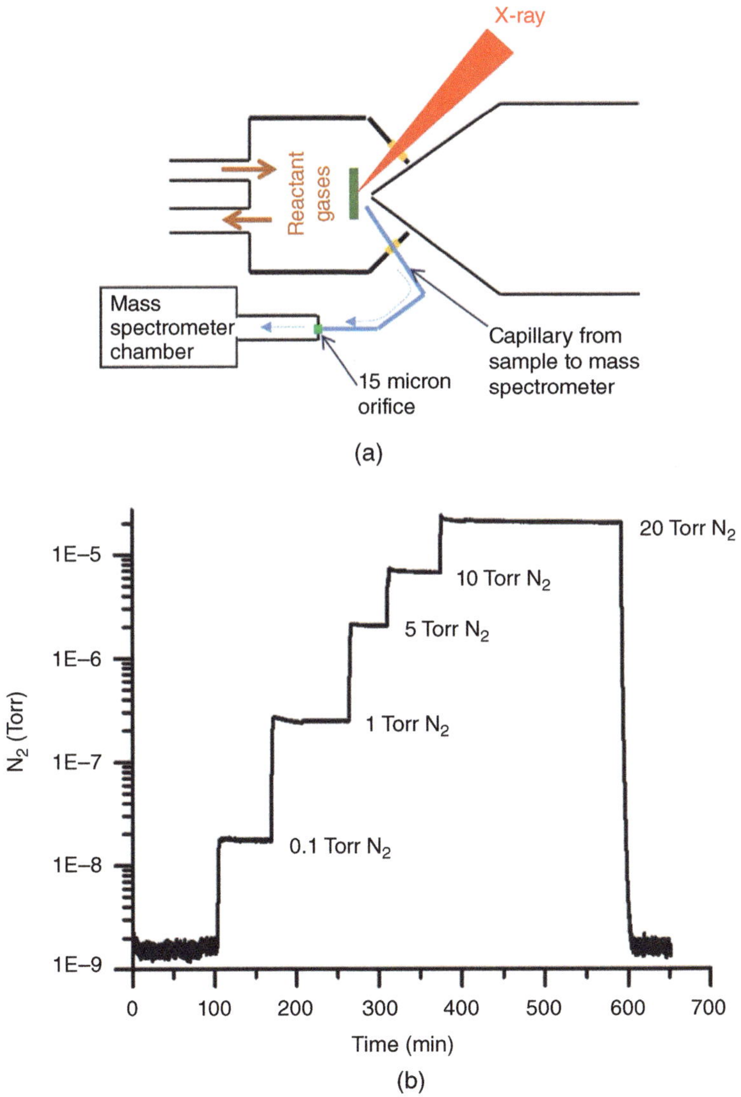

Figure 7.4 (a) Diagram showing the setup for online gas products analysis. A small capillary was installed on the reaction cell. This capillary protrudes into the reaction and situates close to the sample surface. This allows the capillary to sample reaction gas product responsively into a mass spectrometer chamber for analysis with high sensitivity. In (b), an example of N_2 mass spectrometer data at different N_2 pressure in reaction cell. *Source:* Reproduced with permission from Tao et al.[1]/ AIP Publishing.

have boiling points higher than room temperature, it is necessary to keep the nozzle and introduction tubing (between the reaction cell and the additional HV chamber) at a temperature higher than their boiling points to make sure such reactant(s) or product(s) can be delivered to the mass spectrometer. There is no significant difference between the two extraction methods. A notable difference is that the second method (Figure 7.4a) allows a readily warming to gas tubing in contrast to the first method. This is because remaining the two stages at a relatively high temperature in the first extraction method could damage the electronics in the two stages of the differential pumping energy analyzer.

With the assembled mass spectrometer, change of gas composition or variation of partial pressure of a specific reactant or products upon any perturbations such as an elevation of catalyst temperature can be tracked. As shown in Figure 7.4b, the change of N_2 pressure in the reaction cell can be readily sensed through the online mass spectrometer. A working curve can be established by plotting the actual pressure in the gas cell (measured with a capacitance gauge) as a function of the pressure read from the mass spectrometer, generating a sensitive factor. With the sensitive factor, the gas pressure in the reaction cell can be evaluated upon a measurement of mass spectrometry.

The online mass spectrometer can be used to track how partial pressure of a specific gas changes due to a specific perturbation. For instance, it can track the partial pressure of a specific product along the increase of temperature of a catalyst, by which the onset temperature of a catalysis on a specific catalyst can be identified. Through measuring the partial pressure or concentration of a specific product at a specific temperature, the reaction rate at this temperature in terms of turnover rate under a kinetics-controlled regime can be evaluated. By measuring reaction rates in an interested temperature range, an Arrhenius plot can be built for evaluation of the apparent activation barrier. Another example is that it can readily uncover the specific temperature where a poisonous gas such as CO can deactivate the catalyst through monitoring the evolution of partial pressure or concentration of the specific product such as CO_2 along the increase of catalysis temperature. By integrating the evolution of surface chemistry of a catalyst along the perturbation into the evolution of partial pressure or concentration of a specific product measured by the mass spectrometer, the correlation between a catalyst surface during catalysis and its corresponding catalytic performance can be established. This correlation can aid proposing a catalytic mechanism at a molecular level through computational studies.

7.6 Spectroscopic Titration of Surface Species

Exposure, immersion, or soaking of an active surface structure of a catalyst such as a metal single crystal or metal nanoparticle in static or flowing reactant gas such as O_2 at certain temperature could form a few coexisting surface phases. Here a surface phase is a broadly defined term. A surface phase is not necessarily a catalytic phase. If a surface phase is active for a catalytic reaction, it is called a catalytic phase. To pursue a fundamental understanding of catalytic mechanism, it is crucial to identify the authentic catalytic phase. However, distinguishing an active catalytic phase(s) from a non-active one(s) or differentiating two coexisting active catalytic phases is quite challenging for most surface analytical techniques.

AP-XPS is the right technique to distinguish a catalytic phase from those noncatalytic ones since AP-XPS allows for differentiating surface phases in gas phase and tracking the evolution of the quantity of each surface phase when the catalyst surface is being consumed or transformed by a titrant such as CO. In addition, based on the titration rate, it can differentiate a highly active phase from a moderately active phase since the highly active phase has a higher titration rate than a moderately active phase. The following gives one example how AP-XPS can be used for this purpose. Here the identification of an active phase for CO oxidation on Pt model catalyst is briefed.[3]

In the case of CO oxidation on a Pt catalyst, a surface phase can be a surface oxide thin film, surface oxide, or chemisorbed oxygen atoms on a metal surface. O 1s of each surface phase has a different binding energy; thus, collection of O 1s peak allows for distinguishing these surface phases and thereby identifying catalytic phases. AP-XPS was used in tracking the evolution of the quantity of each surface phase in terms of analyte when the surface is being titrated by the titrant, CO. The raw quantity gained through a titration is the intensity of O 1s of each surface phase. In addition, the raw data of each surface phase can be represented by the coverage of O atoms of each surface phase on Pt(111). Thus, the coverage of O atoms of each surface phase, similar to the concentration of an analyte, was plotted as a function of the amount of the CO dose, similar to the amount of titrant (Figure 7.5). If a containing-O surface phase is active for CO oxidation, CO can consume oxygen atoms of this containing-O surface phase to form CO_2. The titration happens. Then, the coverage of the oxygen atoms of this surface phase must decrease. The evolution of the surface O

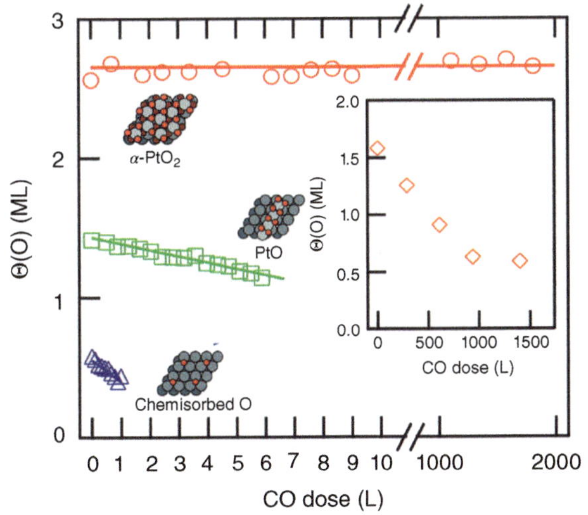

Figure 7.5 The coverage of surface-oxygen species [Θ(O)] while dosing CO, as determined using in situ O 1s XPS, for three oxygen layers on Pt(111): (red circles) an α-PtO_2 film with an initial coverage of 2.8 ML; (green squares) a 4O/α-PtO_2 mixture with an initial coverage of 1.4 ML; (blue triangles) an O_{ad}/α-PtO_2 mixture with an initial coverage of 0.47 ML. The solid lines are guides to the eye. P(CO) was 1×10^{-8} Torr in all measurements except that of the 2.8 ML α-PtO_2 film, for which a pressure of 1×10^{-5} Torr was employed. Inset: CO dose-dependent change of the surface-oxygen species from α-PtO_2 partially covering the surface. Note that Θ(O) excludes the coverage of CO_{ad}, i.e. Θ(O) = Θ(Ototal) − Θ(CO). *Source:* Reproduced with permission from Miller et al.[3]/American Chemical Society.

coverage of this containing-O surface phase as a function of carbon monoxide dose introduced to the surface gives the titration lines here in Figure 7.5. How fast the oxygen atom coverage of this containing-O surface phase decreases is the indicator of the activity of this containing-O surface phase. A larger absolute value of the slope of a surface phase in Figure 7.5 means this containing-O surface phase is more active for CO oxidation.

Figure 7.5 is the titration result.[3] As shown in Figure 7.5, exposure of a clean Pt(111) surface to different amount of O_2 can produce three different types of containing-oxygen surface phases that are α-PtO_2, PtO, and chemisorbed oxygen (also called O_{ad}/α-PtO_2 mixture with an initial coverage of 0.47 ML).[3] Their reactivity to CO was identified by exposing the three phases to CO. α-PtO_2 is not active for CO oxidation even though it was exposed to over 2000 Langmuir dose of CO. Figure 7.5 presents how the amounts of different surface phases decrease along the reaction with the titrant CO. Distinctly different from the inertness of α-PtO_2, 1 Langmuir dose of CO was consumed to react with the chemisorbed oxygen atoms (O_{ad}/α-PtO_2). The low consumption of CO is due to the low concentration of the phase of the chemisorbed oxygen as seen at the left-bottom corner of Figure 7.5. The PtO (also called a 4O/α-PtO_2 mixture with an initial coverage of 1.4 ML) requests about 7 Langmuir dose of CO for reaction with all O atoms of the surface PtO. The three titrations performed with AP-XPS shown with red, green, and blue symbols suggest that α-PtO_2 is not active for CO oxidation at all (red circles in Figure 7.5); the surface adsorbed O (O_{ad}/α-PtO_2) is the most active phase for CO oxidation (blue triangle in Figure 7.5) as the absolute value of its slope in Figure 7.5 is highest among the three surface phases. The PtO surface phase (green square in Figure 7.5) is moderately active for CO oxidation.

References

1 Nguyen, L. and Tao, F. 2016. "Development of a reaction cell for in-situ/operando studies of surface of a catalyst under a reaction condition and during catalysis." *Rev. Sci. Instrum.* 87, 064101.

2 Tao, F., Grass, M. E., Zhang, Y. *et al.* 2008. "Reaction-driven restructuring of Rh-Pd and Pt-Pdcore-shell nanoparticles." *Science* 322, 932–934.

3 Miller, D., Casalongue, H. S., Bluhm, H. *et al.* 2014. "Different reactivity of the various platinum oxides and chemisorbed oxygen in CO oxidation on Pt(111)." *J. Am. Chem. Soc.* 136, 6340–6347.

8

Difference in Data Analysis Between AP-XPS and High Vacuum XPS

8.1 Potential Difference in Measuring Atomic Ratio of Two Elements on Catalyst Surface

The basic principle of data analysis in AP-XPS is the same as high vacuum XPS as they follow the same concept. However, strictly speaking in some case the data collected in AP-XPS cannot be analyzed directly. This is due to a potential difference in the decay factor of photoelectrons intensity among different subshells. Here the decay factor, F_p, is defined to the extent of reducing the intensity of photoelectrons leaving the catalyst surface by the gas phase between the sample surface and aperture, $\frac{I_p}{I_s}$, through the equation, $F_p = \frac{I_p}{I_s}$. As shown in Figure 8.1a, in high vacuum XPS the intensity of photoelectrons entering an energy analyzer (I_s) is approximately the same as that leaving surface of a catalyst (I_s). However, the intensity of photoelectrons entering the energy analyzer (I_p) in an AP-XPS study is only a portion of the intensity of photoelectrons leaving surface of a catalyst (I_s). The decay factor for photoelectron intensity, F_p, is important when atomic ratio of two elements of a catalyst is analyzed. Only when the F_p of element A under the experimental condition is the same as that of element Z under the same condition, the ratio of photoelectron intensity of elements A to Z entering an energy analyzer, $\frac{I_{p(A)}}{I_{p(Z)}}$, is the same as the ratio of photoelectron intensity of elements A to Z leaving the catalyst surface, $\frac{I_{s(A)}}{I_{s(Z)}}$. Only under this assumption [$F_{p(A)} = F_{p(Z)}$], the photoelectron intensities collected at the aperture (Figure 8.1b) can be directly used for the calculation of atomic ratio.[1]

As suggested by Figure 5.8b, the intensity of photoelectrons after travelling through a gas phase of 0.5 or 1.0 mm and then being collected at aperture, I_p, is definitely lower than that of photoelectrons just leaving sample surface, I_s. Whether the decay factor, F_p, is the same for different photoelectrons is the first question to be addressed before we discuss whether the scattering of photoelectrons by the gas phase could distort the original information of surface catalyst. Detailed discussion can be found in Sections 8.5–8.7.

Application of Ambient Pressure X-ray Photoelectron Spectroscopy to Catalysis, First Edition. Franklin Tao.

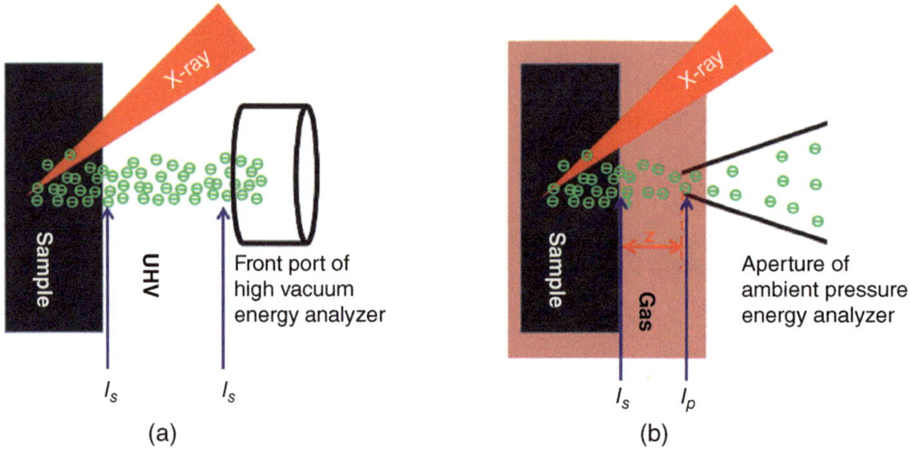

Figure 8.1 Schematic showing the intensity of photoelectrons entering the energy analyzer of AP-XPS (I_p) is only a portion of the intensity of photoelectrons leaving the catalyst surface (I_s). (a) XPS in high vacuum. (b) XPS in gas phase. What matters in determining surface composition is the intensity of photoelectron leaving surface of catalyst (I_s). Unfortunately, the intensity of photoelectron leaving the catalyst surface cannot be measured directly. Note: the energy analyzer of AP-XPS is called ambient pressure energy analyzer. The main difference between the energy analyzer of AP-XPS and the one in high vacuum XPS is the differential pump stage. Thus, it is called differential pumping energy analyzer sometime. The prototype of differential pumping energy analyzer was made in 2002. *Source:* Reproduced with permission from Tao and Nguyen[1]/Royal Society of Chemistry.

8.2 Difference in Intensity of Photoelectrons Collected by Energy Analyzer

In high vacuum XPS studies, the intensity of photoelectrons entering its energy analyzer is the same as that of photoelectrons leaving surface of the catalyst. In AP-XPS, the intensity of photoelectrons entering the energy analyzer (or aperture) of AP-XPS strongly depends on the pressure and thickness of the gas between catalyst surface and aperture. The attenuation can be evaluated through the equation, $F_p = \frac{I_p}{I_0} = e^{-\frac{z\sigma_{KE}p}{kT}}$. Figure 5.8a presents how a travel distance and gas pressure make the intensity of photoelectrons decay. A notable feature is that when the gas pressure is lower than 0.1 Torr, the intensity of photoelectrons with energy of 500 eV does not decay obviously after these photoelectrons travel through the 0.1 Torr gas region with a thickness of submillimeters.[2] This feature could justify the phenomenon that majority of reported AP-XPS studies in literature were performed at 0.1 Torr or lower.

8.3 Difference in Resolution and Baseline of Spectrum

Compared to high vacuum XPS, the scattering of photoelectrons by gas molecules in AP-XPS does not broaden the XPS peak. Based on the formation mechanism of an XPS peak discussed in Chapter 3, there is no base to support a claim that inelastic scattering

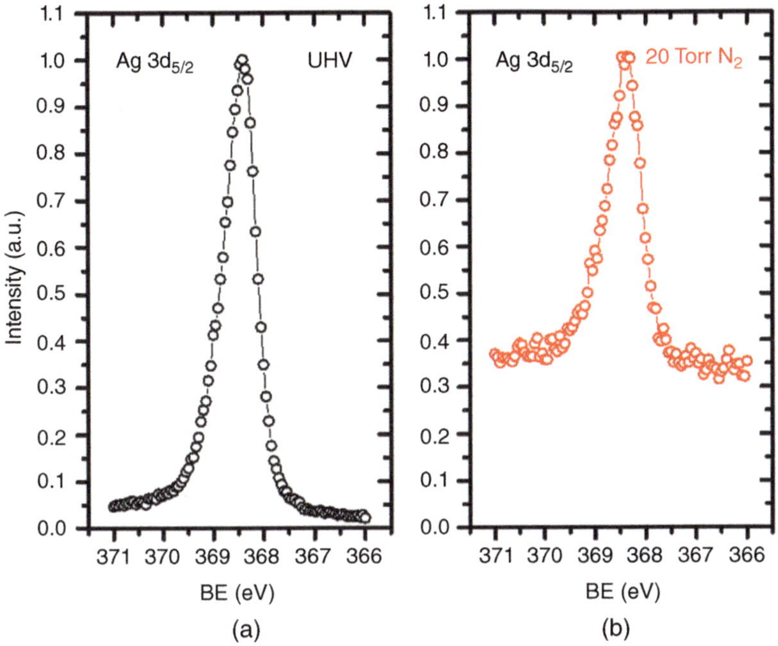

Figure 8.2 Ag $3d_{5/2}$ spectra of Ag foil under UHV environment (a) and 20 Torr N_2 (b) collected on the AP-XPS system of the Tao group. The spectra were normalized to peak height of 1 for easy comparison. It is noted that the intensity of the baseline for the Ag $3d_{5/2}$ spectrum collected under 20 Torr N_2 is much higher than that collected in UHV. *Source:* Reproduced with permission from Tao and Nguyen[1]/Royal Society of Chemistry.

between gas molecules and photoelectrons makes peak of AP-XPS broaden. In fact, these inelastic scatterings generate certain number of photoelectrons with energies lower than their original ones and thus increase the background of the XPS peak collected in AP-XPS. For instance, Ag 3d collected in 25 Torr N_2 through AP-XPS (Figure 8.2b) exhibits the same FWHM as that in ultrahigh vacuum environment collected with the same AP-XPS instrument (Figure 8.2a). An obvious difference between them is the baseline of the spectrum collected at 20 Torr (Figure 8.2b) is much higher than that collected in ultrahigh vacuum (Figure 8.2a). The high baseline exactly shows that a significant portion of photoelectrons leaving solid surface are inelastically scattered by gas molecules in the gas region between the sample surface and aperture. They contributed to the baseline of AP-XPS spectrum at $BE' = h\nu - KE' - \phi$. These photoelectrons that lost some kinetic energy due to inelastic scatterings in gas region appear as noise in terms of baseline at BE'. BE' is higher than BE since KE' is lower than KE due to the inelastic scattering by molecules in the gas phase. Here KE is the kinetic energy of photoelectrons leaving the catalyst surface or before entering the gas phase. Thus, in general, the gas phase between the sample surface and aperture does not vary the spectral resolution but creates a high baseline at the higher BE side of a peak in AP-XPS studies.

8.4 Difference in Spectrum between Free Molecules in Gas and Adsorbed Molecules on Surface

Since the 1970s people have known that X-ray can excite photoelectrons from molecules in the gas phase.[3,4] In many cases, XPS peak of gas phase can be observed in AP-XPS studies (Figure 8.3) because X-ray can photoionize gas phase molecules before it reaches the catalyst surface.[5] A reasonable concern is whether the XPS peak of gas phase overlaps the peak of the molecules adsorbed on the catalyst surface. Kinetic energies of photoelectrons from gas molecules such as CO in gas phase are different from those of adsorbed CO by several eVs since their kinetic energies are referenced to vacuum level instead of Fermi level of the spectrometer.[6] Therefore, the XPS peak of free molecules in gas and that of molecules adsorbed on a catalyst surface do not overlap in an XPS spectrum collected with AP-XPS.

The binding energy of photoelectrons generated from the molecules adsorbed on a surface is referenced to the Fermi level (E_F) of the same surface that is also the same E_F of the energy analyzer used in AP-XPS system since the sample electrically contacts with the energy analyzer. Thus, the kinetic energies of these photoelectrons generated from the molecules adsorbed on the surface are referenced to the Fermi level of the energy analyzer

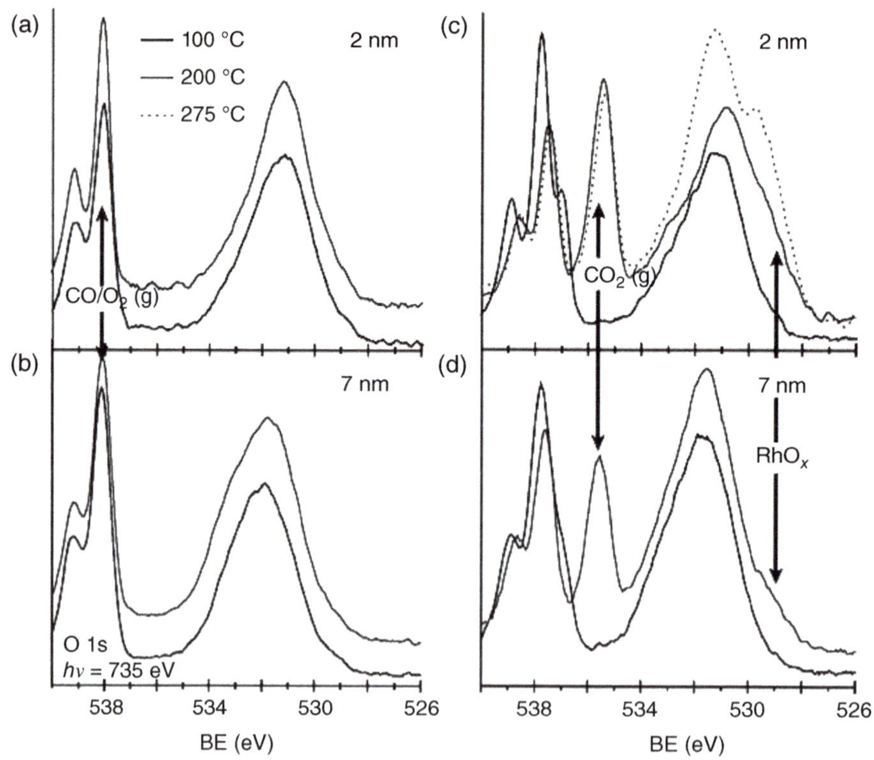

Figure 8.3 O 1s spectra of 2 and 7 nm Rh nanoparticles in 500 mTorr O_2 (a and c) and during catalysis in the mixture of reactants (410 mTorr O_2 and 20 mTorr CO) (b and d). When exposed to both O_2 and CO, a peak at higher binding energy forms, which is attributed to a RhO_x species. *Source:* Reproduced with permission from Grass et al.[5]/John Wiley & Sons.

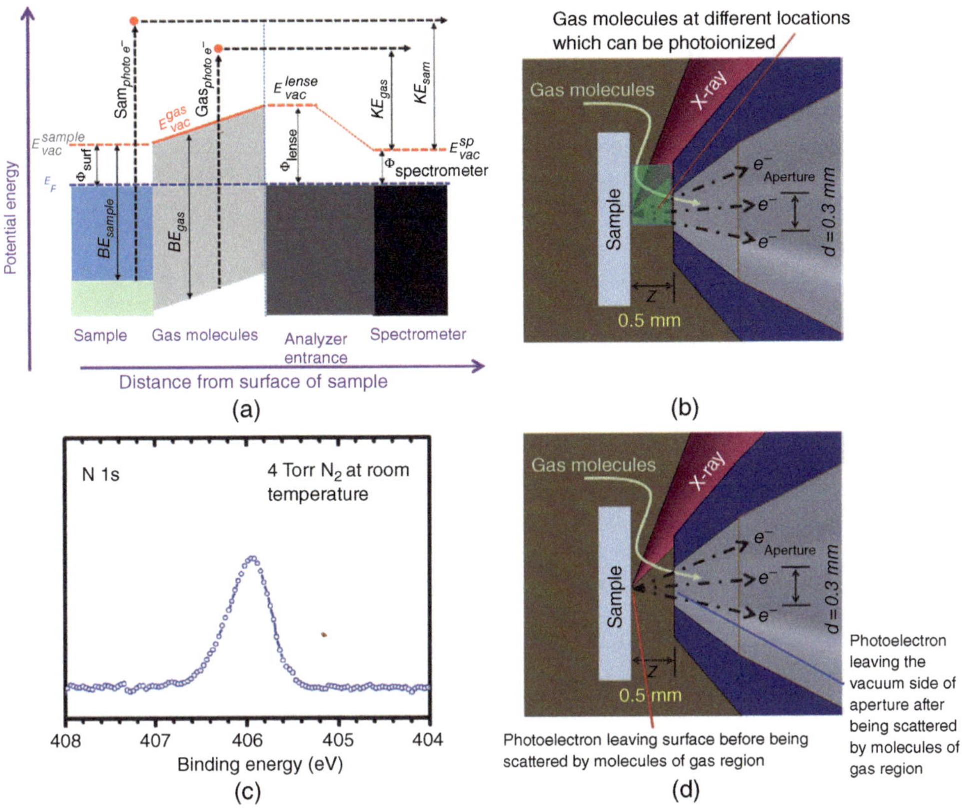

Figure 8.4 Energy diagram of the catalyst surface, gas molecules, and the energy analyzer of the AP-XPS system and its correlation to the peak position of the photoelectrons generated from molecules adsorbed on the surface and molecules traveling in the gas region between the catalyst surface and aperture. (a) Schematic showing the energy alignment of the sample, gas region, and energy analyzer, reproduced from reference.[6] (b) Schematic showing the gas region (between the catalyst surface and the aperture) that is photoionized. (c) Broadened N 1s peak collected from the gas phase of N_2 at 4 Torr from lab-based AP-XPS. (d) Schematic showing the location of photoelectrons leaving the surface of the catalyst before entering the gas phase (marked with red line) and the location of photoelectrons collected at the vacuum side of the aperture (marked with the blue line). The atomic ratio measured through photoelectrons leaving the surface of the catalyst before entering the gas phase is the authentic atomic ratio; the atomic ratio measured through photoelectrons leaving the gas region is the nominal atomic ratio. *Source:* Reproduced with permission from Tao and Nguyen[1]/Royal Society of Chemistry.

as shown in Figure 8.4a.[1] Compared to these molecules adsorbed on surface, these free molecules in gas being photoionized do not have any electrical contact with either surface or the energy analyzer (Figure 8.4a). Thus, it is inappropriate to reference kinetic energies of photoelectrons generated from gas molecules to Fermi level of the energy analyzer. In fact, they are referenced to vacuum level. As shown in Figure 8.4a, the vacuum level of gas molecules, E_{vac}^{gas} (the solid red line) is above the Fermi level of the energy analyzer by certain values. Consequently, the kinetic energies of photoelectrons generated from gas phase molecules are lower than that of molecules adsorbed on the sample surface by certain values. In fact, this difference allows for readily distinguishing these free molecules such as CO in gas phase (Figure 8.3a and b) from these adsorbed molecules such as CO adsorbed on Pt(557). For instance, O 1s peaks of *free* CO and O_2 molecules in gas can be clearly observed in Figure 8.3a and b; In addition, O 1s peaks of free CO, O_2 and CO_2 (product of CO oxidation) in gas phase can be readily identified in Figure 8.3c and d. The gas phase molecules can be readily identified through AP-XPS since their peaks are typically located at the higher BE side of the peaks of molecules adsorbed on a catalyst surface by a few eV.

8.5 Calibration of Nominal Atomic Ratio A/Z of a Catalyst Surface in a Pure Gas

Kinetic energy is one of the many factors contributing to σ_{KE}. σ_{KE} varies as a function of KE. Thus, the decay factor of photoelectron intensity varies with KE based on the equation, $F_p = \frac{I_p}{I_0} = \frac{I_p}{I_s} = e^{-\frac{z\sigma_{KE}p}{kT}}$. Here I_s and I_p are photoelectron intensities collected at catalyst surface and aperture, respectively. Photoelectrons from two subshells of two elements typically have different KEs. Thus, their σ_{KE} [$\sigma_{KE\,(A)}$ versus $\sigma_{KE\,(Z)}$] are different. This difference further leads to different decay factors for photoelectrons of elements A and Z, [$F_{p(A)}$ versus $F_{p(Z)}$]. Then, $\frac{I_{p(A)}}{I_{s(A)}}$ is different from $\frac{I_{p(Z)}}{I_{s(Z)}}$ since the decay factor of A is different from Z, $\frac{I_{p(A)}}{I_{s(A)}} \neq \frac{I_{p(X)}}{I_{s(X)}}$. Furthermore, the measured ratio of intensity of photoelectrons of element A to Z at the aperture, $\frac{I_{p(A)}}{I_{p(Z)}}$, is different from the ratio of intensity of photoelectrons of element *A* to *Z* leaving the catalyst surface or before entering the gas phase, $\frac{I_{p(A)}}{I_{p(Z)}} \neq \frac{I_{s(A)}}{I_{s(Z)}}$. The locations for measuring $\boldsymbol{I_s}$ and $\boldsymbol{I_p}$ are marked in Figure 8.1b. In fact, only the intensity of photoelectrons before entering the gas phase represents the authentic catlayst surface in the gas phase. Thus, $\frac{I_{p(A)}}{I_{p(Z)}}$ should be calibrated to $\frac{I_{s(A)}}{I_{s(Z)}}$ since $\frac{I_{s(A)}}{I_{s(Z)}}$ reflects the authentic atomic ratio $r_{A/Z}$.

As $I_{s\,(A)}$ and $I_{s\,(Z)}$ cannot be measured directly, we can only measure $I_{p\,(A)}$ and $I_{p\,(Z)}$. Thus, calibrations of $I_{p(A)}$ and $I_{p(Z)}$ to $I_{s\,(A)}$ and $I_{s\,(Z)}$ are necessary, respectively if $KE_{(A)}$ is obviously different from $KE_{(Z)}$. A simple way to avoid a calibration is to choose two subshells of elements A and Z which have similar binding energy such as Rh 3d and Pd 3d. However, this is not the case in many studies because one of the interested subshells could have a much lower photoionization cross section in gas phase. Thus, checking photoionization cross section at the energy of the available X-ray is necessary. Reference 9 is a collection of photoionization cross sections of most subshells of most elements to be excited by X-rays with different energies.

The difference between $\frac{I_{p(A)}}{I_{p(Z)}}$ and $\frac{I_{s(A)}}{I_{s(Z)}}$ in AP-XPS studies was demonstrated by Tao et al. in the measurement of atomic ratio of Rh to Ce on surface of a catalyst, 1.0wt%Rh/CeO_2.

As shown in Figure 8.5a and b, 1.0 wt% Rh/CeO_2 is a catalyst consisting of CeO_2 and the loaded Rh nanoparticles. The average size of Rh nanoparticles is about 5.0 ± 3.0 nm. Rh 3d and Ce 3d spectra were collected from surface of the catalyst in UHV, in 1 Torr N_2,

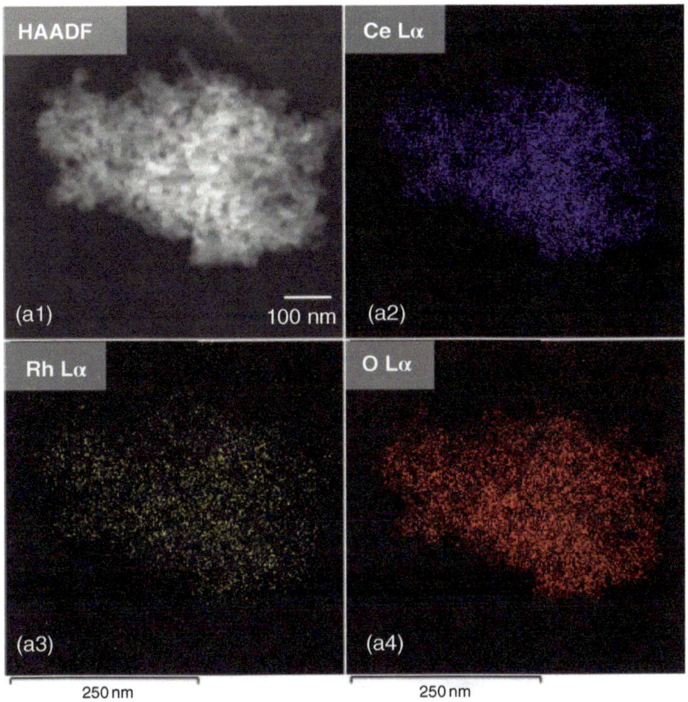

(a)

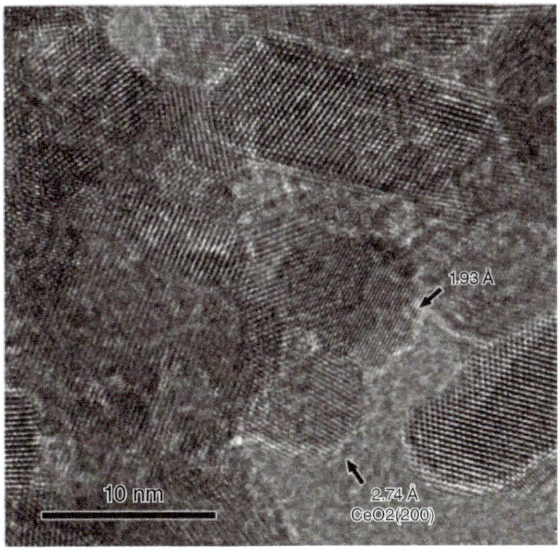

(b)

Figure 8.5 Demonstration of the calibration of the atomic ratio of the elements of a sample when photoelectrons of their subshells have different kinetic energies. Here Rh atoms doped in the surface region of CeO_2 of 1.0 wt% Rh/CeO_2; 1.0 wt% Rh/CeO_2 is the sample of these studies. (a) Energy dispersive X-ray analysis of elemental mapping of 1.0 wt% Rh/CeO_2; (a1) HAADF image of a large area of 1.0 wt% Rh/CeO_2; (a2) mapping of Ce La; (a3) mapping of Rh La; (a4) mapping of O La. (b) High resolution TEM image of nanoparticles of 1.0 wt% Rh/CeO_2. (c) Photoemission feature of Rh 3d in UHV and 1, 3, and 5 Torr of N_2. (d) Photoemission feature of Ce 3d in UHV and 1, 3, and 5 Torr of N_2. (e) Atomic ratios of Rh/Ce of the surface of the same catalyst at different pressures without any correction (line of black squares), with correction by factor, λ_{IMFP}, in gas (line of red circles), and with correction by transmission factor, $F_{transmission}$ (0.8 mm) (line of blue triangles). *Source:* Reproduced with permission from Tao and Nguyen[1]/Royal Society of Chemistry.

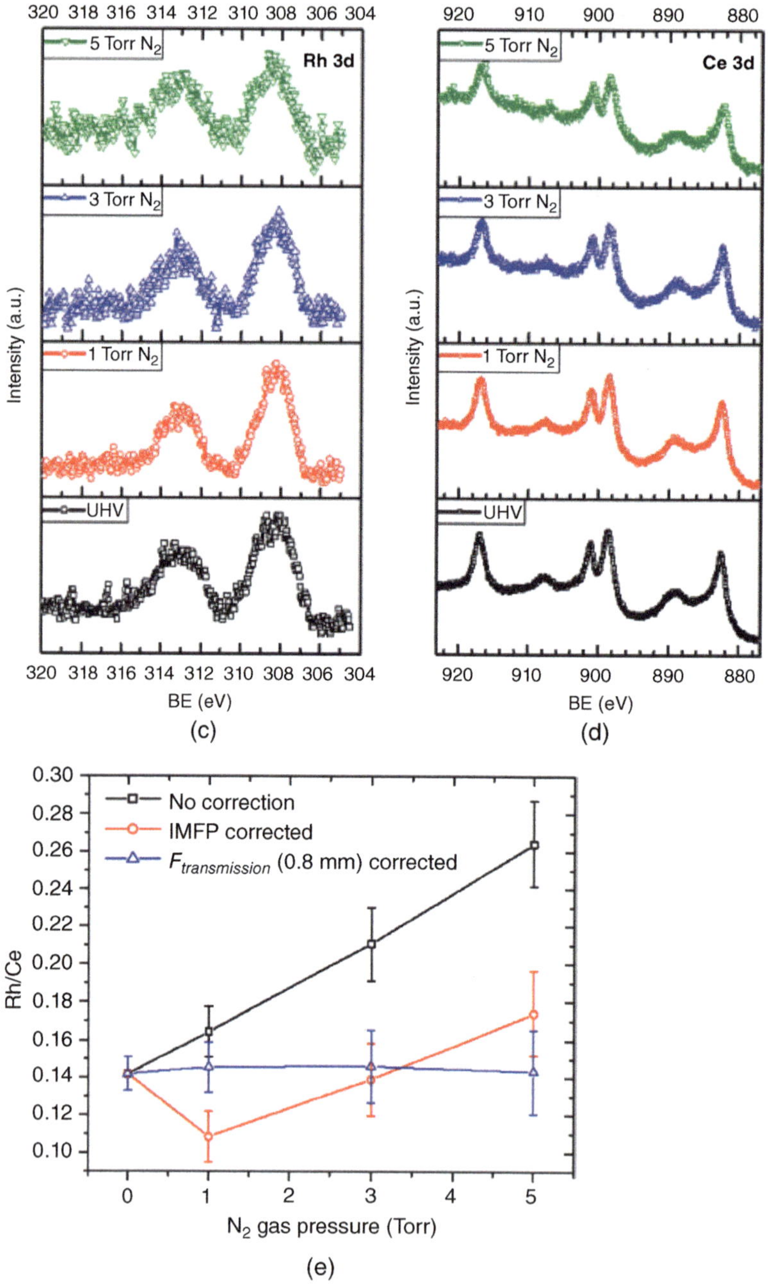

Figure 8.5 (Continued)

in 3 Torr N_2, and in 5 Torr N_2 (Figure 8.5c and d) with the same AP-XPS instrument. Atomic ratios of Rh to Ce, $r_{Rh/Ce}$ were calculated through $r_{Rh/Ce} = \frac{I_{p(Rh3d)}/S_{Rh3d}}{I_{p(Ce3d)}/S_{Ce3d}}$, where $S_{Rh\ 3d}$ and $S_{Ce\ 3d}$ are the sensitive factors of Rh 3d and Ce 3d, respectively. The calculated atomic ratio, $r_{Rh/Ce}$, was plotted in Figure 8.5e. Without calibration of the decay effect, the atomic ratio, $r_{Rh/Ce}$, measured with $\frac{I_{p(Rh)}}{I_{p(Ce)}}$ at 1 Torr N_2, 3 Torr N_2, and 5 Torr N_2 are obviously different from the $r_{Rh/Ce}$ measured in UHV. Here $\frac{I_{s(Rh)}}{I_{s(Ce)}}$ is the ratio of photoelectron intensity collected from the true surface of 1.0 wt% Rh/CeO_2 before Rh 3d and Ce 3d photoelectrons enters N_2 gas phase. It must be the same as that in UHV since N_2 is inert to this catalyst at room temperature. Thus, the $r_{Rh/Ce}$ of the catalyst surface in N_2 should be the same as the $r_{Rh/Ce}$ in UHV at room temperature. In other words, the $r_{Rh/Ce}$ should be 0.14 no matter it is in UHV, 1 Torr N_2, 3 Torr N_2, or 5 Torr N_2. However, the nominal Rh/Ce atomic ratios, $r_{Rh/Ce}$ measured at 1, 3, or 5 Torr N_2 with AP-XPS marked with squares in Figure 8.5e are obviously offset from the true ratio, 0.14 measured in UHV. This difference confirmed the $r_{Rh/Ce}$ measured by collecting photoelectrons at the aperture is different from that measured by collecting photoelectrons directly from the catalyst surface. Thus, it proved the following inequalities.

$$A1: \frac{I_{p(Rh3d)}}{I_{p(Ce3d)}} \neq \frac{I_{s(Rh3d)}}{I_{s(Ce3d)}}$$

$$A2: \frac{I_{p(Rh3d)}}{I_{s(Rh3d)}} \neq \frac{I_{p(Ce3d)}}{I_{s(Ce3d)}}$$

Based on the definition of decay factor, F, the inequality A2 can be rewritten into $F_{p(Rh)} \neq F_{p(Ce)}$. The inequality A2 is confirmed since A1 is proved by the experimental results. Thus, the σ_{KE}–driven difference in decay factor between $F_{p(Rh)}$ and $F_{p(ce)}$ is confirmed in Figure 8.5e. As shown in Figure 8.5e, the difference between $r_{Rh/Ce}$ measured in N_2 without correction and $r_{Rh/Ce}$ measured in UHV increases along the increase of pressure of N_2. It suggests a calibration to retrieve the true $r_{Rh/Ce}$ of the catalyst surface is necessary at high gas pressure since the nominal $r_{Rh/Ce}$ offsets more from the true $r_{Rh/Ce}$ more at a higher pressure of N_2.

Compared to the atomic ratio of Rh/Ce, $r_{Rh/Ce}$ at 0.14 in UHV, it is 0.17 in 1 Torr N_2, 0.21 in 3 Torr N_2, and 0.26 in 5 Torr N_2, respectively, although all the Rh/Ce ratios, $r_{Rh/Ce}$ in UHV or in N_2 at different pressure should have the same value, 0.14. Thus, definitely a calibration for the nominal atomic ratio is necessary in order to obtain the true atomic ratio of catalyst surface in a gas phase.

To calibrate the nominal atomic ratio to its authentic atomic ratio, a calibration factor, C defined by the equation, $C = \frac{\beta_A}{\beta_Z}$. Thus, the calibration can be done with the following equation, $r_{\frac{A}{Z}(authentic)} = r_{\frac{A}{Z}(nominal)} \times \frac{\beta_A}{\beta_Z}$. β can be represented with or replaced by the decay factor, F_p, which can be calculated with the equation, $F_p = \frac{I_p}{I_0} = \frac{I_p}{I_s} = e^{-\frac{z\sigma_{KE}p}{kT}}$; thus, the calibration factor, C, can be expressed with the equation, $C = \frac{\beta_A}{\beta_Z} = \frac{F_{p(A)}}{F_{p(Z)}} = e^{-\frac{zp}{kT}(\sigma_{KE,Z}-\sigma_{KE,A})}$. Then, the authentic atomic ratio can be obtained with the following equation, $r_{\frac{A}{Z}(authentic)} = r_{\frac{A}{Z}(nominal)} \times e^{-\frac{zp}{kT}(\sigma_{KE,Z}-\sigma_{KE,A})}$. In other words, the authentic atomic ratio, $r_{\frac{A}{Z}(authentic)}$ can be obtained through multiplying the nominal atomic ratio, $r_{\frac{A}{Z}(nominal)}$ by the calibration factor, C. Again, the nominal atomic ratio,

Table 8.1 Cross sections of inelastic scattering of photoelectrons with kinetic energy of 200, 300, 600, or 1000 eV in different gas phases of N_2, CO, CO_2, or H_2O. Data were taken from reported values in literature.[7,8]

KE_{e^-} (eV)	σ_{N_2} (cm^2) ($\times10^{-16}$)	σ_{CO}(cm^2) ($\times10^{-16}$)	σ_{CO_2} (cm^2) ($\times10^{-16}$)	σ_{H_2O}(cm^2) ($\times10^{-16}$)
200	2.93	4	3.9	2.7
300	2.34	3	3.5	2.1
600	1.54	1.7	2.0	1.2
1000	1.10	1.1	1.1	0.85

Source: Adapted from Itikawa[7] and Shirai et al.[8]

$r_{\frac{A}{Z}(nominal)}$ is the one measured in AP-XPS without considering the difference in decay factor of photoelectrons of subshells of elements A and Z, $F_{p(A)}$ and $F_{p(Z)}$.

σ is determined by gas molecules and the kinetic energy of photoelectrons. Table 8.1 lists the cross sections of inelastic scattering of photoelectrons (200, 300, 600, and 1000 eV) in N_2, CO, CO_2 and H_2O. For a pure gas such as N_2 here, the $\sigma_{KE\,=\,200\text{ eV}}$, $\sigma_{KE\,=\,300\text{ eV}}$, $\sigma_{KE\,=\,600\text{ eV}}$, and $\sigma_{KE\,=\,1000\text{ eV}}$ are 2.93, 2.34, 1.54, and 1.10, respectively, based on references.[7–9] In pure N_2, it is ~1.10 for photoelectrons with *KE* at ~1000 eV and ~1.54 for photoelectrons with *KE* at 600 eV. When Al Kα is used, the KEs of Rh 3d and Ce 3d photoelectrons are approximately 1180 eV and 560–600 eV, respectively based on literature.[7–9] Thus, approximately $\sigma_{Rh\ 3d}$ and $\sigma_{Ce\ 3d}$ are 1.10 and 1.54, respectively. Thus, the calibration factor, *C*, can be calculated with the $C = e^{\frac{-zp}{RT}\left(\sigma_{KE(Z)}-\sigma_{KE(A)}\right)}$. Then, the $r_{\frac{Rh}{Ce}(nominal)}$ can be readily calibrated to $r_{\frac{Rh}{Ce}(authentic)}$. The blue line in Figure 8.5e is the $r_{\frac{Rh}{Ce}(authentic)}$ obtained through calibration. In summary, these analyses demonstrated how a calibration for nominal atomic ratio of a catalyst surface obtained in a pure gas at certain pressure such as 1 Torr N_2 can be done. Notably, such a calibration is not necessary if the kinetic energies of the photoelectrons of the two subshells such as Rh 3d and Pd 3d are very close. In addition, such a calibration can be avoided by choosing different X-rays to generate photoelectrons with the same kinetics energy for subshells of two interested elements when synchrotron radiations such as hν=500 eV for Rh 3d and hν=1050 eV for Ce 3d are available. Alternatively, in some cases, we could try to choose two subshells which have similar binding energies and reasonably strong photoionization cross sections for the two interested elements to avoid a complicate calibration if synchrotron radiations are unavailable. Notably, of the above calibration method needs the cross-section of inelastic scattering of photoelectrons with a specific KE in a specific gas, σ. However, in many cases, such parameters are not available.

8.6 Calibration of Nominal Atomic Ratio A/Z of a Catalyst Surface in a Mixture of Reactants

Section 8.5 describes how a nominal atomic ratio of two elements of a catalyst surface in a pure gas was calibrated to the authentic ratio. However, in terms of heterogeneous catalysis, most catalytic reactions have two or even more gaseous reactants. In other words, the gas where these photoelectrons have to travel through is a mixture of two or more types of molecules. As shown in Table 8.1, different molecules have quite different cross sections of inelastic scattering even if these photoelectrons have the same or similar kinetic energy. For instance, σ_{N_2}, σ_{CO}, σ_{CO_2}, and σ_{H_2O} for photoelectrons with kinetic energy of 600 eV are 1.54, 1.7, 2.0, and 1.2, respectively. Thus, for a mixture of gases, σ_{KE} of photoelectrons with a specific kinetic energy to gas need to be calculated with the following equation, $\sigma_{KE,\ mix} = \sum_1^n \frac{p_i \sigma_i^o}{p}$, which takes all constituting gases into account.[1] p_i and p are partial pressure of a constituting gas and total pressure of the gas mixture, respectively. σ_i^o is the cross section of a pure gas, i. The summation of the contributions of all gases gives the cross section of the gas mixture for photoelectrons with a specific kinetic energy (KE). For 1.0 wt% Rh/CeO_2 in flowing mixture of CH_4 (2 Torr) and O_2 (0.5 Torr), the $\sigma_{KE,\ mix}$ for Rh 3d photoelectrons (KE= ~1000 eV) and $\sigma_{KE,\ mix}$ for Ce 3d photoelectrons (KE=~ 600 eV) can be calculated on the basis of the data listed in Table 8.1. Then the calibration factor, C, can be calculated through the equation, $C = e^{-\frac{zp}{kT}\left(\sigma_{KE,\ Rh3d} - \sigma_{KE,\ Ce3d}\right)}$. With this calibration factor, C, the authentic atomic ratio Rh/Ce in the mixture of 2 Torr CH_4 and 0.5 Torr O_2, $r_{\frac{A}{Z}(authentic)}$ can be obtained by multiplying the nominal atomic ratio, $r_{\frac{A}{Z}(nominal)}$ with calibration factor, C. The nominal ratio $r_{A/Z\ (nominal)}$ is the ratio measured with the data of AP-XPS without a calibration.

8.7 Calibration of Nominal Atomic Ratio A/Z of a Catalyst Surface in a Pure Gas Obtained at Different Temperature for Fair Comparison

Temperature is an important factor for catalysis. Most catalytic reactions are performed in the temperature range of 200–800 °C. In some cases, it is necessary to explore the evolution of surface chemistry of a catalyst such as surface composition as a function of catalysis temperature. To uncover the authentic evolution of atomic ratio along the increase of catalysis temperature, it is necessary to make sure the comparison of atomic ratio at different catalysis temperatures is meaningful. Thus, the nominal atomic ratio at each temperature obtained from AP-XPS should be calibrated to its authentic atomic ratio.

As found in the decay factor, $F_p = \frac{I_p}{I_0} = e^{-\frac{z\sigma_{KE}p}{kT}}$, the temperature of a gas where the photoelectrons travel through is a parameter contributing to the decay factors. Thus, a comparison of atomic ratio of surface at different temperatures is meaningful only after the temperature factor is calibrated. The variation can be evaluated by checking how a catalysis temperature influences the decay factor of photoelectrons, F_p. The influence of the temperature on the decay factor, F_p can be expressed with a term, θ.[1] The temperature calibration term, θ_A for element A can be defined to the ratio of transmission rate at T_2 to T_1. It can be calculated with the equation: $\theta_A = \frac{F_{transmission\ of\ A\ at\ T_2}}{F_{transmission\ of\ A\ at\ T_1}} = \frac{e^{\frac{-zp\sigma_{A\ at\ T_2}}{kT_2}}}{e^{\frac{-zp\sigma_{A\ at\ T_1}}{kT_1}}}$. For data of a series of temperature, we can set the lowest temperature as T_1 and take T_1 as a reference temperatures. Similarly, the temperature calibration term θ_Z for Z element can be calculated with a similar equation: $\theta_Z = \frac{F_{transmission\ of\ Z\ at\ T_2}}{F_{transmission\ of\ Z\ at\ T_1}} = \frac{e^{\frac{-zp\sigma_{Z\ at\ T_2}}{kT_2}}}{e^{\frac{-zp\sigma_{Z\ at\ T_1}}{kT_1}}}$. Then, the authentic ratio of elements A to Z at temperature T_2, $r_{\frac{A}{Z}at\ T_2(authentic)}$ can be calibrated through the equation, $r_{\frac{A}{Z}at\ T_2(authentic)} = r_{\frac{A}{Z}at\ T_2(nominal)} \times \frac{\theta_A}{\theta_Z}$. By introducing the above equations of θ_A and θ_Z, the authentic atomic ratio, $r_{\frac{A}{Z}at\ T_2(authentic)}$ can be calibrated with the equation, $r_{\frac{A}{Z}at\ T_2(authentic)} = r_{\frac{A}{Z}at\ T_2(nominal)} \times \frac{e^{\frac{zp\sigma_{Z\ at\ T_2} - zp\sigma_{A\ at\ T_2}}{kT_2}}}{e^{\frac{zp\sigma_{Z\ at\ T_1} - zp\sigma_{A\ at\ T_1}}{kT_1}}}$. If σ_Z and σ_A can be assumed to be independent of temperature, the above equation can be much simpler; then, $r_{\frac{A}{Z}at\ T_2(authentic)}$ can be calibrated with the following simple equation for a fair comparison: $r_{\frac{A}{Z}at\ T_2(authentic)} = r_{\frac{A}{Z}at\ T_2(nominal)} \times e^{zp(\sigma_Z - \sigma_A)} e^{\frac{1}{kT_2} - \frac{1}{kT_1}}$. Strictly speaking, temperature could influence σ_A and σ_Z and thus make the calibration to $r_{\frac{A}{Z}at\ T_2(nominal)}$ quite complicated. If the dependence of σ on temperature is not negligible, the calculation of the temperature calibration term requests σ of a gas at a specific temperature when the gas molecules scatter photoelectrons with a specific kinetic energy. Depending on the requested accuracy in this calibration, probably more assumptions for approximation and simplification have to make.

Overall, whether a calibration is necessary and how it is calibrated, depend on both the level of requested accuracy for information and whether the cross sections of inelastic scattering of photoelectrons with certain kinetic energy in a specific gas at a specific temperature, σ are available or not. In addition, it is worth mentioning that such calibrations discussed in Sections 8.5, 8.6, and 8.7 are not necessary if tunable X-rays are available to excite atoms of the two elements (Rh and Ce) to generate photoelectrons that have similar kinetic energies.

References

1 Tao, F. and Nguyen, L. 2018. "Interactions of gaseous molecules with X-ray photons and photoelectrons in AP-XPS study of solid surface in gas phase." *Phys. Chem. Chem. Phys.* 20, 9812–9823.

2 Ogletree, D. F., Bluhm, H., Lebedev, G. *et al.* 2002. "A differentially pumped electrostatic lens system for photoemission studies in the millibar range." *Rev. Sci. Instrum.* 73, 3872.

3 Siegbahn, K. 1970. *ESCA Applied to Free Molecules.* North-Holland Pub. Co., Amsterdam.

4 Hamrin, K., Johansson, G., Gelius, U. *et al.* 1968. "Ionization energies in methane and ethanemeasured by means of ESCA." *Chem. Phys. Lett.* 1, 613–615.

5 Grass, M. E., Zhang, Y., Butcher, D. R. *et al.* 2008. "A reactive oxide overlayer on rhodium nanoparticles during CO oxidation and its size dependence studied by in situ ambient-pressure X-ray photoelectron spectroscopy." *Angew. Chem. Int. Ed.* 47, 8893–8896.

6 Bluhm, H. 2010 "Photoelectron spectroscopy of surfaces under humid conditions." *J. Electron Spectrosc. Relat. Phenom.* 177, 71–84.

7 Itikawa, Y. 2006. "Cross sections for electron collisions with nitrogen molecules." *J. Phys. Chem. Ref. Data* 35, 31–53.

8 Shirai, T., Tabata, T., and Tawara, H. 2001. "Analytic cross sections for electron collisions with CO, CO_2 , and H_2O relevant to edge plasma impurities." *Atomic Data Nuclear Data Tables* 79, 143.

9 Majeed, T. and Strickland, D. J. 1997. "New survey of electron impact cross sections for photoelectron and auroral electron energy loss calculations." *J. Phys. Chem. Ref. Data* 26, 335–349.

9

Significance of Using AP-XPS in Studies of Catalysis

9.1 Fundamental of Catalyst Surface

Over 80% of chemical transformations in the world are performed through one or more catalytic processes. Fundamental studies of catalysis are to provide insights in terms of understanding of catalytic mechanism at a molecular level toward offering guidance for developing catalysts with better catalytic performance. One of the most important components in this fundamental study is characterization of the catalyst surface during catalysis.

A great challenge in fundamental understanding of catalytic mechanism of a heterogeneous catalyst is characterization of the authentic surface of a catalyst under a working condition. Compared to a heterogeneous catalyst working at solid–gas interface, it is relatively easy to identify structure of a homogeneous catalyst. This is because a molecular catalyst typically catalyzes a reaction at a mild condition and thus in many cases its molecular structure during catalysis is very similar or even the same as that before and after catalysis. However, in general a heterogeneous catalyst works at a relatively high temperature or even quite high temperature up to 1000 °C in gas phase at 1 atm or a higher pressure up to tens or even hundreds of atm. In addition, a nanoparticle catalyst could perform catalysis at a solid–liquid interface or even at a solid–liquid–gas interface.[1–7] The real player of a heterogeneous catalyst is one or a couple of atomic layers of the topmost surface of a catalyst particle. Driven by the harsh condition in terms of high temperature of a catalyst, high pressure of gas reactants, or complex environment in liquid, in many cases the surface of a catalyst during catalysis is different from that before or after catalysis. The difference between a catalyst during catalysis and that before or after catalysis has been reported in literature.[7–12] Compelling studies have justified the restructuring of a catalyst surface driven by temperature or/and pressure of the reactant gases. The latest discovery reported by Sautet and Tao groups first found that the surface of a catalyst can be restructured at atomic or nano scale by a gaseous product of a catalytic reaction.[13] It is found that a product such as CO of methane steam reforming is crucial for remaining highly dispersed catalytic sites. Under a condition of a low catalysis temperature or a low pressure of the product CO, the active catalytic clusters, $Rh_m(CO)_n$, readily sintered to form CO-free Rh metal nanoclusters; however, when the pressure of the product CO is increased, the sintered Rh nanoclusters are broken into subnanometer clusters bonded with more CO molecules, forming $Rh_m(CO)_n$. Clearly, the catalyst surface structure is truly determined by its catalytic condition.

9.2 Significance of Characterization of Surface of a Catalyst in Gas Phase

Development of catalysts with high activity, high selectivity, and long durability in chemical transformations has continuously driven fundamental studies of catalysis. It is generally acknowledged that the fundamental understanding of catalytic mechanisms of catalysts have fallen behind the expectation for guiding the development of efficient catalysts; One of the main reasons is the lack of an intrinsic correlation between an observed catalytic performance and its corresponding catalyst surface. A deeper reason is the unavailable information on the authentic catalyst surface during catalysis.

Why is information on the surface chemistry and structure of a catalyst during catalysis not available? The obvious reason is the lack of in situ or/and operando characterization of the catalyst surface during catalysis. This is definitely one of the most challenging part in fundamental studies of catalysis. It is similar to the study of behavior of a deep-sea fish, such as lantern fishes. A simple observation of a dead lantern fish on a beach could never tell us how lantern fishes can survive in deep ocean in terms of the waters far below the epipelagic zone where the hydrostatic pressure is as high as a few hundreds of atm and there is no natural illumination. A simple observation of a dead lantern fish never makes us finding out that they catch foods by their high sensitivity to small changes of local pressure and never let us being aware of the fact that their eyes are much more sensitive to light than human eyes by hundreds of times. Similarly, by simply checking a fresh or used catalyst, one has not necessarily known the authentic surface structure of a catalyst during catalysis.

For simplicity, characterization of a fresh or used catalyst surface in vacuum or in air at room temperature is a practice in the past decades. These historical practices are based on two assumptions. One is that it is extremely challenging technically if not impossible to characterize surface of a catalyst during catalysis since the working condition is high temperature high pressure; the second assumption is that the surface of a catalyst before or after catalysis can represent the surface of a working catalyst under high temperature or/and high pressure. In fact, the advances in instrumentation in the last decades have offered the catalysis community capabilities of characterizing catalysts under a working condition or semi-working condition although a lot of characterization techniques could not study a catalyst yet under working conditions. In terms of the second assumption, numerous studies in the past a couple of decades have confirmed that the surface at room temperature in high vacuum or air cannot represent the surface of a catalyst under working conditions.[7–13] Obviously, the second assumption must have misled the fundamental understanding of catalysis as it interpreted catalysis by taking surface of a catalyst in vacuum or air at room temperature as the surface of the catalyst during catalysis. Simply, the second assumption considers a catalyst surface structure remains static. In fact, catalyst surface structure is evolving and dynamic along with a change of one or more parameters of a catalytic reaction. Similar to a biologist who must observe and track alive fishes in sea instead of observing dead fishes on beach or in kitchen, a researcher in the community of catalysis must closely observe a catalyst during catalysis under a working condition or at least a semi-working condition if possible, for gaining a fundamental understanding of the catalytic mechanism at a molecular level.

9.3 Significance of Using AP-XPS in Fundamental Studies of Catalysis

A catalyst is a highly complicated system although in general only the shallow layers of atoms in the surface region of catalyst particles play roles in catalysis. When it functions in gas phase, the complexity of the surface increases by a magnitude. The essential part of a catalyst is the surface region of a catalyst particle in gas or/and liquid environment during catalysis. AP-XPS is the right characterization technique for observing the essential part of a catalyst during catalysis. The following chapters discussed how these valuable AP-XPS studies can provide essential information for understanding catalytic mechanisms at a molecular/atomic level.

References

1 Zhang, X., Sun, Z., Wang, B. *et al.* 2018. "C-C coupling on single-atom-based heterogeneous catalyst." *J. Am. Chem. Soc.* 140, 954–962.

2 Tang, Y., Li, Y., Fung, V. *et al.* 2018. "Single rhodium atoms anchored in micropores for efficient transformation of methane under mild conditions." *Nat. Commun.* 9, 1231.

3 Tang, Y., Li, Y., and Tao, F. 2022. "Activation and catalytic transformation of methane under mild conditions." *Chem. Soc. Rev.* 51, 376–423.

4 Li, Y., Tang, Y., Nguyen, L. *et al.* 2019. "Catalytic oxidation of ethane to carboxylic acids in the liquid phase at near room temperature at ambient pressure." *ACS Sustainable Chem. Eng.* 7, 4707–4715.

5 Li, Y., Khivantsev, K., Tang, Y. *et al.* 2019. "Synthesis of Na@nanoFAU zeolite catalyst and catalysis for production of formic acid with Na@nanoFAU." *Catal. Lett.* 149, 1965–1964.

6 Huang, W., Zhang, S., Tang, Y. *et al.* 2016. "Low-temperature transformation of methane to methanol on Pd1O4 single sites anchored on the internal surface of microporous silicate." *Angew. Chem. Int. Ed. Engl.* 55, 13441–13445.

7 Tao, F., Grass, M. E., Zhang, Y. W. *et al.* 2008. "Reaction-driven restructuring of Rh-Pd and Pt-Pd core-shell nanoparticles." *Science* 322, 932–934.

8 Tao, F., Dag, S., Wang, L. W. *et al.* 2010. "Break-up of stepped platinum catalyst surfaces by high CO coverage." *Science* 327, 850–953.

9 Tao, F. and Salmeron, M. 2011. "In situ studies of chemistry and structure of materials in reactive environments." *Science* 331, 171–174.

10 Dou, J., Tao, F., Tang, Y. *et al.* 2017. "Operando chemistry of catalyst surfaces during catalysis." *Chem. Soc. Rev.* 46, 4747–4765.

11 Tao, F., Zhang, S., Nguyen, L. *et al.* 2012. "Action of bimetallic nanocatalysts under reaction conditions and during catalysis: evolution of chemistry from high vacuum conditions to reaction conditions." *Chem. Soc. Rev.* 41, 7980–7993.

12 Zhang, S., Nguyen, L., Zhu, Y. *et al.* 2013. "In-situ studies of nanocatalysis." *Acc. Chem. Res.* 46, 1731–1739.

13 Yan, G., Tang, Y., Li, Y. *et al.* 2022. "Reaction product-driven restructuring and assisted stabilization of a highly dispersed Rh-on-ceria catalyst." *Nat. Catal.* 5, 119–127.

10

CO Oxidation on Single Crystal Model Catalysts

Starting from this chapter to the end of this book, AP-XPS studies of single crystal model catalysts and nanoparticle catalysts during catalysis of over ten different reactions will be discussed. The theme of these chapters is to demonstrate how AP-XPS can be used to study catalyst surfaces, what insights these studies have offered, and what challenges scientists still face.

Scientists in the field of ultrahigh vacuum surface science have extensively studied well-defined single crystal surfaces and adsorption of a great number of reactants on those surfaces through a high vacuum surface science approach. There are good reasons that scientists started from single crystal surfaces. Single crystal surface has the least number of irregular sites and thus allows for drawing a reliable conclusion on reactivity of a specific type of catalytic sites. Another reason is the high repeatability in preparation of surfaces of these single crystal model catalysts in contrast to the crystal surface of nanoparticles that are typically prepared through kinetics control process of colloidal chemistry. The macroscopic single crystal surface at the scale of 5–10 mm allows for repetitive preparations. In addition, the high concentration of the element such as Pt of Pt(111) or Ti of TiO_2(110) on the surface of a single crystal model catalyst makes the characterization be able to be done easily in contrast to the low surface concentration of metal nanoparticles supported on an oxide. Single crystal surfaces in a reactant gas such as CO or O_2 and during CO oxidation have been extensively investigated with AP-XPS, providing a reliable base for studies of the complex surface of a nanoparticle catalyst. In this chapter, CO oxidations on some of the studied metal model catalysts are discussed. More examples on this topic can be found in literature.

10.1 Pt(557) and Pt(332) in CO

Pt(557) is a preferential choice among single crystal model catalysts since its surface consists of three types of Pt atoms in different coordination environments. For instance, the Pt atoms on terrace, the Pt atoms at edge, and the Pt atoms under an edge have coordination number of 9, 7, and 10, respectively. Coexistence of multiple types of Pt atoms allows for uncovering how these different active sites could respond to a reactant differently. Adsorption of CO on Pt(557) was studied in ultrahigh vacuum (UHV).[1] It is found that the adsorption energies of the three types of Pt atoms to CO are different.

Application of Ambient Pressure X-ray Photoelectron Spectroscopy to Catalysis, First Edition. Franklin Tao.

In surface science, adsorbates also called adsorbed molecules are introduced to the surface of a single crystal model catalyst by exposing the clean surface to the gas at certain pressure, typically 1×10^{-9} to 1×10^{-7} Torr for certain amount of time in the range of 1–1000 seconds. Then, the reactant gas in the UHV chamber is immediately purged to restore the base pressure of the UHV, typically lower than 3×10^{-10} Torr before a characterization of the surface adsorbate is performed. The amount of the introduced molecules is determined by the product of gas pressure in Torr and expsosing time in second other than temperature of the sample. It is also called exposure. This product in terms of p (*in* Torr)$\times t$ (*in* second) can be converted to a quantality in a unit of 10^{-6} Torr·second. The unit 10^{-6} Torr·second is defined to 1 Langmuir (L). The quantity in unit of L is called exposure. It represents the amount of reactant introduced to the surface. However, neither the coverage of the adsorbate on the surface nor the number of molecules adsorbed on surface could be provided from this exposure unless the sticking coefficient of the molecules on the surface is known.

Other than the studies of chemisorption of CO on Pt(557) with UHV surface science approach, surface of Pt(557) in CO at Torr pressure was studied with AP-XPS.[1] The data in terms of XPS spectra including Pt 4f and O 1s were collected by using a synchrotron AP-XPS with X-rays with energies of 340 and 800 eV, by which the kinetic energy of both Pt 4f and O 1s photoelectrons are nearly the same, about 260-280 eV and thus the photoelectrons of Pt 4f and O 1s in gas have the same mean free path. Furthermore, the decay factors of Pt 4f and O 1s photoelectrons are the same in gas phase. Then, the measured nominal atomic ratio, O/Pt in the gas phase is the same as the authentic atomic ratio. Moreover, the atomic ratio of O to Pt can be used for calculations of surface coverage of adsorbed CO molecules in CO gas at different pressure.

The bottom spectra in Figure 10.1a and b are Pt 4f and O 1s of the Pt(557) in UHV with a base pressure of 2×10^{-10} Torr before any CO was introduced.[2] As we can see from Figure 10.1, along the increase of CO pressure surrounding the Pt(557), shoulders at the left side of Pt $4f_{7/2}$ and $4f_{5/2}$ were clearly observed.[2] The dash line in Figure 10.1a marks the growing shoulder of Pt $4f_{7/2}$ at 72 eV. The shoulder becomes larger at a higher pressure. At the highest pressure of this study (0.5 Torr), the shoulder becomes a dominant peak (Figure 10.1a). The difference in the relative intensity of the shoulder between the low pressure at 5×10^{-9} Torr and relative high pressure 5×10^{-1} Torr is quite large. The red arrow in Figure 10.1a marks the major shoulder at 72 eV. Notably, all these data were collected when the catalyst was in the CO gas at a pressure as marked in Figure 10.1a. Immediately after the data acquisition at 0.5 Torr, the CO was purged and reach a pressure at 2×10^{-8} Torr; the Pt 4f and O 1s were collected at the 2×10^{-8} Torr CO. Interestingly, the shoulder became small and even completely disappeared along the decrease of CO pressure. Such an appearance of Pt $4f_{7/2}$ and O_{1s} shoulders along increase of CO pressure and disappearance along decrease of CO pressure were repeatably observed as shown in Figure 10.1a.

Figure 10.1b is the O 1s XPS peak collected at the same conditions as Pt 4f of Figure 10.1a. There is a peak of O 1s at 532.6 eV observed at 5×10^{-9} CO; it then progressively upshifts to 532.95 eV along the increase of CO pressure from 5×10^{-9} to 0.5 Torr. At 0.5 Torr CO, the shoulder O 1s at 532.9 eV marked with a red arrow is obvious (Figure 10.1b). It is consistent with the evolution of the Pt 4f shoulder at 72 eV in Figure 10.1a. The pressure-dependent Pt 4f and O 1s features clearly show chemical state of Pt atoms on surface of Pt(557) is

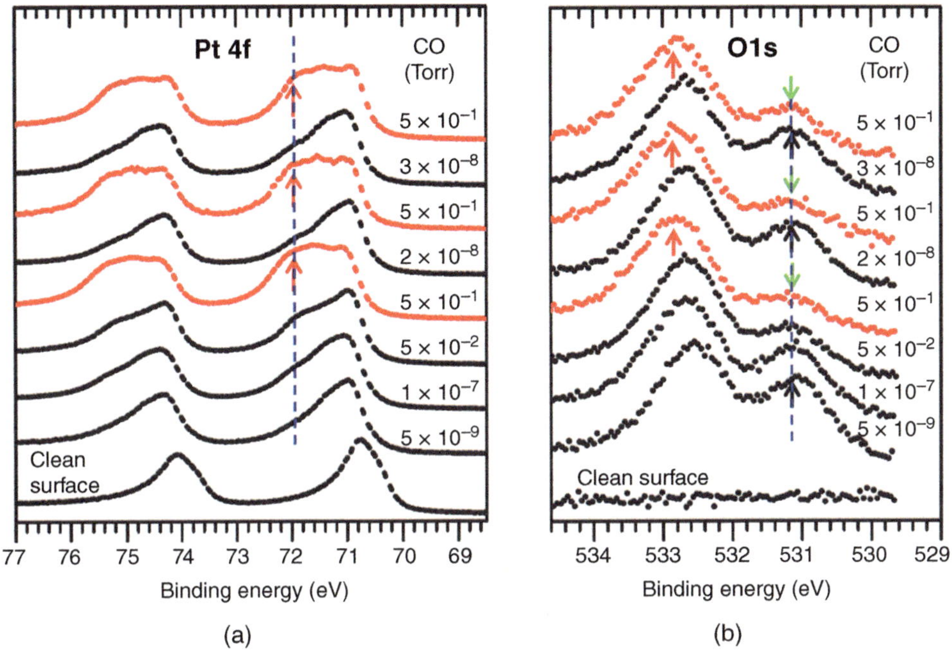

Figure 10.1 Photoemission spectra of the Pt 4f (a) and O 1s core levels (b) acquired with 340- and 800-eV X-rays, respectively, under different CO pressures. Binding energies were referenced to the Fermi level measured under the same conditions. The red spectra were obtained at 5×10^{-1} Torr. The red and green arrows in (a) and (b) mark the increase of photoemission intensity in the high-binding energy side of the Pt $4f_{7/2}$ and the simultaneous decrease of intensity in the low-binding energy side of the O 1s spectra, respectively, at high pressure. The black arrows in (b) mark the relatively larger photoemission intensity in the low-binding energy side at low pressure, compared to that at high pressure. *Source:* Reproduced with permission from Tao et al.[2]/American Association for the Advancement of Science.

determined by the pressure of CO surrounding the Pt(557) surface.[2] Notably, such obvious evolutions of Pt 4f and O 1s features were not observed for Pt(111) in CO. Here the most important message is that such pressure-dependent evolutions of chemical states of Pt atoms and chemisorbed layer of CO could never be observed if AP-XPS was not used. This is because the large shoulder of Pt 4f formed in 0.5 torr CO marked with red arrows in Figure 10.1a could never be observed with a high vacuum XPS when CO with a pressure at 0.05 or 0.5 Torr was purged to restore a UHV or HV for XPS studies.

The pressure-dependent evolutions of Pt 4f and O 1s suggest that the surface structure of a model catalyst could be largely determined by the gaseous environment. The pressure-dependent structural evolution was confirmed by High Pressure Scanning Tunneling Microscopy (HP-STM). The step edges of each terrace of Pt(557) were clearly presented in UHV. In CO, the massive transformation from the surface consisting of nearly straight step edges to a surface of highly curl step edges at 5×10^{-8} Torr has made surface appear largely different (Figure 5.2a versus b). This change clearly shows Pt(557) is restructured to some extent even at a CO pressure as low as 5×10^{-8} Torr (Figure 5.2b). This restructuring in terms of transportation of Pt atoms along step edge was driven by the existing CO gas instead of other nonchemical factors, confirmed by the lack of curl step edge in 5×10^{-8} Torr N_2. At a higher CO pressure, 5×10^{-7} Torr, the surface performed massive restructuring, forming double step edge (Figure 5.2c). The uneven contrast in a double step suggests that the atoms on surface have high mobility although atom-resolved images were not observed. In fact, the incapability of atom-resolved image at 5×10^{-7} Torr CO supports the high mobility of Pt atoms on surface in CO gaseous environment. Surprisingly, in CO at 0.5 Torr, the surface performed a dramatic restructuring. As seen in HP-STM (Figure 5.2d), homogeneous triangular nanoclusters were formed on Pt(557) in CO at 0.5 Torr. The whole surface is covered with Pt nanoclusters with a size of 2.3 nm or so. This is how it is called an entire breakup of Pt(557) surface by Torr pressure CO. Further surprisingly, the broken surface can reunite into a step surface once the CO gas is purged to high vacuum, showing that restructuring of Pt(557) is driven by CO pressure and the formed nanocluster surface is maintained by high-pressure CO. The CO pressure-driven restructuring is affirmed by the consistent observations of the evolution of surface structure by HP-STM and the change of spectral feature of Pt 4f and O 1s by AP-XPS. This is the first observation of the reversible massive surface restructuring modulated by pressure of a reactant gas.[2] The significance of AP-XPS studies (Figure 10.1) in this finding is the first example of the necessity of examining surface of a catalyst in gas phase instead of a study of a catalyst before or after catalysis.

How could a Pt(557) perform restructuring along the change of CO pressure? The quantitative analysis of AP-XPS data collected at different pressure allowed us to uncover the driving force. As shown in Figure 10.2, the CO coverage increases along the increase of CO pressure surrounding the Pt(557).[2] Here the coverage is defined as the ratio of the chemisorbed molecules to the number of atoms in the topmost layer of the surface in the same surface area. It is a vacuum surface science term. Here we borrow it to parameterize the surface density of adsorbed CO molecules in CO gas at different CO pressure. At a CO pressure of 0.5 Torr, the CO coverage reaches nearly 100%, obviously higher than the saturated coverage of CO molecules chemisorbed on Pt(557) achieved in high vacuum environment. The coverage of CO on Pt(557) in CO gas was calibrated by using the coverage, 0.5 reported

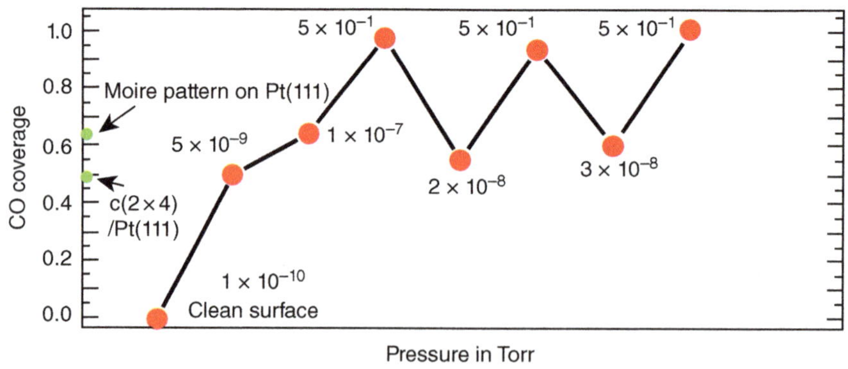

Figure 10.2 Coverage of CO on Pt(557) in CO at different pressures as marked in the graph. The CO coverage was determined by the XPS peak areas calibrated with the published coverage of CO in the c(2 × 4) CO adsorbate structure found at 5×10^{-9} Torr. *Source:* Reproduced with permission from Tao et al.[2]/American Association for the Advancement of Science.

in literature for CO in the c(2×4) adsorption layer on Pt(332), Pt(355), and Pt(111) at room temperature formed upon exposed to 5×10^{-7} Torr CO.[3–5]

The correlation between the extent of surface restructuring in CO gas (Figure 5.2) and CO coverage listed in Figure 10.2 shows that the restructuring of stepped surface is driven by the increase of CO coverage.[2] It suggests that a high CO coverage is the driving force for the restructuring of Pt(557) surface. How could the increase of CO coverage make the surface of Pt(557) broken? The increase of CO coverage results in CO molecules repelling each other because the distance between two adjacent adsorbed CO molecules is only 3 Å or so. The intermolecular repulsion between two CO molecules chemisorbed on two adjacent Pt atoms can break Pt–Pt bonds of step edges since CO are polar molecules, the densely packed CO must repel its adjacent CO molecules, and binding of CO to Pt atoms at step edges is strong. Once the Pt–Pt bonds at the step edge are broken and curly step edges are formed, the length of the step edge increases and the portion of Pt atoms at step edge increases. The atomic-scale mechanism of the restructuring of the stepped Pt surface was uncovered through an in situ time-dependent HP-STM observation and machine-learning accelerated first principles atomistic simulations. At a higher pressure of CO at 0.5 Torr, as CO can bind to Pt atoms at edges of a cluster through a fan-out geometry. More CO can pack on a Pt atom on average. Thus, the coverage of CO on the Pt(557) increased after the surface is broken. Notably, although the Pt(557) surface is broken into nanoclusters, the total number of Pt atoms on surface does not increase. The "fan-out" geometry of CO molecules at step edges of nanoclusters makes the surface bind more CO molecules. Why CO–CO repulsion-based breakup of the Pt surface typically starts at step edges instead of the middle of a terrace? A Pt atom at a step edge of Pt(557) only coordinates with other 7 atoms; however, a Pt atom in the middle of a terrace of Pt(557) or Pt(111) bonds with 9 Pt atoms.

Why could the chemisorbed layer having high coverage of CO formed at 0.5 Torr not survive at a lower pressure? In other words, why could the nanocluster surface observed at 0.5 Torr in Figure 5.2d not be formed at low pressure such as 5×10^{-8} Torr? A similar question is why such a surprising restructuring could not be observed when CO was introduced to the surface through surface science exposure. Notably, the CO coverage in the chemisorbed CO layer on Pt(557) in CO gas, nearly 100%, is definitely higher than the saturated coverage of CO in the chemisorbed CO layer on Pt(557) in UHV.

As reported in literature,[2] Pt(332) also exhibits similar reversible restructuring along the increase of CO pressure. Similar restructuring was observed on Cu(111) in CO[6] since the cohesive energy of Cu is definitely lower than Pt. Notably, such a change on Pt(557) was not observed in the H_2 gas. Likely, the main reason is the weak repulsion between two adjacent H atoms on Pt(557) cannot force Pt–Pt bond to break. In fact, the lower adsorption energy of H atoms to Pt atoms allows the adsorbed H atoms readily "slide" from step-edge Pt atoms to terrace Pt atoms. Compared to the strong repulsion of two adjacent CO molecules and strong binding of CO to Pt atoms, the adjacent, nonpolar H atoms weakly repel and thus such a weak repulsion cannot force their underlying Pt–Pt bond to break. This study demonstrates that such a quantitative analysis through AP-XPS studies of pressure-dependent surface can provide unique information for elucidating surface structure of a catalyst under a reaction condition or during catalysis.

10.2 CO Oxidation on Pd(100), Pd(111), and Pd(110)

10.2.1 CO Oxidation on Pd(100)

CO oxidation on Pd model catalysts was studied with AP-XPS and its online mass spectrometer. The Pd(100) is not active for CO oxidation until 175 °C.[7] Surface of Pd(100) in the mixture of 0.2 Torr O_2 and 0.02 Torr CO was tracked with AP-XPS along the increase of catalysis temperature. The onset catalytic oxidation of CO at 175 °C was rationalized with the observed evolution of surface chemistry of Pd(100) in AP-XPS. When the temperature of Pd(100) is higher than 175 °C, Pd(100) is partially oxidized as suggested by the observed new peaks at 334.6, 335.4, and 336.2 eV (Figure 10.3b). These three peaks are contributed from Pd atoms at the interface of surface oxide layer and metallic Pd, twofold Pd atoms of surface palladium oxide, and fourfold Pd atoms of surface palladium oxide, respectively. Another significant difference between the temperature of inactive surface (<175 °C) and the temperature of active surface (>175 °C) is the lack of C 1s XPS peak when the temperature is higher than 175 °C. This difference suggests that the lack of activity of Pd(100) at low temperature (<175 °C) is due to the strong chemisorption of CO on metallic surface of Pd(100) at <175 °C; in other words, the strong chemisorption of CO deprives O_2 of adsorption sites on Pd(100). Furthermore, the confirmed activity in CO oxidation at a temperature >175 °C and the observed formation of surface oxide on Pd(100) at this temperature regime (>175 °C) suggest that the active phase of Pd(100) for CO oxidation at >175 °C is surface oxide instead of metallic Pd. Thus, the authentic active phase of Pd(100) is surface oxide of palladium oxide instead of metallic Pd(100). From this point of view, we can say the Pd(100) is a nominal catalyst as it does not give information on the authentic catalytic phase during CO oxidation.

We can imagine that the active phase of Pd(100) for CO oxidation could have been wrongly assigned to the metallic Pd(100) surface if there were a lack of AP-XPS study.[7] Another surprising observation is that the surface turns to metallic Pd at >370 °C (Figure 10.3b). This transformation from surface palladium oxide to metallic Pd at >370 °C was supported by the decrease of activity of CO oxidation at the high temperature (>370 °C) (Figure 10.3a). By establishing the correlation between catalytic activity for CO oxidation and the corresponding authentic surface chemistry observed with AP-XPS, the authentic catalytic phase of nominal Pd(100) can be identified.

10.2.2 CO Oxidation on Pd(111)

Formation of different types of chemisorption states of oxygen atoms on Pd(111) in 0.2 Torr O_2 in the temperature range of 200–500 °C was investigated with AP-XPS. Figure 10.4a and b are Pd $3p_{3/2}$ and Pd $3d_{5/2}$ at different temperatures, respectively. At 200 °C, a broad O 1s peak at 530 eV was observed. It can be deconvoluted into three chemisorbed species O(I), O(II), and chemisorbed O atoms on a surface oxide although Pd $3p_{3/2}$ overlaps O 1s. The areas of O(I) and O(II) at 200 are similar to those at 300 °C. O(I) and O(II) in Figure 10.4a were assigned to surface oxide Pd_5O_4 based on the reference.[9] Surface structure of Pd_5O_4 was presented in Figure 10.4c. Notably, the O(I) and O(II) observed at 200–300 °C are threefold and fourfold coordinated oxygen atoms as shown in Figure 10.4c. Corresponding to the

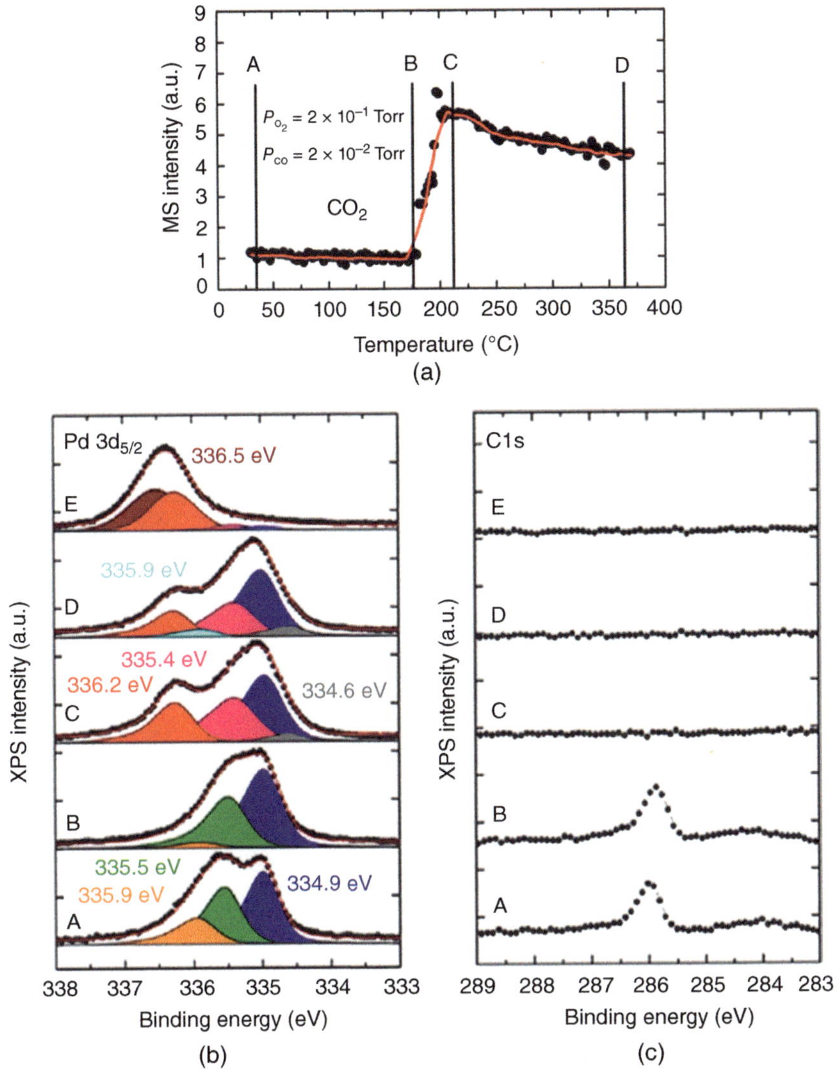

Figure 10.3 AP-XPS studies of CO oxidation on Pd(100) facet. (a) Mass spectra of CO_2 under exposure to 200 mTorr O_2 and 20 mTorr CO at 30–370 °C. (b) Pd $3d_{5/2}$ and (c) C 1s XPS spectra of Pd(100) exposed to 200 mTorr O_2 and 20 mTorr CO at 30 °C (A), 175 °C (B), 210 °C (C), and 370 °C (D), and 200 mTorr O_2 at 370 °C (E). *Source:* Reproduced with permission from Toyoshima et al.[7]/ American Chemical Society.

O 1s spectrum at 300 °C, it is found that the Pd $3d_{5/2}$ at 300 °C (Figure 10.4b) consists of three peaks including 334.9 eV for metallic bulk Pd, 335.4 eV for two fold coordinated Pd of oxide, and 336.2 eV for fourfold coordinated Pd of oxide based on literature.[8–10]

Similar to the above CO oxidation on Pd(100) in Section 10.2.1,[7] AP-XPS studies of CO oxidation on nominal catalyst Pd(111) suggests that the active phase for CO oxidation on Pd(111) is surface oxide instead of metallic Pd.[8] As shown in Figure 10.5c, Pd(111) surface at 200 °C remains metallic in the mixture of 0.02 Torr CO and 0.2 Torr O_2. α, β, and γ are three

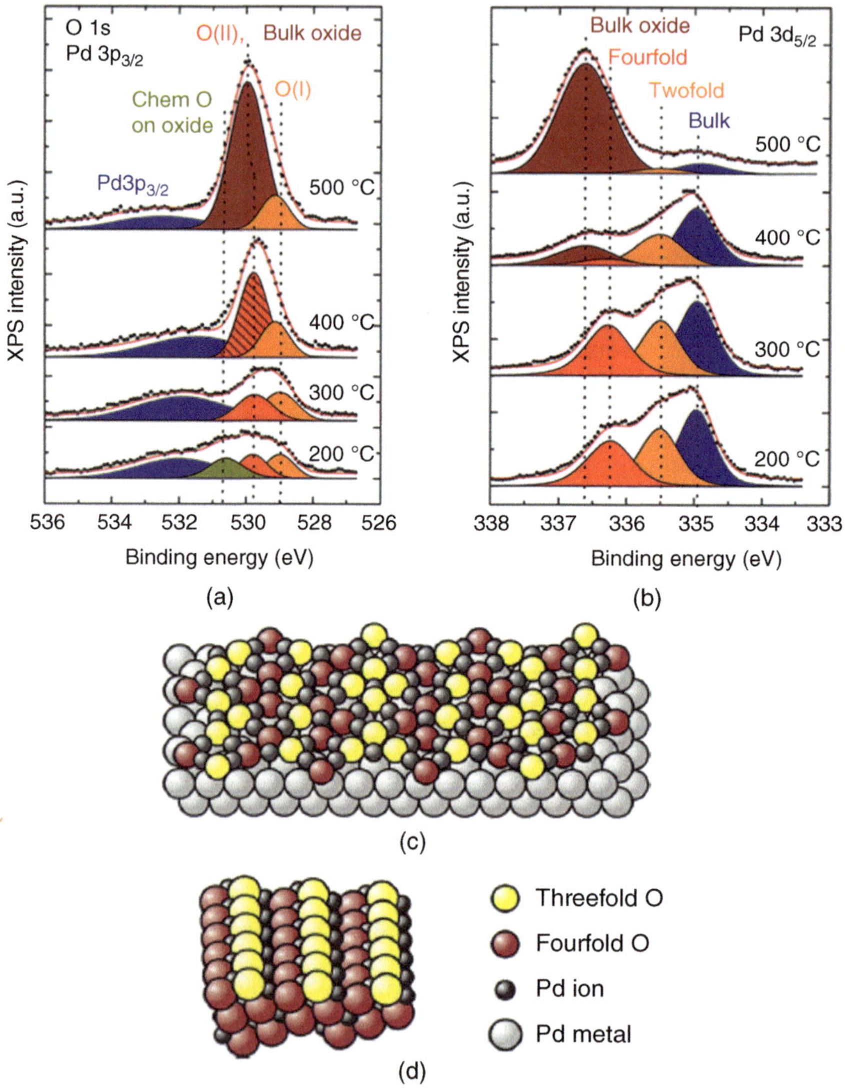

Figure 10.4 AP-XPS studies of Pd(111) surface in 0.2 Torr O_2 in the temperature range of 200–500 °C and the related structure models. (a) XPS spectra of O1s and Pd $3p_{3/2}$ and (b) Pd $3d_{5/2}$ regions for a Pd(111) surface taken at various temperatures under the presence of O_2 gas at 200 mTorr. The XPS spectra are deconvoluted into several components, as indicated by different colors. (c) Structure models of the Pd_5O_4 surface oxide. (d) Structure model of PdO bulk oxide with the (101) surface orientation. *Source:* Reproduced with permission from Toyoshima et al.[8]/American Chemical Society.

different time zones in the mixture of CO and O_2 as defined in Figure 10.5a. Four peaks of Pd $3d_{5/2}$ at 334.9, 335,4, 335.6, and 336.2 eV are assigned to bulk Pd atoms, and Pd atoms with adsorbed CO on hollow, bridge, and on-top sites of the metallic Pd surface, respectively (Figure 10.5c). At 200 °C, no CO_2 was formed in the mixture of 0.02 Torr CO and 0.2 Torr O_2 (region α in Figure 10.5a). It is understandable that the CO molecules cover the surface of Pd(111) and thus block the access of O_2 to Pd(111) based on the collected C 1s spectrum.

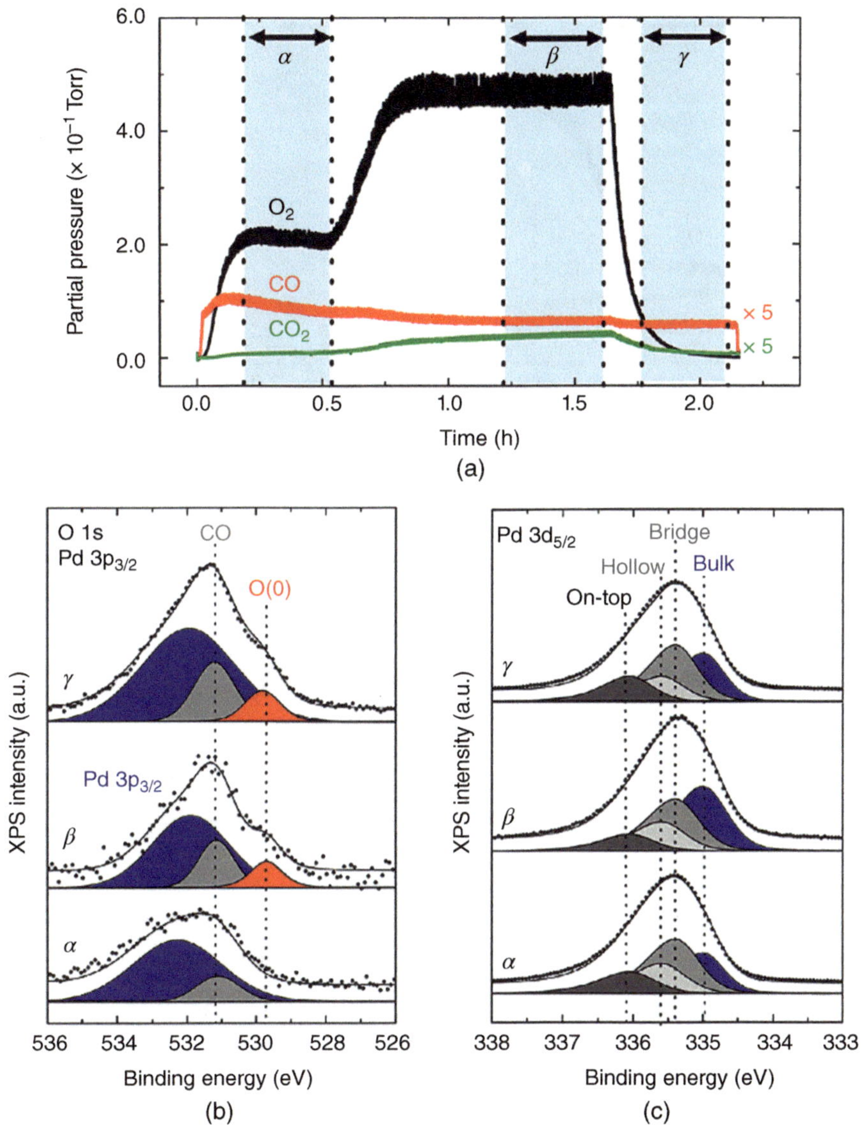

Figure 10.5 AP-XPS studies of Pd(111) surface CO oxidation reaction on Pd(111) at 200 °C in the mixture of CO and O_2 at α, β, and γ phrases. (a) Partial pressures of O_2, CO, and CO_2 monitored by mass spectroscopy; At β phrase, the partial pressure of O_2 was obviously increased. (b) O 1s and Pd $3p_{3/2}$; and (c) Pd $3d_{5/2}$ AP-XP spectra of a Pd(111) surface under exposure to O_2 and CO gases, up to 500 and 20 mTorr, respectively. *Source:* Reproduced with permission from Toyoshima et al.[8]/ American Chemical Society.

The O 1s peaks of adsorbed CO or O on Pd(111) in the mixture of CO and O_2 are presented in Figure 10.5b. There are no chemisorbed oxygen atoms formed at the α phrase as suggested by Figure 10.5b. By increasing O_2 partial pressure to 0.5 Torr at 200 °C, a new peak in O 1s spectrum at 529.8 eV was observed in the β zone of Figure 10.5b, suggesting that subsurface O atoms were formed at 200 °C when the partial pressure of O_2 is relatively high (β phrase in Figure 10.5a).

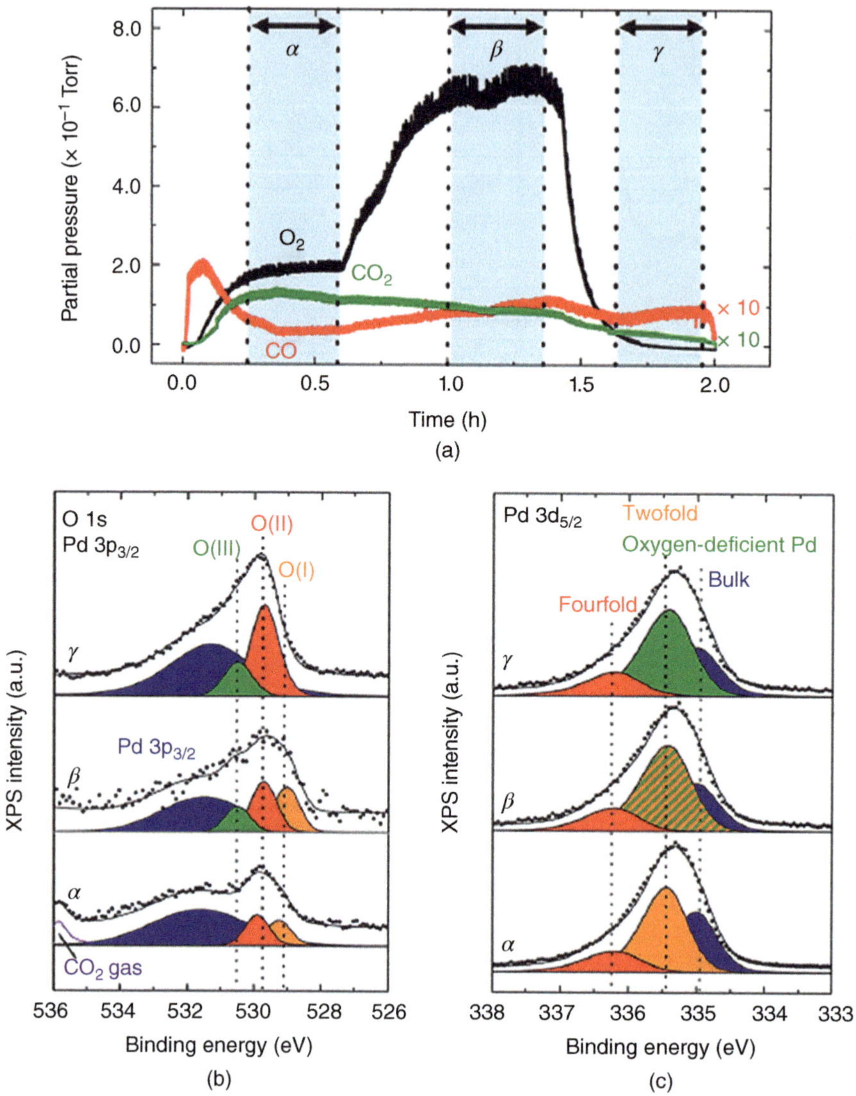

Figure 10.6 AP-XPS studies of Pd(111) surface at 300 °C in the mixture of CO and O_2. (a) Partial pressures of O_2, CO, and CO_2 monitored by mass spectroscopy, (b) O 1s/Pd $3p_{3/2}$, and (c) Pd $3d_{5/2}$ AP-XP spectra of a Pd(111) surface under exposure to O_2 and CO gases, up to 600 and 20 mTorr, respectively. *Source:* Reproduced with permission from Toyoshima et al.[8]/American Chemical Society.

At 300 °C in the same gas mixture, no CO molecules are adsorbed on Pd(111) although the dominant adsorbates on Pd(111) are CO molecules at 200 °C. At 300 °C, Pd(111) was partially oxidized to surface oxide, Pd_5O_4 (Figure 10.6b). The two deconvoluted O 1s peaks at 529.1 [O(I)] and 529.8 eV [O(II)] are assigned to threefold and fourfold O atoms of surface Pd_5O_4;[7] accordingly, the two peaks of Pd $3d_{5/2}$ at 335.8 and 336.2 eV in Figure 10.6c were due to threefold and fourfold Pd atoms of Pd_5O_4. The simultaneous observation using a mass spectrometer shows that the rate of CO_2 formation at 300 °C in this mixture obviously increases while pressure of CO clearly decreased under the condition that both CO and O_2 were constantly flowing through Pd(111) at 300 °C. It shows that the surface palladium oxide Pd_5O_4 formed at 300 °C is active for CO oxidation on metallic Pd(111) (Figure 10.6a) although Pd(111) at 200 °C is nearly inactive for CO oxidation (Figure 10.5a). Compared to the α phrase, the partial pressure of O_2 in the β phrase was increased to 0.5 Torr (Figure 10.6a). Compared to the α phrase at 300 °C, the activity for CO oxidation in the β phrase decreased to some extent (Figure 10.6a).

At the β phrase in terms of the condition (0.02 Torr CO and 0.5 Torr O_2) at 300 °C, oxygen-deficient Pd oxide was formed, evidenced by the observed new O 1s component at 530.5 eV (O(III) in Figure 10.6b) and Pd $3d_{5/2}$ at 335.8 eV (Figure 10.6c). The decrease of activity for CO oxidation in β region and the formation of a new chemical state of Pd–O labeled with O(III) show the newly formed O(III) is not active for CO oxidation. Now the question is which of O(I) and O(II) is active for CO oxidation. By turning off the flowing O_2 (γ region of Figure 10.6a), only CO flows through the catalyst. Surprisingly, the component O(I) disappeared immediately, suggesting I(I) is active for CO oxidation. As the relative intensities of both O(II) and O(III) showed no decrease in the γ phrase at 300 °C while only CO was flowing, they are not active phases for CO oxidation. AP-XPS clearly identified the three surface O-containing phases and distinguished the active phase and two spectator phases. If surface of Pd(111) used for CO oxidation at 300 °C were examined with high vacuum XPS at room temperature, the high vacuum XPS must find the remained O(II) and O(III) other than O(I) and thereby incorrectly assign O(I), O(II), and O(III) to active phases of CO oxidation.

At 400 °C, in the mixture of 0.02 Torr CO and 0.2 Torr O_2, CO_2 was obviously formed (Figure 10.7a). As shown in Figure 10.7b, both O(I) and O(II) were observed at 400 °C, similar to surface at 300 °C in the mixture of 0.02 Torr CO and 0.2 Torr O_2. The O(I) surface phase is active for CO oxidation since CO_2 was observed in the α phrase in Figure 10.7a. However, the formation rate of CO_2 subsequently decayed rapidly when the partial pressure of O_2 was increased in the β phrase in Figure 10.7a. Meanwhile, the O(III) formed rapidly in Figure 10.7b along the decay of CO_2 production (β to γ in Figure 10.7a). In the γ phrase, the O_2 partial pressure further decreased. Correspondingly, the production of CO_2 further decreased. In γ region of Figure 10.7b O(II) significantly increased but O(I) disappeared along the decrease of production rate of CO_2 (γ phrase in Figure 10.7). These evolutions of O(I), O(II), and O(III) at 400 °C in Figure 10.7b suggest that only O(I) is the active phase for CO oxidation. Thus, O(I) phase is the active phase of Pd(111) at 400 °C for CO oxidation.

During CO oxidation on Pd(111), the common feature of O 1s at 300 and 400 °C is the coexistence of O(I) and O(II) with roughly equivalent intensity; they are assigned to the surface oxide of Pd_5O_4. This surface oxide contains two oxygen species, O(I) and O(II), that have different coordination numbers as show in Figure 10.4c. O(I) atoms shown in yellow and O(II) in pink in Figure 10.4c coordinate with three and four Pd atoms, respectively. Titration of O atoms of surface of Pd_5O_4 with CO dosing technique suggests that only O(I) is active for CO oxidation. This is understandable since the low coordination of O(I) allows the access of CO for CO oxidation. Compared to Pd_5O_4, the bulk PdO (Figure 10.4d) is not active for CO oxidation as it is spatially separated from CO and O_2 of gas phase.[8]

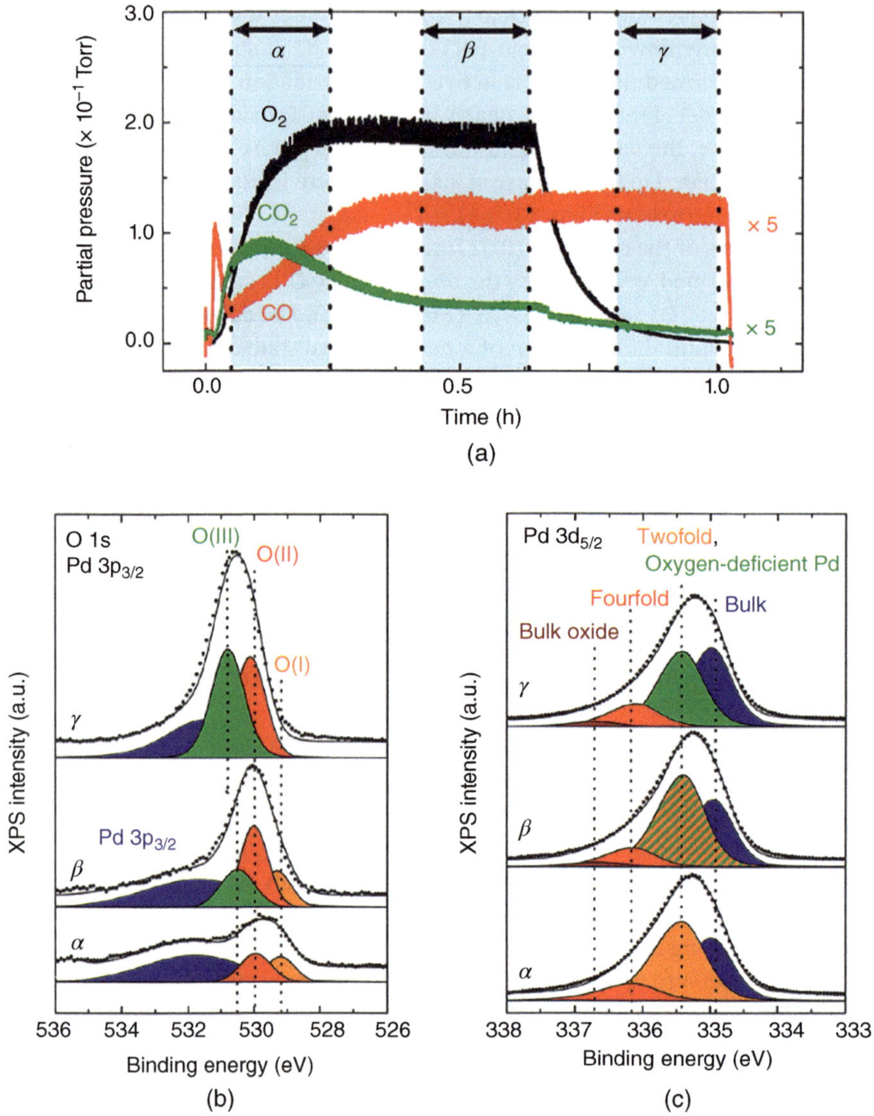

Figure 10.7 CO oxidation reaction at 400 °C. (a) Partial pressures of O_2, CO, and CO_2 monitored by mass spectrometry; (b) O1s and Pd $3p_{3/2}$; and (c) Pd $3d_{5/2}$ AP-XP spectra of a Pd(111) surface under exposure to O_2 and CO gases, up to 200 and 20 mTorr, respectively. *Source:* Reproduced with permission from Toyoshima et al.[8]/American Chemical Society.

10.2.3 CO Oxidation on Pd(110)

CO oxidation on Pd(110) was studied with AP-XPS and mass spectrometry during catalysis.[11] Different from Pd(100) and Pd(111), no surface oxide was formed on Pd(110) up to 290 °C in the mixture of 0.02 Torr CO and 0.2 Torr O_2. At a temperature lower than 158 °C, surface in the mixture of CO and O_2 is covered with two types of chemisorbed CO, rationalizing the lack of activity of CO oxidation in the temperature range of 25–158 °C. At 150 °C or so, the

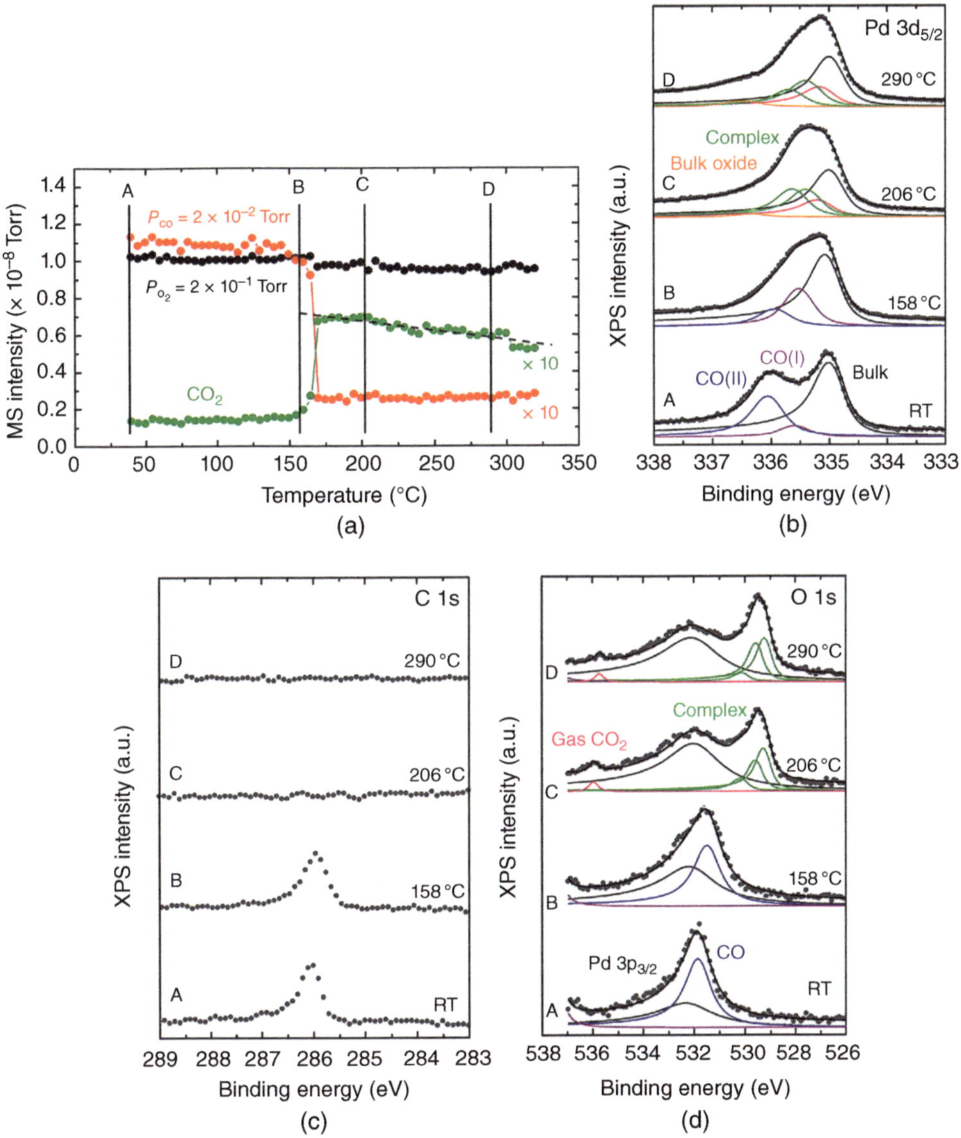

Figure 10.8 AP-XPS and mass spectrometry studies of CO oxidation Pd(110). (a) Temperature dependence of mass spectral intensity of CO_2 under exposure of $p_{O_2} = 2 \times 10^{-1}$ Torr and $p_{CO} = 2 \times 10^{-2}$ Torr. Above the critical temperature (150–160 °C), the CO_2 intensity increased drastically along the obvious consumption of CO; however, further heating to 290 °C causes a gradual decay of the CO_2 signal indicated by the dashed line. Each label (A, B, C, and D) indicates temperature where XPS spectra were measured. (b) Pd $3d_{5/2}$, (c) C 1s, and (d) O 1s. In (c) and (d), gas phase CO_2 peak is observed at 536 eV. *Source:* Reproduced with permission from Toyoshima et al.[11]/American Chemical Society.

intensity of Pd $3d_{5/2}$ of Pd atoms adsorbing CO molecules is obviously weak in contrast to that at 25 °C (spectrum B versus A in Figure 10.8b). At 206 °C, the formation rate of CO_2 increased along the disappearance of chemisorbed CO (C in Figure 10.8a and C in Figure 10.8c). As seen in Figure 10.8b, at 206 °C three major peaks in Pd $3d_{5/2}$ were observed; they are peaks at 335.0 eV contributed from metal Pd atoms of deep atomic layers in the surface region, 335.4–335.8 eV from Pd atoms binding to the chemisorbed O atoms on surface, and 335.3 eV from Pd atoms of bulk palladium oxide. Along the increase of temperature to 290 °C, the formation rate of CO_2 starts to decrease (D in Figure 10.8a). Corresponding to the slight decrease of formation rate of CO_2 at 290 °C, the intensity of chemisorbed oxygen atoms on Pd(110) decreases on the basis of the decrease of Pd $3d_{5/2}$ at 335.4–335.8 eV in D of Figure 10.8b. This activity–structure correlation between the decrease of formation rate of CO_2 observed with mass spectrometry and the decrease of intensity of chemisorbed oxygen atoms on Pd(110) tracked with AP-XPS suggest that the chemisorbed oxygen atoms layer is the catalytically active phase for CO oxidation on Pd(110).[11] In addition, consistent with the observation of CO_2 in mass spectrometry when Pd(110) is at 206 °C and 290 °C (phrase C and D in Figure 10.8a), gaseous CO_2 was observed with AP-XPS at the two temperatures (spectrum C and D in Figure 10.8d).

10.3 CO Oxidation on Pt(110) and Pt(111)

10.3.1 CO Oxidation on Pt(110)

Surface of Pt(110) during CO oxidation was investigated with AP-XPS studies.[12] In 0.2 Torr O_2 at 150 °C, surface roughening was observed in the STM image at the left-bottom corner in Figure 10.9. The roughening of surface at 150 °C in 0.2 Torr O_2 observed with HP-STM corresponds to the formation of a surface oxide based on AP-XPS studies performed under the same condition.[12] The roughened surface is formed in an O_2-rich environment; it restores to a flat surface under an O_2-deficient condition.

When Pt(110) was in 0.05 Torr O_2, a shoulder at 72.0 eV was observed by AP-XPS (Figure 10.9b). This peak results from the chemisorbed oxygen atoms on the surface. A new peak appeared at 73.4 eV (Figure 10.9c) when Pt(110) was exposed to 0.5 Torr O_2. This new Pt $4f_{7/2}$ peak was marked by a red dash line and labeled "Surf O" in Figure 10.9. It is attributed to the formation of surface oxide. In terms of O 1s spectrum collected at 0.5 Torr O_2, two O 1s peaks at 529.7 and 530.8 eV were observed and they are the adsorbed oxygen atoms and surface oxide formed in 0.5 Torr O_2.

To identify the nature of surface-oxygen species, surface titration using CO as a titrant was performed while the AP-XPS was simultaneously tracking surface of the catalyst. With AP-XPS, surface coverages of chemisorbed oxygen atoms (Chem-O) and surface oxide (Surf-O), total coverage of O atoms and total coverage of CO were measured as a function of time after the two steps including (i) exposing Pt(110) to 0.5 Torr O_2 and (ii) then exposing to 10^{-6} Torr CO at 270 K. During the exposure to 10^{-6} Torr CO (the step (ii) or called titration step), the surface chemistry of Pt(110) was tracked by AP-XPS as a function of titration time when the surface oxygen-containing species was being titrated. O 1s peaks of O-containing surface phases upon titration for 36 and 276 seconds are presented in Figure 10.10a and 10.10b, respectively. Figure 10.10c presents the coverage of these oxygen-containing surface phases as a function of time of exposing to CO.

As shown in Figure 10.10a, three O 1s peaks at 529.7, 530.8, and 532.6 eV were still observed upon exposing Pt(110) to 0.5 Torr O_2 and then titrated with 10^{-6} Torr CO for

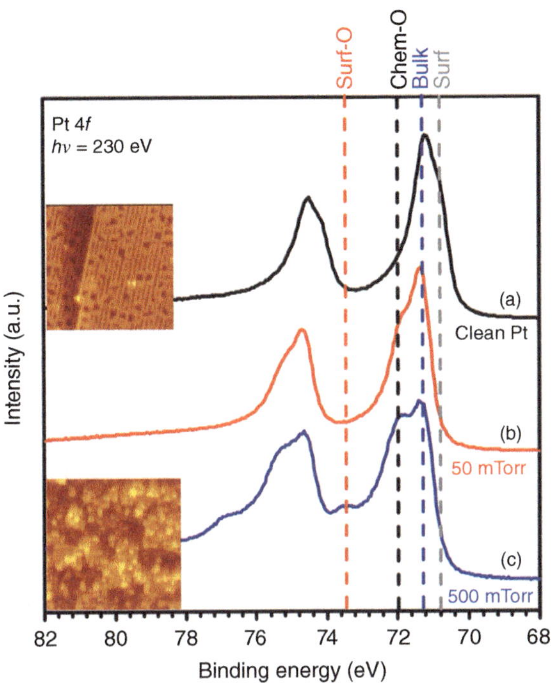

Figure 10.9 Surface structure and chemistry of Pt(110) in UHV and O_2 at 0.05 and 0.5 Torr at room temperature. (a) The Pt 4f spectrum taken after sputtering and annealing in UHV is composed of a peak from bulk Pt at 71.2 eV and a surface peak at lower binding energy (70.8 eV, CLS = 0.4 eV). (b) Under 0.05 Torr of O_2 an additional peak at 71.9 eV (CLS = 0.7 eV) appears. This peak corresponds to surface Pt atoms bonded to the oxygen atom. (c) When the O_2 pressure is increased to 0.5 Torr, a third peak at 73.4 eV is observed, corresponding to the formation of a surface oxide. STM images taken under similar conditions (clean surface and under 0.2 Torr of O_2 at 423 K) are shown as inserts. The high-pressure image shows a roughened surface with nanometer islands. *Source:* Reproduced with permission from Butcher et al.[12]/American Chemical Society.

36 seconds; they are assigned to the chemisorbed O atoms (Chem-O in black), surface oxide (Surf-O in red), and adsorbed on-top CO molecules (in gray), respectively (Figure 10.10a). Notably, upon a longer period of exposing to CO (276 seconds) as suggested by Figure 10.10b, the atomic fractions of the three chemical states of O atoms changed in contrast to the short exposure as shown in Figure 10.10a. After a period of 276 seconds, the intensities of two types of oxygen-containing surface species including chemisorbed O atoms at 529.7 eV and surface oxide at 530.8 eV decreased because the longer exposure to CO consumed more oxygen atoms. The difference between Figure 10.10a and b suggests the two types of oxygen-containing surface species are active phases for CO oxidation. Quantitative analyses of AP-XPS to the two types of surface oxygen species (Chem-O at 529.7 eV and Surf-O at 530.8 eV) and the adsorbed CO provided information on how fast these surface phases can change as a function of time during CO oxidation (Figure 10.10c). As the surface coverage of chemisorbed O atoms (Chem-O) decreased faster than surface oxide (Surf-O) in the titration of CO, the chemisorbed O atoms are more active for CO oxidation (Figure 10.10c). As the formed surface oxide is active for CO

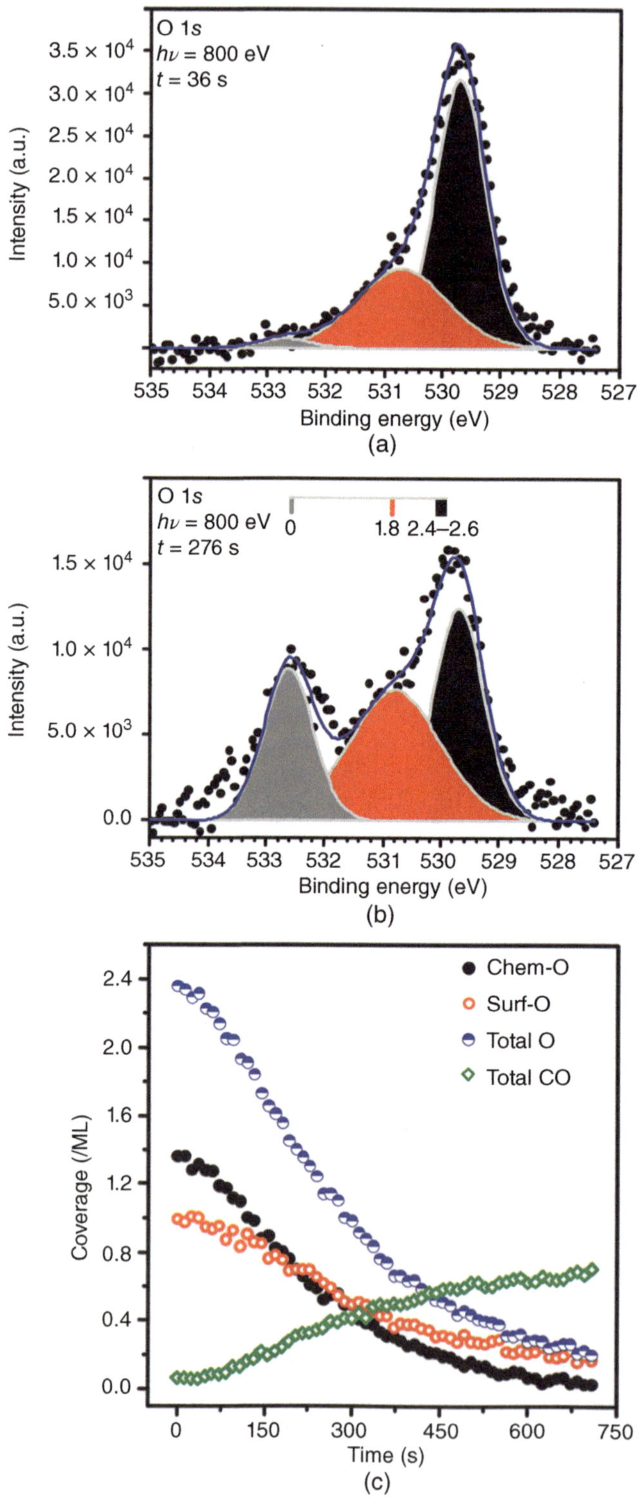

Figure 10.10 AP-XPS studies of O 1s spectrum of Pt(110) surface exposed to 0.5 Torr O_2 and then exposed to 10^{-6} Torr CO at 270 K for 36 seconds (a) and 276 seconds (b). (c) Coverage of surface CO and O species on Pt(110) surface exposed to 0.5 Torr O_2 and then exposed to 10^{-6} Torr CO at 270 K. *Source:* Reproduced with permission from Butcher et al.[12]/American Chemical Society.

oxidation as well, both Langmuir–Hinshelwood and Mars–van Krevelen mechanisms are responsible for CO oxidation on Pt(110).[12]

These studies demonstrate the unique function of AP-XPS in tracking surface chemistry of a catalyst under reaction condition (in gas of one reactant) or during catalysis (in gase of all reactants of a chemical reaction). In addition, the function of real-time track of the evolution of the surface species of the catalyst is a unique function in fundamental understanding of catalytic mechanisms. For instance, the AP-XPS observation of titration of surface (Figure 10.10) allows for identifying catalytic phases from surface spectator phases and distinguishing coexisting multiple catalytic phases with different activities.

10.3.2 CO Oxidation on Pt(111)

CO oxidation on Pt(111) was investigated with AP-XPS.[13] Based on literature, surface oxide α-PtO_2 on Pt(111) can be formed upon exposing Pt(111) to 10 Torr O_2 while the Pt(111) was repeatedly heated to 800 K and then cooled to 300 K for four times.[13] The formation of α-PtO_2 was confirmed by the observed XPS peak of Pt 4f (red spectrum in Figure 10.11). Although the XPS peak of Pt 4f (red spectrum in Figure 10.11) was not deconvoluted, it is readily deduced that the Pt 4f peak is contributed from metallic Pt and α-PtO_2. Upon the formation of α-PtO_2 islands on Pt(111), oxygen atoms can be still chemisorbed on patches of bare Pt(111) where no α-PtO_2 islands were formed. Based on the information gained from Figure 10.11, Pt(111) surface consisting of three different surface structures including 2.8 ML α-PtO_2 (green in Figure 10.12), 1.1 ML 4O film (red in Figure 10.12), and 0.5 ML c(4×2)-2CO_{ad} adlayer (blue in Figure 10.12).

These containing-oxygen surface phases formed on Pt(111) exhibit different catalytic activity for CO oxidation. The difference in their activity was identified through the AP-XPS spectroscopic titration using CO. As shown in Figure 10.13, the amounts of the three phases

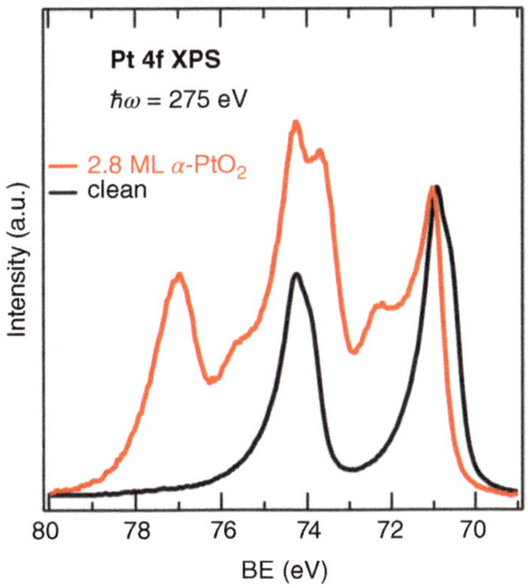

Figure 10.11 Pt 4f XPS spectra of a 2.8 ML α-PtO_2 film (red line) and of an adsorbate-free Pt(111) surface (black line). *Source:* Reproduced with permission from Miller et al.[13]/American Chemical Society.

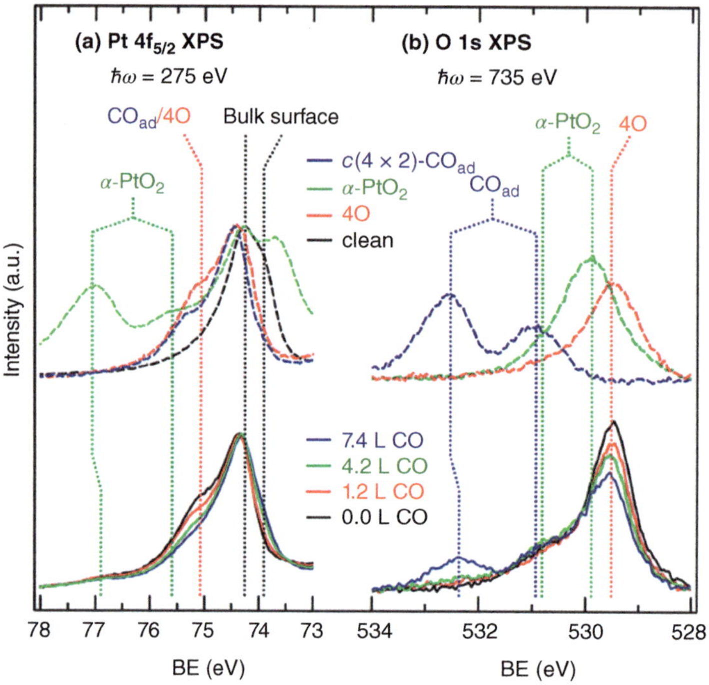

Figure 10.12 AP-XPS titration of the evolution of the mixed PtO/α-PtO_2 formed on Pt(111) by exposing to different amount of CO. (a) (solid lines at bottom) Pt $4f_{5/2}$ XP spectra of a mixed PtO/α-PtO_2 film after exposure to 0.0 L (black), 1.2 L (red), 4.2 L (green), and 7.4 L (blue) CO; (dashed lines at top) reference Pt $4f_{5/2}$ XP spectra of an adsorbate-free Pt(111) surface (black), a 1.1 ML 4O film (red), a 2.8 ML α-PtO_2 film (green), and a 0.5 ML c(4 × 2)–$2CO_{ad}$ adlayer (blue). (b) The corresponding O 1s XP spectra during these titrations. *Source:* Reproduced with permission from Miller et al.[13]/American Chemical Society.

exposed to different doses of CO were measured. For the chemisorbed oxygen atoms, denoted as O_{ad} (blue triangles in Figure 10.13) in terms of 0.5 ML c(4×2)-$2CO_{ad}$ adlayer, it is highly active for reaction with CO to form CO_2. Only a small amount of CO (1 L) can remove these chemisorbed oxygen atoms. Regarding the PtO (green squares in Figure 10.13), its requests much more CO (7L) to titrate it. However, the titration suggests that α-PtO_2 is not active at all as its surface coverage represented by the coverage of oxygen atoms of α-PtO_2 phase never decreases regardless of how much CO it is exposed to (see red circle in Figure 10.13). The titration suggests that both O_{ad} and PtO are active for CO oxidation since their coverages decrease along exposing to more CO. As the absolute value of the slope of titrating O_{ad} is larger than that of titrating PtO, O_{ad} is more active than PtO phase for CO oxidation (Figure 10.13). Thus, these AP-XPS titrations uncovered that the activity of these surface O-containing phases formed on Pt(111) increases along the ordering: α-$PtO_2 << PtO < O_{ad}$.[13]

These AP-XPS studies of CO oxidation on Pt(110) and Pt(111) have suggested very rich surface chemistries of Pt model catalysts during CO oxidation even at sub-Torr pressure range of CO and O_2.

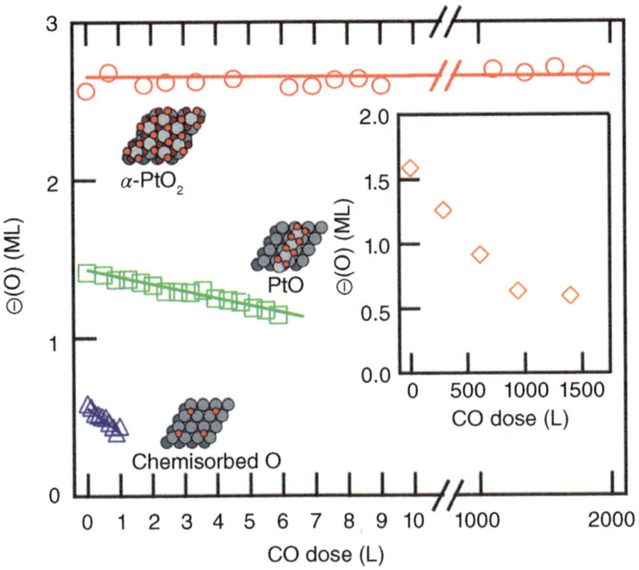

Figure 10.13 The coverage of surface-oxygen species [Θ(O)] while dosing CO, as determined using AP-XPS, for three oxygen layers on Pt(111): (red circles) an α-PtO_2 film with an initial coverage of 2.8 ML; (green squares) a 4O/α-PtO_2 mixture with an initial coverage of 1.4 ML; (blue triangles) an O_{ad}/α-PtO_2 mixture with an initial coverage of 0.47 ML. The solid lines are guides to the eye. P(CO) was 1×10^{-8} Torr in all measurements except that of the 2.8 ML α-PtO_2 film, for which a pressure of 1×10^{-5} Torr was employed. Inset: CO dose-dependent change of the surface-oxygen species from α-PtO_2 partially covering the surface. Note that Θ(O) excludes the coverage of CO_{ad}, i.e. Θ(O) = Θ(Ototal) − Θ(CO). *Source:* Reproduced with permission from Miller et al.[13]/American Chemical Society.

10.4 CO Oxidation on Rh(110)

Compared to Rh(111), Rh(110) is a much more open surface. Thus, it is readily restructured once it is exposed to a small dosage of O_2.[14] For instance, once a clean Rh(110) is exposed to O_2 at a pressure as low as 2×10^{-8} Torr O_2 for 5 minutes at 500 °C, the Rh(110)-1×1 was immediately restructured to Rh(110)-1×2-O.[14] Rh(110)-1×2-O is called a missing row structure.[14] On this missing-row structure, these oxygen atoms are evenly chemisorbed on surface with a coverage of 0.5 for O atoms. Its binding energy is 529.9 eV.[14] Notably, these adsorbed O atoms of Rh(110)-1×2-O were quite active for CO oxidation since their O 1s peak at 529.9 eV in AP-XPS progressively disappears when surface O atoms of the Rh(110)-1×2-O were titrated with 8×10^{-8} Torr CO at 25 °C (Figure 10.14). Other than AP-XPS, this titration using CO was visualized with in situ real-time HP-STM. In the STM image, these O atoms adsorbed on the Rh(110)-1×2 surface appear in blue (Figure 10.14a1). Based on HP-STM studies, the adsorbed surface O atoms of Rh(11)-1×2-O were progressively titrated by CO to form CO_2 by which the surface sites (Rh atoms) are available. In fact, CO molecules are adsorbed on Rh atoms immediately upon O atoms are removed through the formation of CO_2. In the STM images in Figure 10.14, the adsorbed O atoms and CO molecules are represented with blue and pink, respectively. As shown in Figure 10.14a1-a6,

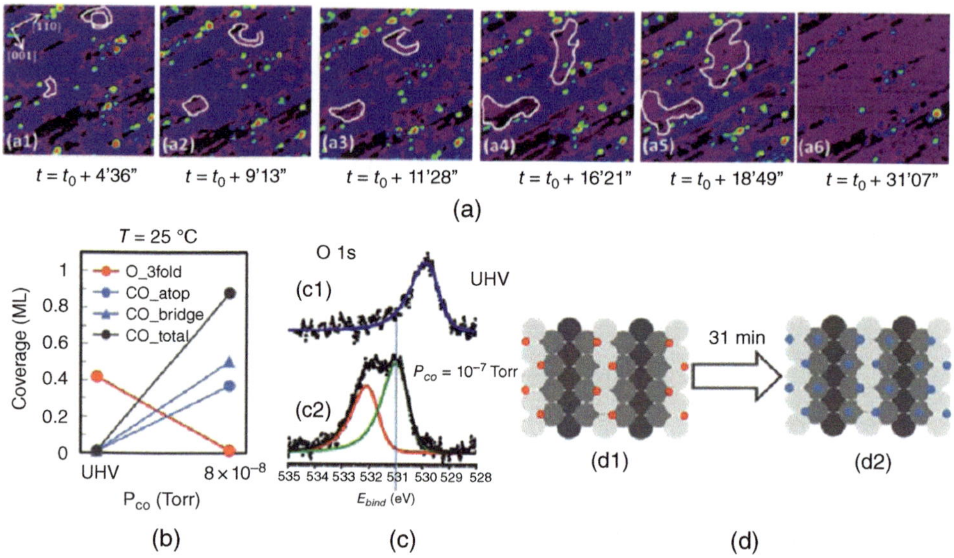

Figure 10.14 In situ observation of CO adsorption on Rh(110)-(1 × 2)-O at 25 °C and $P_{CO} = 8 \times 10^{-8}$ Torr. (a) Sequential STM images of the same area as a function of time while being exposed to CO gas at 8×10^{-8} Torr; blue and pink represent adsorbed O and CO species, respectively. (b) AP-XPS studies of O and CO coverages based on O 1s and Rh 3d photoemission intensities. (c) O 1s XPS peak of Rh(1 × 2)-O surface in UHV and CO at $P_{CO} = 8 \times 10^{-8}$ Torr (c1 and c2, respectively). (d) Adsorption models showing the replacement by (1 × 2)-O with CO. *Source:* Reproduced with permission from Nguyen et al.[14]/American Chemical Society.

the blue representing O_{ad} was progressively transferred to pink representing the adsorbed CO molecules. The titration of surface O atoms using CO visualized with STM (Figure 10.14a) is consistent with the decrease of coverage of O atoms adsorbed at three-fold Rh atom sites and the simultaneous increase of coverage of CO molecules adsorbed on surface (Figure 10.14b). Upon titratiation of these surface O atoms with CO, the CO molecules strongly chemisorbed on Rh(110)-2×1 through on-top and bridge binding configurations, evidenced by the observed 531.0 and 532.1 eV in AP-XPS studies (Figure 10.14c). In addition, HP-STM studies show that Rh(110) still remained Rh(110)-1×2 structure at 25 °C (Figure 10.14a6) upon the replacement of the adsorbed O atoms by CO molecules.

The above titration transformed Rh(110)-1×2-O to the Rh(110)-1×2-CO. Rh(110)-1×2-CO is stable at 25 °C in 8×10^{-8} Torr CO. However, the increase of CO pressure to 0.08 Torr and elevation of the temperature of Rh(110)-1×2-CO to 55 °C triggered a significant restructuring for surface from Rh(110)-1×2-CO to Rh(110)-1×1-CO. This restructuring was clearly observed by HP-STM (Figure 10.15). As shown in Figure 10.15c2, the inter-row distances of Rh(110)-1×2 and Rh(110)-1×1 are 7.6 and 3.8 Å, respectively. Thus, the newly formed Rh(110)-1×1 can be readily identified through its inter-row distance of 3.8 Å in STM images. It is marked with blue circles in Figures 10.15a1, b1 and c1. This structural transformation from Rh(110)-1×2 to Rh(110)-1×1 at 55 °C in 0.08 Torr CO was observed by the time-stream HP-STM (Figure 10.15). Figure 10.15a1, b1, and c1 are STM images of the same area at $t = t_0$, $t = t_0 + 2$ minutes, and $t = t_0 + 4$ minutes, respectively. In the region marked with green ellipses, more areas of Rh(110)-1×2 in Figure 10.15 were transformed into Rh(110)-1×1 while Rh(110)-1×2 was remained in CO for longer time.[14]

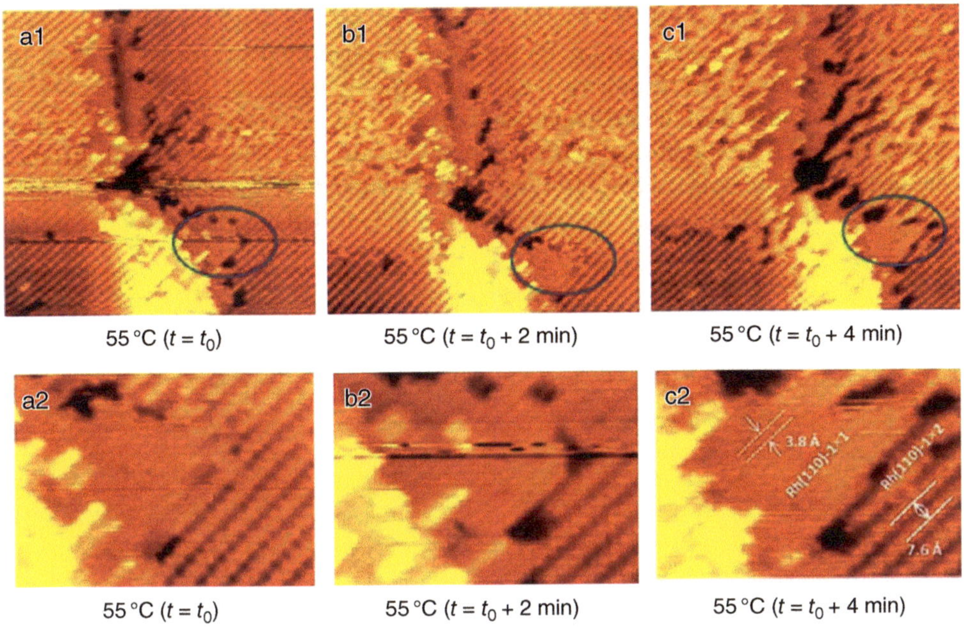

Figure 10.15 Time lapse in situ HP-STM images of the same area at P_{CO} =0.08 Torr and T=55 °C. (a1–c1) Sequential STM images of the same area in 2 minutes intervals; (a2–c2) enlarged images of areas marked by blue ovals. The inter-row distances in the [001] direction on two distinct regions of the sample are marked in (c2). *Source:* Reproduced with permission from Nguyen et al.[14]/American Chemical Society.

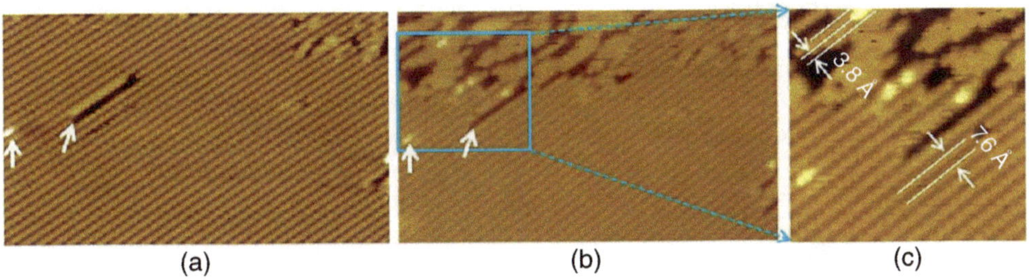

Figure 10.16 In situ STM images of Rh(110)-(1 × 2)-O in a gas environment of a mixture of CO and O_2 at P_{tot} =0.1 Torr in a 4/1 (CO/O_2) mixture and T=25 °C. (a) STM image at $t = t_0$ with the white box representing a largely unreconstructed Rh(110)-(1 × 2)-CO surface with 7.6 Å inter-row distance. (b) Time lapse image of the same region marked in (a) at $t = t_0 + 25$ minutes; the green box marks the region of image to be enlarged. (c) Enlarged image of region in the blue box in (b). The pairs of white arrows in (a) and (b) are mark landmarks used to identify the same region in the STM images taken at different times. *Source:* Reproduced with permission from Nguyen et al.[14]/American Chemical Society.

Other than an external heating to Rh(110)-1×2-O in terms of the elevation of temperature to 55 °C, a self-driven restructuring from Rh(110)-1×2-O to Rh(110)-1×1-CO without any external heating was uncovered with AP-XPS and HP-STM. As shown in Figure 10.16, in the mixture of 0.08 Torr CO and 0.02 Torr O_2, the Rh(110)-1×2-O was progressively restructured to Rh(110)-1×1-CO at 25 °C instead of 55 °C.[14] In this restructuring at 25 °C, the formation of CO_2 was observed through mass spectrometry. Obviously, the restructuring of surface from Rh(110)-1×2 to Rh(110)-1×1 was driven by the CO oxidation since no external heating was provided to the catalyst. As CO oxidation at 25 °C is exothermic, the CO oxidation event on the surface heated the local surface to 55 °C or higher since 55 °C is the threshold temperature for transition of Rh(110)-1×2-O to Rh(110)-1×1-CO based on the finding in Figure 10.15.

Surface of Rh(110)-1×2 during CO oxidation in a mixture of 0.08 Torr CO and 0.02 Torr O_2 in the temperature range of 25–200 °C was tracked with AP-XPS (Figure 10.17). AP-XPS studies show that CO coverage decreased during CO oxidation along the increase of

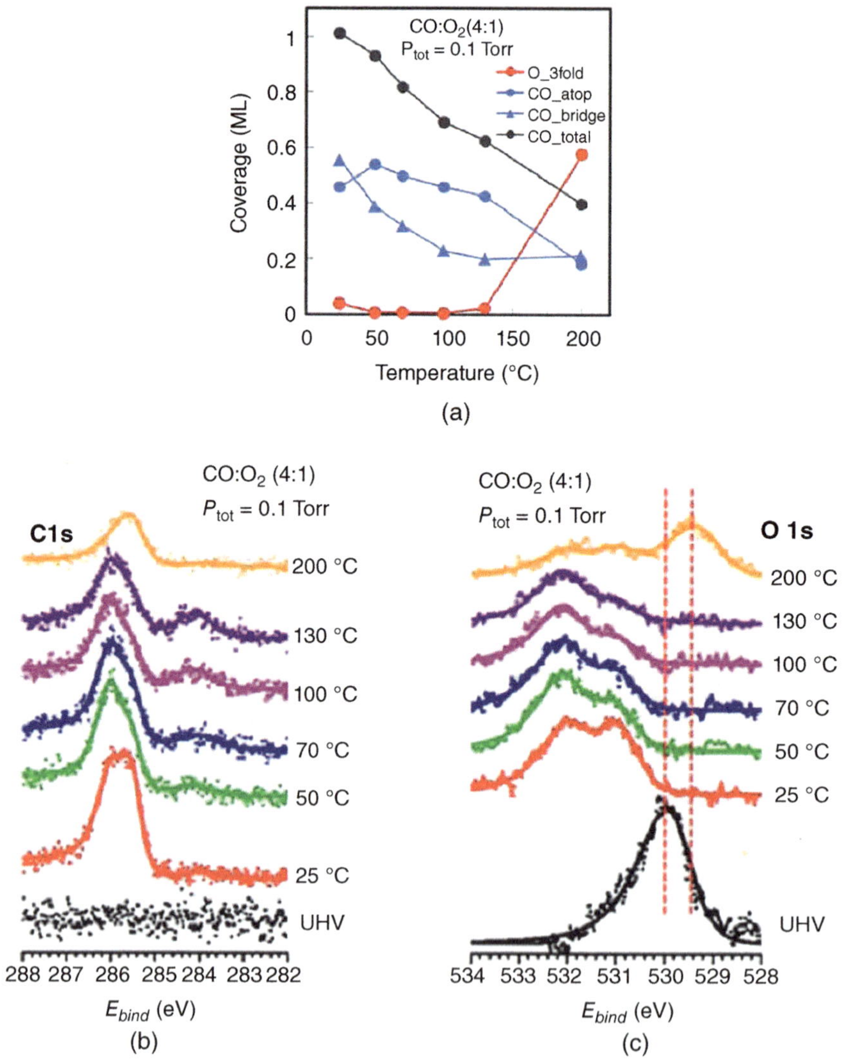

Figure 10.17 In situ AP-XPS studies of CO oxidation on Rh(110)-(1 × 2)-O as a function of temperature at P_{tot} = 0.1 Torr for a 4/1 (CO/O_2) mixture. (a) Calculated surface coverage for CO and O adsorbates. (b, c) C 1s and O 1s photoemission spectra of adsorbates stacked vertically as a function of temperature. Scatter points and black lines at the bottom of (b) and (c) indicate respective spectra under UHV conditions. *Source:* Reproduced with permission from Nguyen et al.[14]/American Chemical Society.

catalysis temperature from 25 to 200 °C. The trend of decrease of CO coverage and intensity of C 1s and O 1s XPS peak of adsorbed CO along the increase of catalysis temperature was seen from Figure 10.17. Notably, as shown by the red line in Figure 10.17a and the O 1s spectrum at 200 °C in Figure 10.17c, surface oxide was formed at 200 °C in the mixture of 0.08 Torr CO and 0.02 Torr O_2. The simultaneous online mass spectrometry measurement shows that CO_2 product was formed at 200 °C. The AP-XPS studies of the surface and the on-line mass spectrometry suggest the surface oxide formed on Rh(110) at 200 °C is an active phase for CO oxidation.

This complexity of relatively simple Rh single-crystal model catalysts suggests that the surface chemistry of a model catalyst during CO oxidation at sub-Torr could be never simply extrapolated to the surface chemistry of a nanoparticle catalyst during CO oxidation at ambient pressure. It makes us humbly aware of the challenge of in situ/operando studies of catalysis under a catalytic condition. Furthermore, it is not unreasonable to predict that we have only had a glimpse of an extremely complicated real world under a real catalytic condition.

10.5 CO Oxidation on Cu(111)

Cu(111) model catalyst is active for CO oxidation. Its surface chemistries in CO gas and during CO oxidation were studied with AP-XPS.[15] As shown in Figure 10.18a, CO molecules chemisorb on Cu(111) at room temperature in 0.3 Torr CO, evidenced by the observed weak peaks at 531.4 eV for O 1s and 286.1 eV for C 1s.[15] When the pure CO with a pressure of 0.2 Torr was switched to a mixture of 0.03 Torr O_2 and 0.3 Torr CO, two new O 1s peaks at 529.4 and 530.2 eV were observed. They were contributed from the chemisorbed O atoms on Cu(111) and O atoms of Cu_2O phase formed on Cu(111). Notably, the Cu_2O phase can be readily formed on Cu(111) in O_2 even at room temperature (Figure 10.18a). The fraction of Cu_2O increases along the increase of O_2 to CO ratio in the mixture of O_2 and CO as shown in Figure 10.18c. Other than the two types of O atoms, O atoms of CO chemisorbed on Cu_2O contributed to the peak at 534.2 eV, supported by the observed C 1s peak at 287.9 eV (Figure 10.18b).[15] In addition, the reaction intermediate $CO_2^{\delta-}$ was identified at 531.5 eV for O 1s and 289.0 eV for C 1s (Figure 10.18a and b).

AP-XPS studies suggest that the coverages of CO and $CO_2^{\delta-}$ increase as a function of O_2/CO partial pressure ratio in the mixture of O_2 and CO (Figure 10.19b and c).[15] The increase of coverage of intermediate $CO_2^{\delta-}$ as a function of O_2/CO ratio in Figure 10.19c is consistent with the evolution of the fraction of Cu_2O phase formed on Cu(111) as a function of O_2/CO ratio (Figure 10.18c). This consistence suggests that the production rate of CO_2 is higher when there is a large fraction of Cu_2O formed on surface. In other words, this consistence suggests that Cu_2O is the active phase for CO oxidation. These AP-XPS studies of CO oxidation on Cu(111) model catalyst revealed that the active phase of Cu(111) for CO oxidation is in fact the Cu_2O in the mixture of O_2 and CO at Torr pressure range. This finding demonstrated the significance of AP-XPS in fundamental studies of catalysis on metal catalysts since the active phases of metal catalysts in oxidative catalysis such as CO oxidation are some types of containing-oxygen surface phases in many cases, instead of the original metallic surface in many cases.

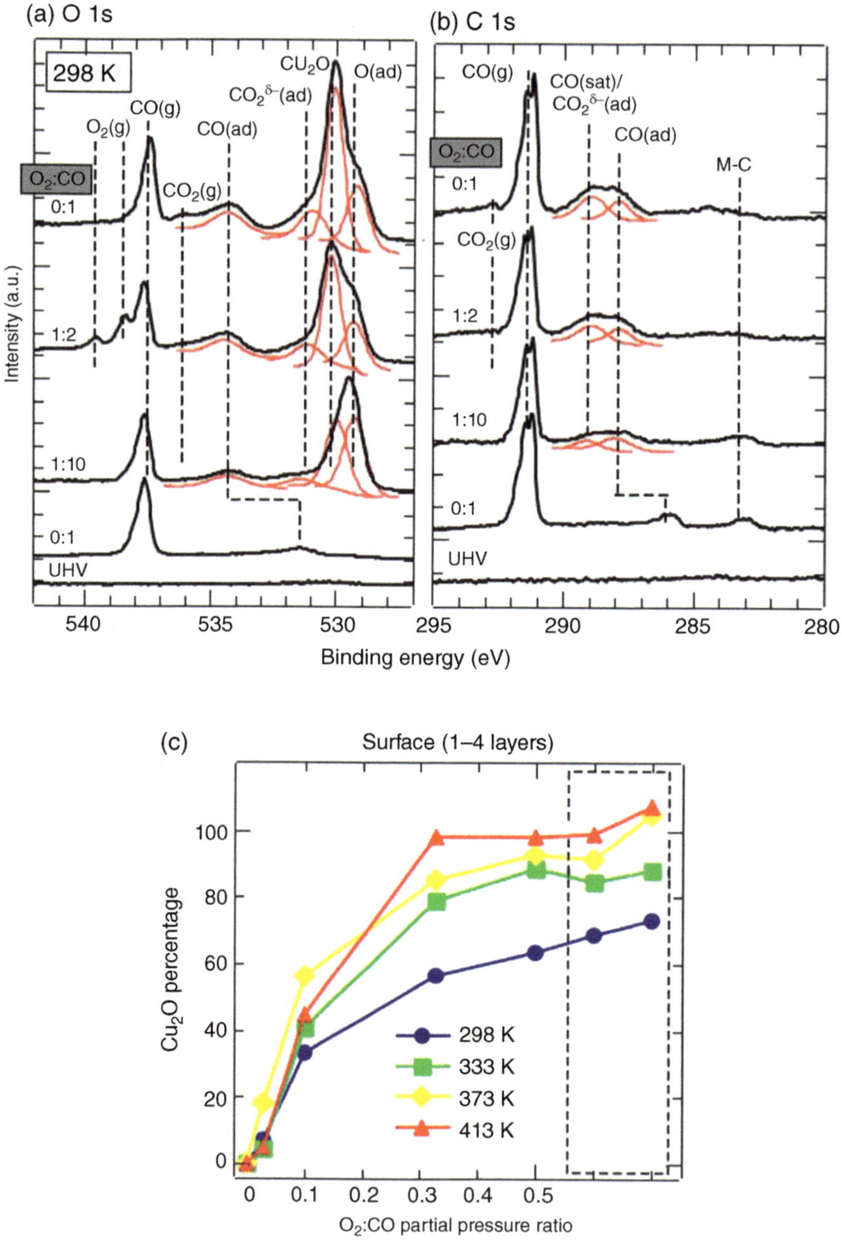

Figure 10.18 AP-XPS studies of (a) O 1s and (b) C 1s spectra of a Cu(111) sample at 298 K. From bottom to top: in UHV, under 0.3 Torr of CO, under 0.03 Torr of O_2 +0.3 Torr of CO, under 0.15 Torr of O_2 +0.3 Torr of CO, and after pumping the O_2. The gas phase CO, O_2, and CO_2 peaks appear at binding energies higher than 536 eV in the O 1s region and higher than 290 eV at the C 1s region. Peaks arising from molecularly adsorbed CO are observed at 531.5 and 286.1 eV on metallic Cu and around 534.2, 287.9, and 289 eV (satellite) on the Cu_2O covered region. The chemisorbed O peak produced a peak at 529.4 eV both on the metallic Cu and on Cu_2O. The lattice O from Cu_2O appears at 530.2 eV. Adsorbed $CO_2^{\delta-}$, intermediate of the CO oxidation reaction, occurs at 531.5 and 289 eV. Small amounts of CH_x (284.4 eV) and carbon (M-C, 283.1 eV) are observed sometimes when O_2 is absent from the gas phase, due to beam-induced dissociation of CO. (c) Cu_2O percentage (rest is metallic Cu) of the surface estimated from the intensity ratio of O 1s (530.2 eV) to Cu 2p XPS peaks. *Source:* Reproduced with permission from Eren et al.[15]/American Chemical Society.

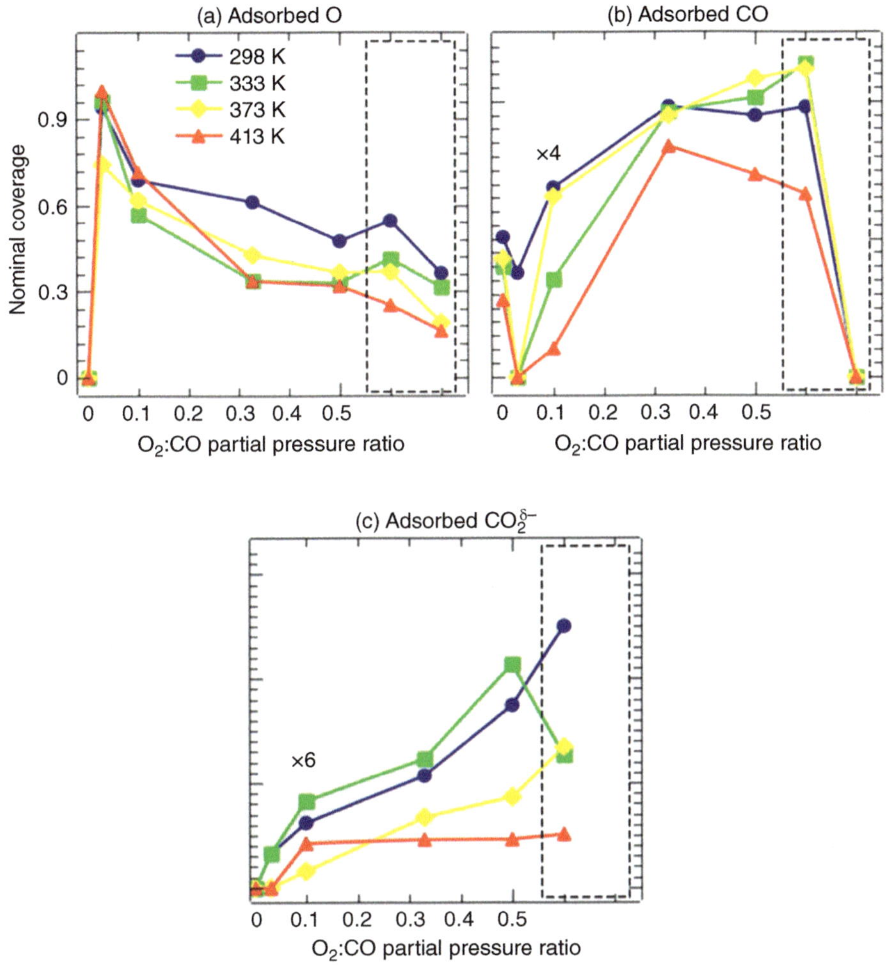

Figure 10.19 Coverage (±20%) of adsorbed species with respect to a flat Cu(111) surface as a function of O_2:CO partial pressure and temperature calculated from the corresponding O 1s AP-XPS peaks. (a) Dissociatively adsorbed oxygen, (b) molecularly adsorbed CO, and (c) $CO_2\delta^-$ (reaction intermediate). The peak intensity ratios are calibrated using intensity ratios of samples with known coverages. The last two data points in the dashed box are taken, respectively, after O_2 is pumped out (with CO remaining in the chamber) and after CO is pumped out. *Source:* Reproduced with permission from Eren et al.[15]/American Chemical Society.

References

1 Sumaria, V., Nguyen, L., Tao, F. *et al.* 2023. "Atomic-scale mechanism of platinum catalyst restructuring under a pressure of reactant gas." *J. Am. Chem. Soc.* 145, 392–401.

2 Tao, F. Dag, S. Wang, L. W., Liu, Z. *et al.* 2010. "Break-up of stepped platinum catalyst surfaces by high CO coverage." *Science* 327, 850–853.

3 Tränkenschuh, B., Fritsche, N., Fuhrmann, T. *et al.* 2006. "A site-selective in situ study of CO adsorption and desorption on Pt(355)." *J. Chem. Phys.* 124, 074712.

4 Tränkenschuh, B., Papp, C., Fuhrmann, T. *et al.* 2007. "The dissimilar twins – a comparative, site-selective in situ study of CO adsorption and desorption on Pt(322) and Pt(355)." *Surf. Sci.* 601, 1108–1117.

5 Longwitz, S. R., Schnadt, J., Vestergaard, E. K. *et al.* 2004. "High-coverage structures of carbon monoxide adsorbed on Pt(111) studied by high-pressure scanning tunneling microscopy." *J. Phys. Chem. B* 108, 14497–14502.

6 Eren, B., Zherebetskyy, D., Patera, L. L. *et al.* 2016. "Activation of Cu(111) surface by decomposition into nanoclusters driven by CO adsorption." *Science* 351, 475–478.

7 Toyoshima, R., Yoshida, M., Monya, Y. *et al.* 2012. "Active surface oxygen for catalytic CO oxidation on Pd(100) proceeding under near ambient pressure conditions." *J. Phys. Chem. Lett.* 3, 3182–3187.

8 Toyoshima, R., Yoshida, M., Monya, Y. *et al.* 2012. "In situ ambient pressure XPS study of CO oxidation reaction on Pd(111) surfaces." *J. Phys. Chem. C* 116, 18691–18697.

9 Ketteler, G., Ogletree, D. F., Bluhm, H. *et al.* 2005. "In situ spectroscopic study of the oxidation and reduction of Pd(111)." *J. Am. Chem. Soc.* 127, 18269–18273.

10 Gabasch, H., Unterberger, W., Hayek, K. *et al.* 2006. "In situ XPS study of Pd(111) oxidation at elevated pressure, part 2: palladium oxidation in the 10^{-1} mbar range." *Surf. Sci.* 600, 2980–2989.

11 Toyoshima, R., Yoshida, M., Monya, Y. *et al.* 2013. "In situ photoemission observation of catalytic CO oxidation reaction on Pd(110) under near-ambient pressure conditions: evidence for the Langmuir–Hinshelwood mechanism." *J. Phys. Chem. C* 117, 20617–20624.

12 Butcher, D. R., Grass, M. E., Zeng, Z. *et al.* 2011. "In situ oxidation study of Pt(110) and its interaction with CO." *J. Am. Chem. Soc.* 133, 20319–20325.

13 Miller, D., Casalongue, H. S., Bluhm, H. *et al.* 2014. "Different reactivity of the various platinum oxides and chemisorbed oxygen in CO oxidation on Pt(111)." *J. Am. Chem. Soc.* 136, 6340–6347.

14 Nguyen, L., Liu, L., Assefa, S. *et al.* 2017. "Atomic-scale structural evolution of Rh(110) during catalysis." *ACS Catal.* 7, 664–674.

15 Eren, B., Heine, C., Bluhm, H. *et al.* 2015. "Catalyst chemical state during CO oxidation reaction on Cu(111) studied with ambient-pressure X-ray photoelectron spectroscopy and near edge X-ray adsorption fine structure spectroscopy." *J. Am. Chem. Soc.* 137, 11186–11190.

11

CO Oxidation on High Surface Area Catalysts

Industrial catalysts are typically high surface area metal or oxide nanoparticles loaded on supports that are particles with a relatively larger size compared to the supported nanoparticles. AP-XPS has been widely used in studying nanoparticle catalysts. One of the earliest reactions studied for nanoparticle catalysts is CO oxidation.

11.1 CO Oxidation on Rh Nanoparticles

Rh has been one of the metals studied extensively in surface science and catalysis community since it is active in catalyzing a great number of reactions. In most of those studies, Rh single crystals were used as model catalysts. Their active phases for CO oxidation were identified. Rh in the form of nanoparticles is active for CO oxidation as well. A clear size-dependent catalytic activity for CO oxidation on 2–11 nm nanoparticles was reported.[1] Turnover frequency (TOF) of producing CO_2 on Rh nanoparticles was calculated by dividing the rate of produced CO_2 molecules measured under kinetic control by the number of Rh atoms on surface of Rh nanoparticles. Figure 11.1a presents the relative TOF of Rh nanoparticles referred to Rh foil.[1] Clearly, TOF of these Rh nanoparticles decreases along the increase of their sizes. Consistent with this size-dependent catalytic activity for CO oxidation, the measured apparent activation barrier of CO oxidation increases along the increase of size of Rh NPs (right axis in Figure 11.1a),[1] suggesting the size dependence of catalytic activity is an intrinsic property of Rh nanoparticle catalysts. Another important finding is that the activity for CO oxidation in terms of TOF for formation of CO_2 on Rh nanoparticles is obviously higher than Rh foil (Figure 11.1a). To achieve fundamental understanding of this size-dependent catalytic performance, surfaces of Rh nanoparticles with different sizes in pure CO, pure O_2 and mixture of CO and O_2 at different temperatures were studied with AP-XPS. Particularly, surfaces of 2 nm Rh NPs and 7 nm Rh NPs were tracked with AP-XPS during catalysis.

Surface phases of these Rh nanoparticles in O_2 or during CO oxidation were studied with AP-XPS.[1] In 0.5 Torr O_2 at 200 °C, 54% of Rh atoms in surface region of 7 nm Rh NP are cationic, suggesting more than half of the surface region of Rh NPs was oxidized at 200 °C in 0.5 Torr O_2

Application of Ambient Pressure X-ray Photoelectron Spectroscopy to Catalysis, First Edition. Franklin Tao.

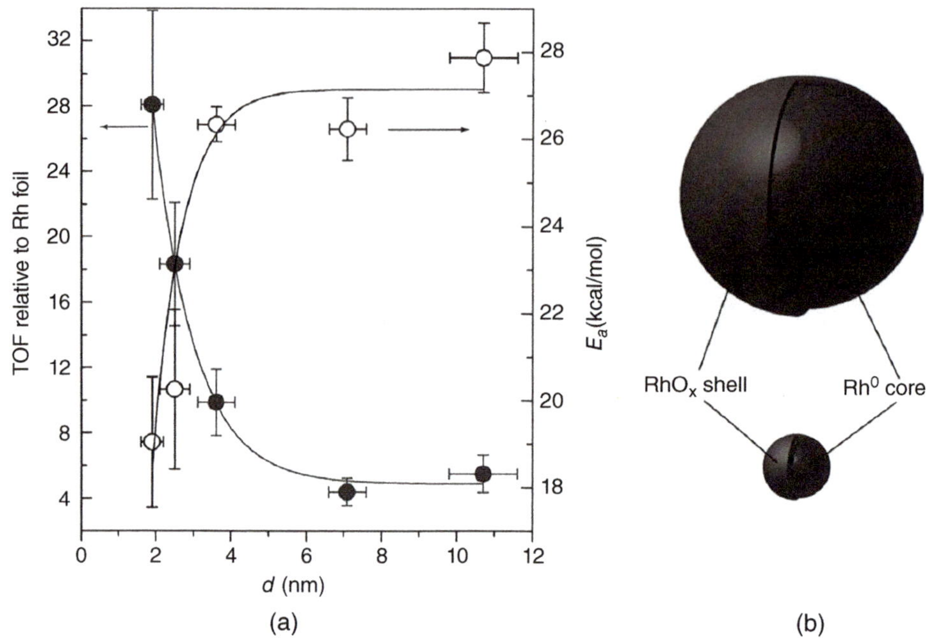

Figure 11.1 Size-dependent catalytic activity in CO oxidation. (a) Turnover frequency relative to rhodium foil at 50 Torr O_2, 20 Torr CO, at 200 °C, and activation energy (150–225 °C) for CO oxidation. (b) Schematic showing the thickness of the oxide shell and metal core of 2 and 7 nm nanoparticles. *Source:* Reproduced with permission from Grass et al.[1]/John Wiley & Sons.

(Figure 11.2d). Here the surface region can be defined to surface layers with a thickness of 0.5 nm, approximately the inelastic mean free path of photoelectrons with KE at 200 eV since photoelectrons in Rh 3d subshell were excited with soft X-ray of 510 eV. Under the same condition, 70% of surface region of 2 nm Rh NP is cationic in terms of Rh^{x+} (Figure 11.2h). In the mixture of 0.2 Torr CO and 0.5 Torr O_2 at 150 °C, the fraction of cationic Rh of 7 nm Rh nanoparticles decreases from 54% in 0.5 Torr O_2 (Figure 11.2c) to 8% (Figure 11.3); In addition, the fraction of cationic Rh on 2 nm Rh nanoparticles decreases from 70% in 0.5 Torr O_2 (Figure 11.2g) to 43% in the mixture of 0.5 Torr O_2 and 0.2 Torr CO for 2 nm Rh NPs (Figure 11.3).

In the same mixture of CO and O_2 at 200 or 275 °C, AP-XPS studies uncovered that the fraction of cationic Rh in terms of Rh^{x+} is size dependent. For instance, the fraction of cationic Rh is 25% for 7 nm Rh nanoparticles and 67% for 2 nm Rh nanoparticles at 200 °C (Figure 11.3). Clearly, the surface of 2 nm Rh NPs has much larger fraction of cationic Rh than 7 nm Rh NPs during CO oxidation. The high concentration of cationic Rh makes 2 nm Rh nanoparticles exhibit higher activity in terms of TOF.[1] More description on this correlation can be found in literature.[1] In addition, quantitation of AP-XPS spectra found out that the thickness in terms of the fraction of cationic rhodium is larger at a higher catalysis temperature.

The O 1s of the RhO_x shell observed in the mixture of O_2 and CO during CO oxidation in Figure 11.4c and d is 529.5 eV; However, this peak was not observed in the pure O_2 along heating these Rh NPs from 100 to 200 °C in pure O_2 (Figure 11.4a and b). It suggests this specific rhodium oxide with O 1s at 529.5 eV can only form during CO oxidation at 200 or

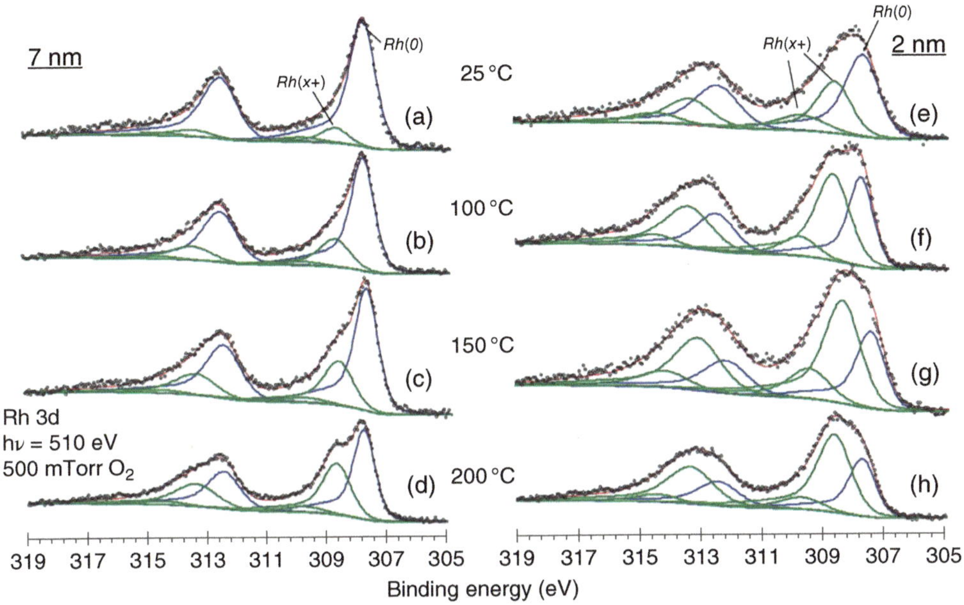

Figure 11.2 AP-XPS spectra of Rh 3d taken under 0.5 Torr O_2 in the temperature of 25–200 °C. Each spectrum was deconvoluted into Rh(0) and Rh(x^+) shown in blue and green, respectively. *Source:* Reproduced with permission from Grass et al[1]/John Wiley & Sons.

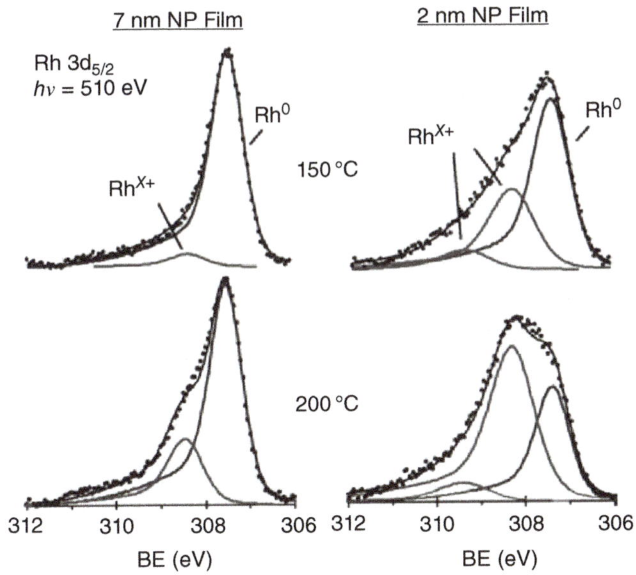

Figure 11.3 AP-XPS spectra of the Rh $3d_{5/2}$ peaks of 2 and 7 nm Rh nanoparticles in the mixture of 0.2 Torr CO and 0.5 Torr O_2 at 150 and 200 °C. They were deconvoluted for reduced rhodium (307.2 eV) and oxidized rhodium (308.2, 309.4 eV). The NPs of both films (2 and 7 nm particles) at higher temperatures are more oxidized compared to their initial state. Compared to the film that is comprised of 7 nm, the 2 nm NPs are more oxidized at all temperatures. The two Rh^{x+} peaks are attributed to rhodium atoms in two different oxidation states or coordinated to a different number of the fractions of Rh^{x+} for 7 nm and 2 nm Rh NPs at 200 °C in the mixture of CO and O_2 are 25% and 67%, respectively. *Source:* Reproduced with permission from Grass et al.[1]/John Wiley & Sons.

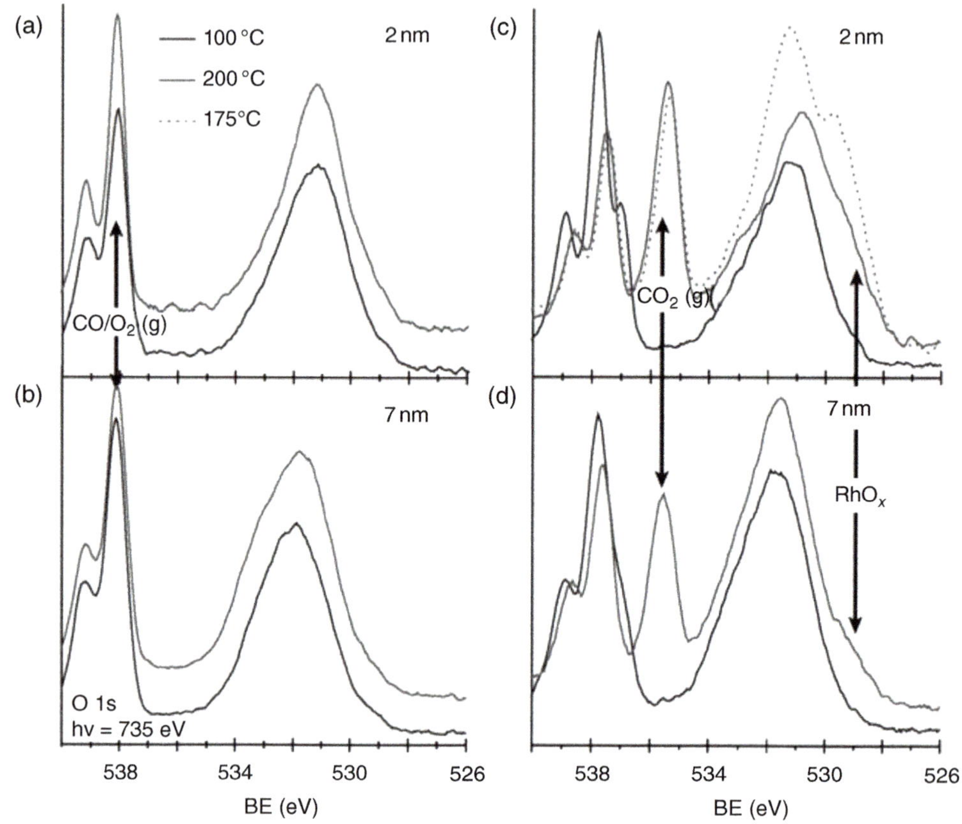

Figure 11.4 O 1s spectra of 2 and 7 nm Rh NPs in 500 mTorr O_2 (a and b) and during reaction (410 mTorr O_2, 20 mTorr CO; c and d). When exposed to both O_2 and CO, a peak at higher binding energy forms, which is attributed to a RhO_x species. *Source:* Reproduced with permission from Grass et al.[1]/John Wiley & Sons.

275 °C but not in pure O_2 at these temperatures. Thus, this specific oxide is stabilized by CO or some intermediate of CO oxidation. The preservation of this reactive oxide by the mixture of reductants in terms of CO and O_2 is supported by the observation that this oxide diminished upon the gas was purged.[1]

As described in Chapter 6, AP-XPS allows us to identify molecules of the gas phase since the binding energy of electrons on subshells of an atom of a molecule in gas phase is typically several eV higher than the molecules adsorbed on a catalyst surface. In terms of 2 and 7 nm Rh nanoparticles at 200 and 275 °C in the mixture of CO and O_2, product CO_2 was readily identified with AP-XPS (Figure 11.4c and d). The formation of gaseous product CO_2 along the simultaneous observation of the rhodium oxide phase with O 1s at 529.5 eV at 200 and 275 °C by AP-XPS (Figure 11.4c and d) clearly shows that the new phase RhO_x is the active phase formed on Rh nanoparticles for CO oxidation.[1]

The Y-axis value in Figure 11.1a is the ratio of TOF of Rh nanoparticles to Rh foil. Clearly, 2 and 7 nm Rh nanoparticles exhibit much higher catalytic activity in terms of TOF than Rh foil. The much higher activity of the reactive RhO_x formed on Rh nanoparticles in contrast to Rh foil is consistent with the computational studies in literature that activation barrier for CO oxidation on RhO_x is lower than that on Rh metal surface since the chemisorption geometry of oxygen atoms on rhodium oxide is different from that on Rh metal.[2]

This work demonstrated the significant function of AP-XPS, uncovering authentic surface of a catalyst during catalysis. The quantitative analysis of AP-XPS to authentic surface phase and track of catalytic activity under the same catalytic condition allowed for successfully establishing a correlation between catalyst surface under a catalytic condition and its corresponding catalytic performance. Through such a correlation, the active catalytic phase is found.

11.2 CO Oxidation on Ru Nanoparticles

AP-XPS was used for studying CO oxidation on Ru NPs with sizes of 2.8 and 6.0 nm synthesized with colloidal chemistry. In the synthesis, Ru precursor, $Ru(acac)_3$, is reduced with ethylene alcohol in a refluxed liquid; aggregation of reduced Ru atoms forms Ru nanoparticles.[3] These synthesized Ru nanoparticles exhibit flexibility in being reversibly reduced and oxidized. As shown in Figure 11.5, Ru nanoparticles were oxidized at 200 °C in 0.2 Torr pure O_2 and then reduced at 200 °C in 0.08 Torr pure CO that were repeated for three times at 200 °C.

Both 2.8 and 6.0 nm Ru nanoparticles are active for CO oxidation as evidenced by the observation O 1s peak of gaseous CO_2 at 534–535 eV in the mixture of CO and O_2. Figure 11.6 is the O 1s spectra of 6 nm Ru nanoparticle catalyst in the mixture of 0.2 Torr O_2 and 0.08 Torr CO at 50–200 °C. Similar studies of O 1s peak were performed on 2.8 nm Ru nanoparticles. Other than O 1s, Rh 3p was tracked with AP-XPS. Both surfaces of 2.8 and 6 nm Ru nanoparticles are covered with a thin layer of RuO_2, evidenced by the distinguishable Ru 3p at 463.2 eV attributed to the RuO_2 surface layer shown in Figure 11.7. Obviously, the intensity of the deconvoluted peak centered at 463.2 eV [Ru(4+)] increases along the increase of catalysis temperature from 50 to 200 °C (Figure 11.7). Quantitative analysis of the deconvoluted spectra collected with AP-XPS during catalysis confirmed that the molar fraction of Ru (4+) formed on 2.8 or 6 nm Rh NPs increases along the increase of catalysis temperature (Figure 11.8a). The trend of increase of

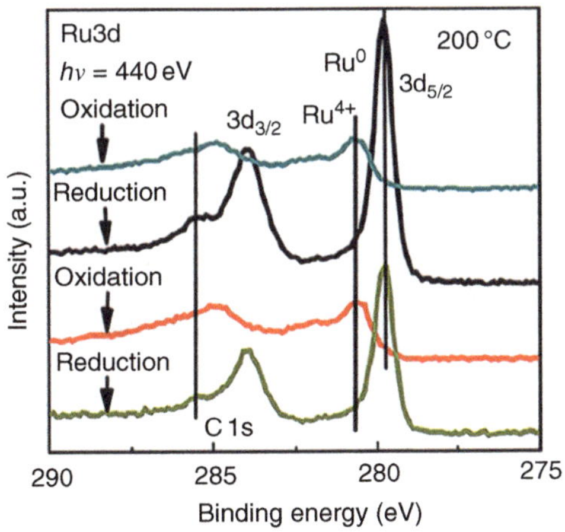

Figure 11.5 Ru 3d spectra showing reversibility of oxidation (under 0.2 Torr O_2) and reduction (under 0.08 Torr CO) of 6 nm Ru nanoparticles at 200 °C during in situ oxidation/reduction cycles. Incident photon energy is 440 eV. *Source:* Reproduced with permission from Qadir et al.[3]/American Chemical Society.

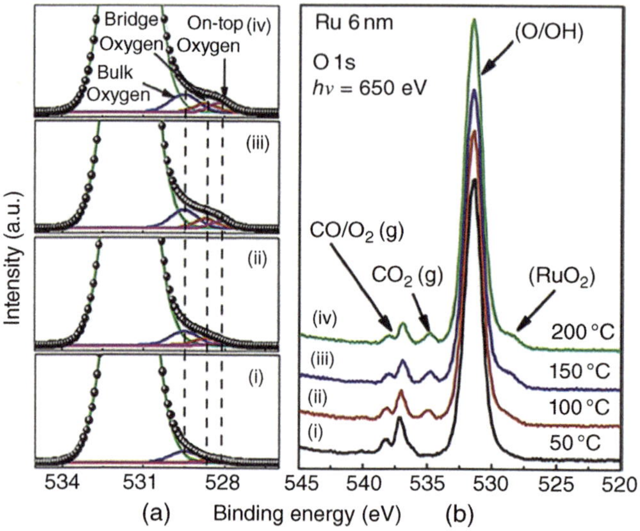

Figure 11.6 AP-XPS spectra of O 1s of 6 nm Ru nanoparticles during CO oxidation (using 0.08 Torr CO and 0.2 Torr O_2) at various temperatures. During CO oxidation, the Ru nanoparticles show progressive formation of well-ordered surface oxide with increasing temperature.
Source: Reproduced with permission from Qadir et al.[3]/American Chemical Society.

thickness of oxide layers on surface of Ru metal nanoparticles with a specific size (2.8 nm or 6 nm) during CO oxidation along increase of catalysis temperature[3] is similar to that observed on Rh nanoparticle during CO oxidation (Figure 11.3).[1]

As shown in Figure 11.8a, atomic ratio of Ru of RuO_2 to metallic Ru increases due to the growth of surface oxide layer along the increase of catalysis temperature. An important feature is that the gaseous peak of CO_2 was clearly observed at 200 °C in the mixture of CO and O_2 (Figure 11.6b). The observation of CO_2 at 200 °C and the formation of dominate RuO_2 at 200 °C suggest that RuO_2 is the active phase for CO oxidation at 200 °C.

The activity for CO oxidation in terms of TOF per Ru site on 6 nm Ru NPs at 240 °C is eight times higher than 2.8 nm Ru NP at the same temperature (Figure 11.8b).[4] However, as shown in Figure 11.8a, the RuO_2/Ru atomic ratio of 6.0 nm Ru NPs at 240 °C is lower than that of 2.8 nm NPs at this temperature. It suggests that the thin RuO_2 layer formed on 6.0 nm Ru NPs is different from the relatively thicker RuO_2 layer formed on 2.8 nm Ru NPs. More importantly, the thin RuO_2 of the 6 nm Ru nanoparticles is more active than the thick RuO_2 of 2 nm Ru nanoparticles. This finding is consistent with early studies of CO oxidation on Ru single crystal model catalyst showing that the RuO_2 layer with a thickness of 1–2 nm formed on Ru single crystal is most active for CO oxidation among those RuO_2 layers with different thicknesses.[5]

Likely, thickness-dependent activity of RuO_2 formed on Ru nanoparticles is driven by the atomic scale surface structure of the RuO_2 which is expected to be quite different along with the increase in the thickness of RuO_2. Thus, the surface structure at the atomic scale for a catalyst during catalysis could be the main factor to decode the size-dependent

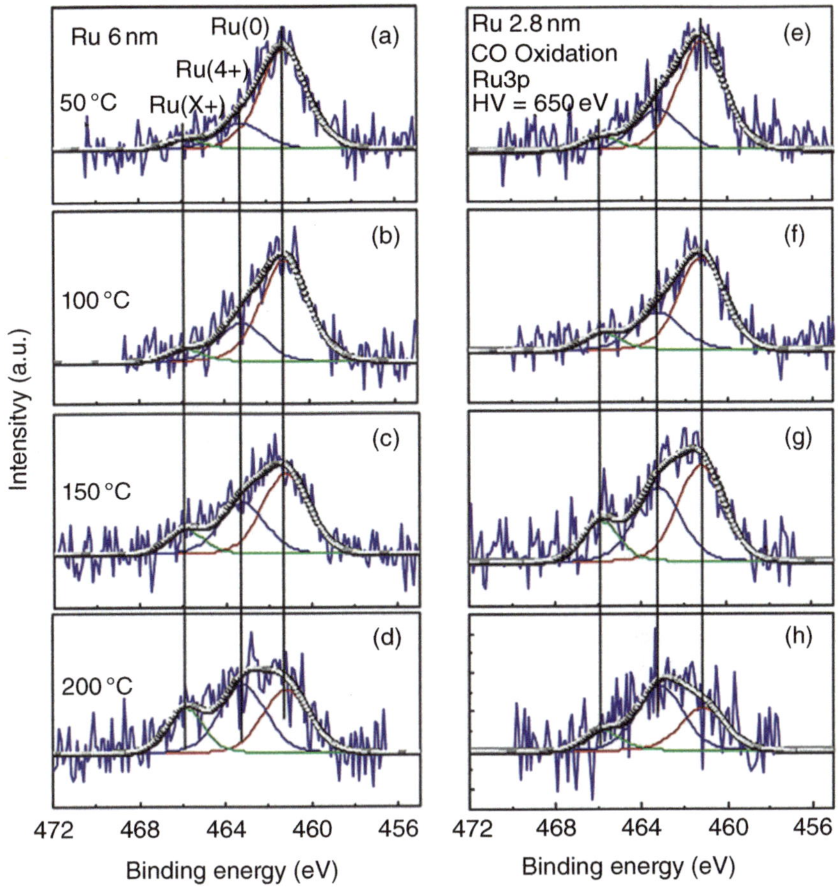

Figure 11.7 AP-XPS spectra of Ru3p of Ru nanoparticles under CO oxidation condition (200 mTorr O_2 and 80 mTorr CO) with increasing temperature (a–d). Spectra of 6 nm Ru nanoparticles acquired at (a) 50, (b) 100, (c) 150, and (d) 200 °C; (e–h) spectra of 2.8 nm Ru nanoparticle acquired at (e) 50, (f) 100, (g) 150, and (h) 200 °C (incident photon energy is 650 eV). The peak at 463.2 eV is attributed to surface layer of RuO_2. Ru3p spectra show a thicker oxide formation around the smaller Ru nanoparticles (2.8 nm). *Source:* Reproduced with permission from Qadir et al.[3]/American Chemical Society.

catalytic performance. However, building a correlation between surface structure at atomic scale and its corresponding catalytic performance for a nanoparticle catalyst is incredibly challenging as having a picture of surface structure of nanoparticles during catalysis has been recognized as one of the extremely challenging tasks in the field of heterogeneous catalysis so far.

In this work,[3,4] AP-XPS identified the active phase for CO oxidation on Ru nanoparticles which is RuO_2 formed on the catalyst. The quantitation function of AP-XPS allows for concluding that the thin RuO_2 layer on 6 nm nanoparticle is more active than the thick RuO_2 layer formed on 2.8 nm Ru nanoparticle. This correlation suggests a picture showing how the atoms of the surface of nanoparticle catalysts pack during catalysis is necessary for understanding the CO oxidation RuO_2 at an atomic scale.

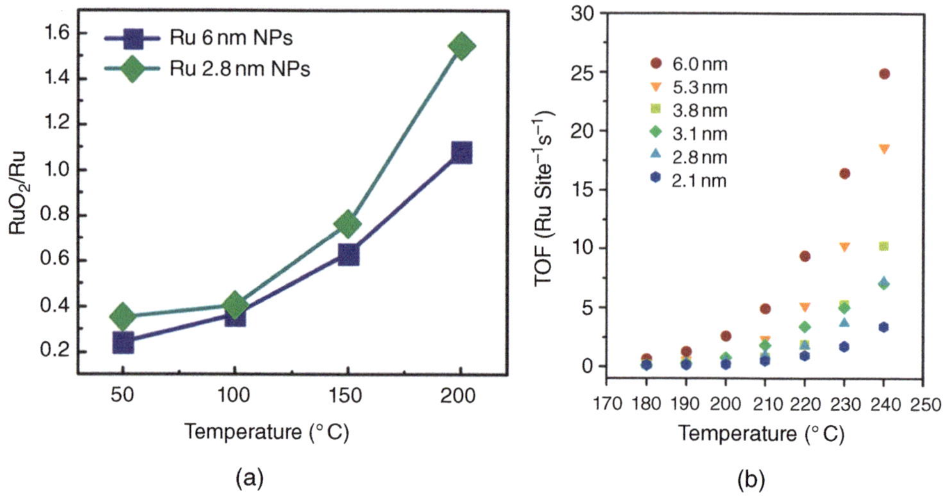

Figure 11.8 AP-XPS of surface of Ru nanoparticles and their catalytic activity in terms of turnover frequency (TOF) measured under kinetics-controlled region. (a) Quantitative analysis of surface phase RuO_2 formed on 2.8 and 6.0 nm Ru nanoparticles during CO oxidation in the temperature range of 50–1200 °C; the *Y*-axis value is the peak area ratio of the RuO_2 peak with respect to the Ru metallic peak (Ru^{4+}/Ru^0), as calculated from AP-XPS spectra of Ru3p for Ru nanoparticles under CO oxidation condition of 0.2 Torr O_2 and 0.08 Torr CO ($CO/O_2 = 1/0.4$). *Source:* Reproduced with permission from Qadir et al.[3]/American Chemical Society. (b) TOFs of nanoparticles of 2.1, 2.8, 3.1, 3.8, 5.3, and 6.0 nm as a function of catalysis temperature. *Source:* Reproduced with permission from Joo et al.[4]/American Chemical Society.

References

1 Grass, M. E., Zhang, Y., Butcher, D. R. *et al.* 2008. "A reactive oxide overlayer on rhodium nanoparticles during CO oxidation and its size dependence studied by in situ ambient-pressure X-ray photoelectron spectroscopy." *Angew. Chem. Int. Ed.* 47, 8893–8896.

2 Gong, X. Q., Liu, Z. P., Raval, R. *et al.* 2004. "A systematic study of CO oxidation on metals and metal oxides: density functional theory calculations." *J. Am. Chem. Soc.* 126, 8–9.

3 Qadir, K., Joo, S. H., Mun, B. S. *et al.* 2012. "Intrinsic relation between catalytic activity of CO oxidation on Ru nanoparticles and Ru oxides uncovered with ambient pressure XPS." *Nano Lett.* 12, 5761–5768.

4 Joo, S. H., Park, J. Y., Renzas, R. *et al.* 2010. "Size effect of ruthenium nanoparticles in catalytic carbon monoxide oxidation." *Nano Lett.* 10, 2709–2913.

5 Assmann, J., Narkhede, V., Breuer, N. A. *et al.* 2008. "Heterogeneous oxidation catalysis on ruthenium: bridging the pressure and materials gaps and beyond." *J. Phys.: Condens. Matter* 20, 184017.

12

Hydrogenation of Carbon Dioxide

Hydrogenation of CO_2 is a promising route to utilize CO_2 to produce value-added chemicals. Ru metal is highly active for this reaction.[1–6] CO_2 hydrogenation on Ru-based high surface area catalyst was studied with AP-XPS. Here AP-XPS study of CO_2 hydrogenation on Ru nanoparticles supported on Co_3O_4, designated as $(Co_{0.95}Ru_{0.05})_3O_4$, was taken as one example for CO_2 hydrogenation on high surface area catalysts.[6]

The reduction of Co_3O_4 without any loaded Ru in the mixture of 0.1 Torr CO_2 and 0.4 Torr H_2 was studied by AP-XPS. Below 200 °C, Co_3O_4 could not be reduced by this mixture. At 200 °C, it was partially reduced, evidenced by the appearance of a shoulder of Co $2p_{3/2}$ at 788.6 eV in Figure 12.1a. At 340 °C, Co cations of surface region were reduced to metallic Co. Notably, before the formation of metal Co, Co_3O_4 was first reduced to CoO, an intermediate phase before a full reduction of Co_3O_4.

Compared to pure Co_3O_4, the reducibility of surface of $(Co_{0.95}Ru_{0.05})_3O_4$ is quite different. In the same mixture of 0.1 Torr CO_2 and 0.4 Torr H_2 as what Co_3O_4 nanoparticles experienced in Figure 12.1a,[6] the Co $2p_{3/2}$ peak of $(Co_{0.95}Ru_{0.05})_3O_4$ at 220 °C obviously downshift, suggesting that Co cations were reduced to metallic at a temperature as low as 220 °C. However, the reduction of Co_3O_4 occurs at 340 °C in the same mixture of CO_2 hydrogenation. In terms of Ru 3p of $(Co_{0.95}Ru_{0.05})_3O_4$ in the mixture of CO_2 and H_2, Ru $3p_{3/2}$ of 220 °C downshift compared to 180 °C, suggesting Ru atoms in surface region were partially reduced to metallic Ru at 220 °C (Figure 12.1c). In addition, the longer inelastic mean free path (λ) of Ru 3p photoelectrons than Co 2p photoelectrons due to the obvious higher kinetic energy of Ru 3p photoelectrons than Co 2p can rationalize (1) the contributions from both metallic Ru atoms on the surface of the catalyst and cationic Ru atoms in the subsurface and (2) the fact that the portion of metallic Co in the Co 2p3/2 intensity of both metallic and cationic cobalt is larger than that of metallic Ru in the Ru 3p intensity. The metallic Co atoms on the surface layers instead of cationic Co atoms in the subsurface are the main contributor to the Co 2p peak intensity of $(Co_{0.95}Ru_{0.05})_3O_4$ (Figure 12.1b).[6]

Based on the downshift of binding energy of Co 2p and Ru 3p (Figures 12.1b and c), both Ru and Co on surface of the $(Co_{0.95}Ru_{0.05})_3O_4$ are metallic at a temperature ≥220 °C. Thus, it is suggested that Co–Ru alloy was formed on surface of the catalyst nanoparticles at a temperature as low as 220 °C. The formation of Co–Ru alloy triggered the syngeneic effect

Application of Ambient Pressure X-ray Photoelectron Spectroscopy to Catalysis, First Edition. Franklin Tao.

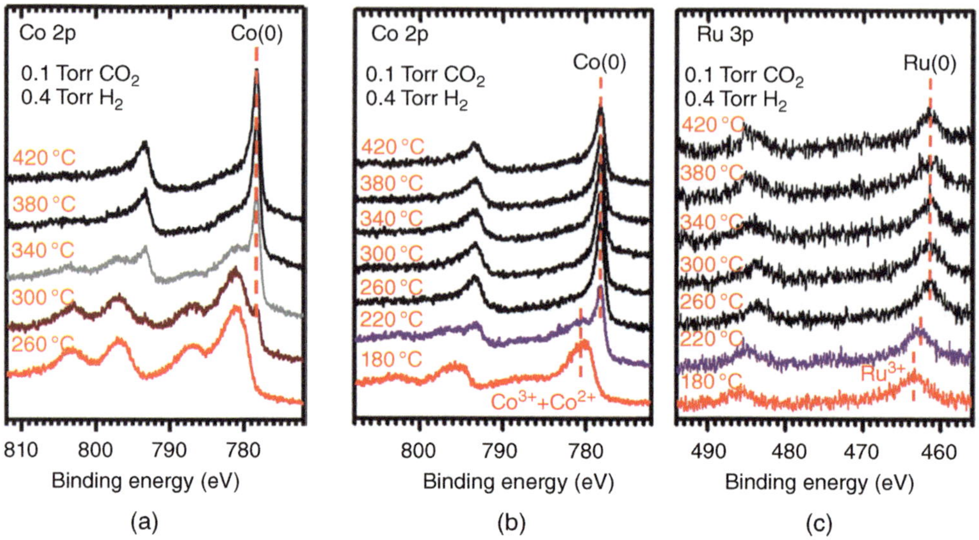

Figure 12.1 Co 2p and Ru 3p XPS peaks of Co_3O_4 and $(Co_{0.95}Ru_{0.05})_3O_4$ under reaction conditions in the temperature of 180–420 °C in the mixture of 0.1 Torr CO_2 and 0.4 Torr H_2. (a) Co 2p of Co_3O_4. (b) Co 2p of $(Co_{0.95}Ru_{0.05})_3O_4$. (c) Ru 2p of $(Co_{0.95}Ru_{0.05})_3O_4$. *Source:* Reproduced with permission from Zhu et al.[6]/American Chemical Society.

on the catalytic performance of metallic Co and Ru as shown in Figure 12.2. For instance, $(Co_{0.95}Ru_{0.05})_3O_4$ exhibits conversion of 3–4% of CO_2 at 220 °C (Figure 12.2c) although cobalt catalyst at this temperature is inert.

Co_3O_4 has the same surface area as $(Co_{0.95}Ru_{0.05})_3O_4$. As shown in Figure 12.2c, $(Co_{0.95}Ru_{0.05})_3O_4$ exhibits higher activity in terms of conversion of CO_2 than Co_3O_4 in the temperature range of 220–420 °C. The much higher activity in hydrogenation of CO_2 on 100 mg $(Co_{0.95}Ru_{0.05})_3O_4$ compared to 100 mg Co_3O_4 must result from the critical role of the Ru on the surface. The significant role of Ru in CO_2 hydrogenation is further supported by the much higher selectivity for production of CH_4 on $(Co_{0.95}Ru_{0.05})_3O_4$ (red line in Figure 12.2b) in contrast to Co_3O_4.

Other than the confirmed role of Ru in the significant promotion of activity and selectivity for production of CH_4 through hydrogenation of CO_2, Co is a necessary element in having the high conversion of CO_2. The important role of Co was confirmed through a comparison of catalytic performance between 5%Ru/SiO_2 and 5%/Co_3O_4 in terms of $(Co_{0.95}Ru_{0.05})_3O_4$ in Figure 12.2. The conversion of CO_2 on 100 mg of 5% Ru/SiO_2 for the hydrogenation of CO_2 was plotted as pink line in Figure 12.2a. It is much lower than the conversion of 100 mg of 5% Ru/Co_3O_4 in terms of $(Co_{0.95}Ru_{0.05})_3O_4$ (red line in Figure 12.2c). This difference shows that Co plays a significant role in the activity of CO_2 hydrogenation on $(Co_{0.95}Ru_{0.05})_3O_4$.

These AP-XPS studies uncovered the active phase of $(Co_{0.95}Ru_{0.05})_3O_4$ for CO_2 hydrogenation is the Co–Ru alloy formed on $(Co_{0.95}Ru_{0.05})_3O_4$ under a catalytic condition. They suggested the high activity and selectivity of $(Co_{0.95}Ru_{0.05})_3O_4$ result from the synergic effect of Co and Ru. The synergic effect of Co and Ru is supported by following analysis. If

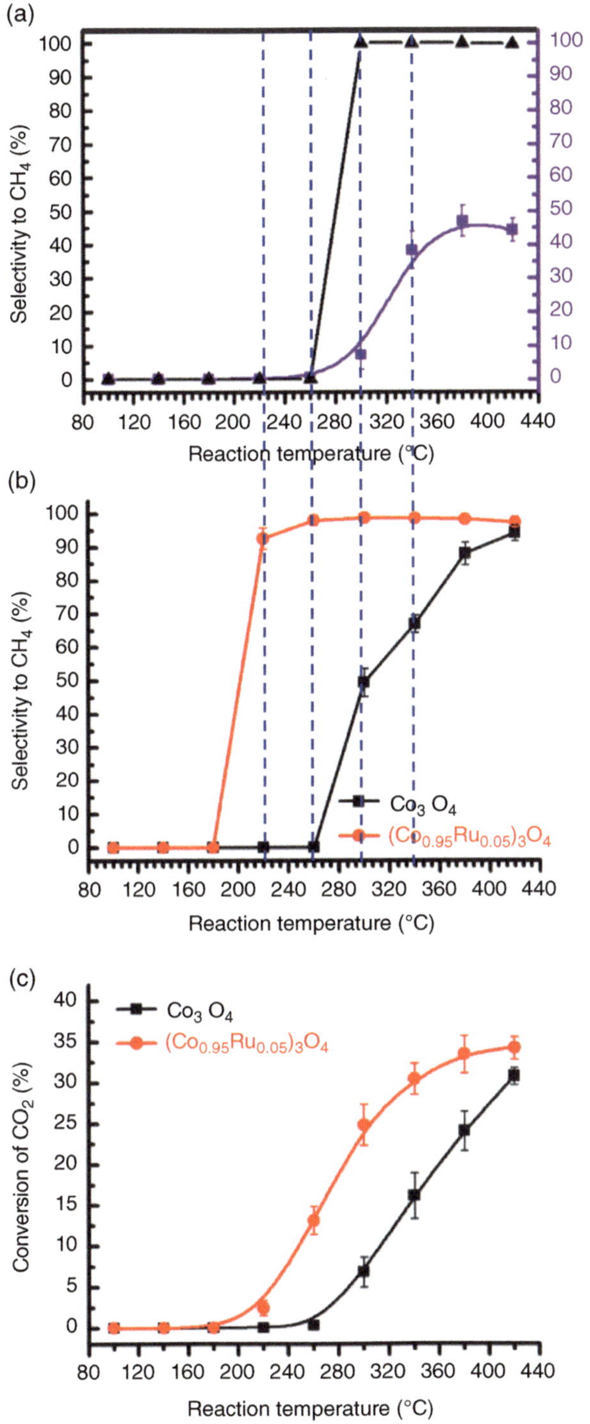

Figure 12.2 Catalytic performances of 5% Ru/SiO_2, Co_3O_4, and $(Co_{0.95}Ru_{0.05})_3O_4$ in the temperature range of 100–420 °C. (a) Catalytic performances of 5%Ru supported on SiO_2. The measurements of catalytic conversion and selectivity were performed in a fixed-bed microreactor at 50 ml/min of the flow rate, 1 bar of the reactant pressure, and the reactant mixing ratio CO_2:H_2 = 1 : 4; weight of catalyst 5% Ru/SiO_2 is 100 mg. The measurements of catalytic performances of 100 mg Co_3O_4 or 100 mg $(Co_{0.95}Ru_{0.05})_3O_4$ were performed under the same catalytic condition of 5% Ru/SiO_2. Notably, the catalytic selectivity for formation of CH_4 was presented with black line in (a). (b) Catalytic selectivity for producing CH_4 on Co_3O_4 (black line) and $(Co_{0.95}Ru_{0.05})_3O_4$ (red line). (c) Catalytic conversion of CO_2 on Co_3O_4 (black line) and $(Co_{0.95}Ru_{0.05})_3O_4$ (red line). *Source:* Reproduced with permission from Zhu et al.[6]/American Chemical Society.

we assume (i) all Ru atoms of $(Co_{0.95}Ru_{0.05})_3O_4$ were reduced to metallic Ru, (ii) all atoms of these reduced Ru atoms are dispersed on surface of the catalyst particles, and (iii) all these metallic Ru atoms participate in the catalysis, then the activity of the $(Co_{0.95}Ru_{0.05})_3O_4$ is 0.0189 CH_4 molecule per Ru atom per second at 260 °C obtained by using the conversion of CO_2 at 260 in Figure 12.2c. Notably, this calculation is a very conservative estimation of turnover frequency (TOF) because lots of Ru atoms could not participate into the catalysis since some metallic Ru must be in the subsurface or core of metallic Ru nanoparticles and some Ru atoms are in cationic state in subsurface or deeper layers. That is, the actual TOFs at 260 is actually definitely higher than 0.0189.

With the conversion of CO_2 (0%) in CO_2 hydrogenation on Co_3O_4 at 260 °C (Figure 12.2c), the activity of Co_3O_4 is 0 CH_4 molecules per Ru per second at 260 °C. Clearly, Ru atoms promoted the activity of Co for CO_2 hydrogenation. Figure 12.3 schematically shows the difference in the catalyst structures between Co_3O_4 and $(Co_{0.95}Ru_{0.05})_3O_4$. The synergic effect of Ru and Co atoms at metallic state on CO_2 hydrogenation is a quite interesting topic which deserves further studies. For instance, a computational study can simulate the reaction pathway of CO_2 hydrogenation on Co–Ru alloy surface and pure Co surface to find out how Ru atoms promote the reaction, achieving a fundamental understanding of this synergistic effect at a molecular level.

Here AP-XPS studies concluded that the active phase is the Co–Ru alloy formed on the surface of the catalyst particles and confirmed the origin of the promotion effect of Co_3O_4 catalyst by doping Ru to Co_3O_4. Based on the XPS spectra of Co 2p and Ru 3p of the catalyst (two bottom spectra in Figure 12.1b and c), both Ru and Co are at cationic state before catalysis. In addition, the active phase, Ru–Co alloy surface must have been oxidized to ruthenium oxide and cobalt oxide once a used catalyst is exposed to air after catalysis. Then an ex situ study of the used catalyst with a high vacuum XPS must have incorrectly suggested that both Co and Ru are at cationic static and the active phase of $(Co_{0.95}Ru_{0.05})_3O_4$ is cobalt oxide or/and ruthenium oxide. Thus, the ex situ studies of either a catalysts before catalysis or a used catalyst after catalysis must give misleading information on the active

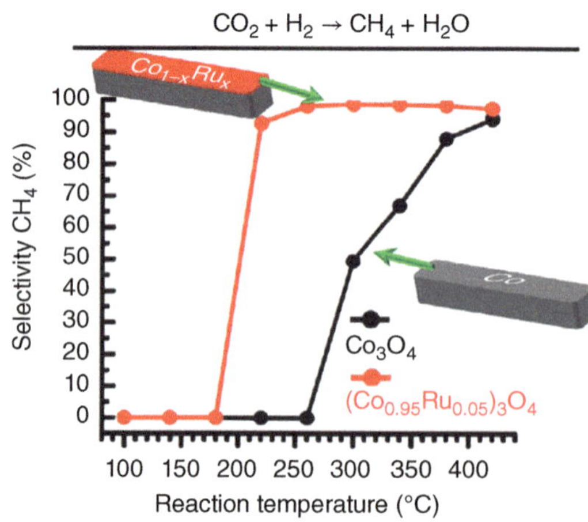

Figure 12.3 Significant promotion of Ru to hydrogenation of CO_2 on Co. *Source:* Reproduced with permission from Zhu et al.[6]/American Chemical Society.

phase of $(Co_{0.95}Ru_{0.05})_3O_4$ for CO_2 hydrogenation. This work demonstrates the necessity of identifying the active phase for reducing reactions such as hydrogenation of CO_2 under the catalytic condition since the reducing condition at high temperature could largely restructure the catalyst surface to form active phase during catalysis. Without such an AP-XPS study, the formed metal or alloy phase on a catalyst surface could not be identified.

References

1 Alayoglu, S., Tao, F., Altoe, V. *et al.* 2011. "Surface composition and catalytic evolution of Au x Pd1−x (x = 0.25, 0.50 and 0.75) nanoparticles under CO/O_2 reaction in torr pressure regime and at 200 °C." *Top. Catal.* 54, 633–640.

2 Melaet, G., Ralston, W. T., Li, C. S. *et al.* 2014. "Evidence of highly active cobalt oxide catalyst for the Fischer–Tropsch synthesis and CO_2 hydrogenation." *J. Am. Chem. Soc.* 136, 2260–2263.

3 Rezayee, N. M., Huff, C. A., and Sanford, M. 2015. "Tandem amine and ruthenium-catalyzed hydrogenation of CO_2 to methanol." *J. Am. Chem. Soc.* 137, 1028–1031.

4 Porosoff, M. D., Yan, B., and Chen, J. G. 2016. "Catalytic reduction of CO_2 by H_2 for synthesis of CO, methanol and hydrocarbons: challenges and opportunities." *Energy Environ. Sci.* 9, 62–73.

5 Carenco, S., Sassoyte, C., Faustini, M. *et al.* 2016. "The active state of supported ruthenium oxide nanoparticles during carbon dioxide methanation." *J. Phys. Chem. C* 120, 15354–15361.

6 Zhu, Y., Zhang, S., Ye, Y. *et al.* 2012. "Catalytic conversion of carbon dioxide to methane on ruthenium–cobalt bimetallic nanocatalysts and correlation between surface chemistry of catalysts under reaction conditions and catalytic performances." *ACS Catal.* 2, 2403–2408.

13

Water–Gas Shift

Water–gas shift (WGS) is an important reaction at the downstream of steam reforming of CH_4 reactor for increasing the ratio of H_2 to CO for producing H_2 with high purity. It is one of the most studied catalytic reactions in the community of fundamental studies of catalysis. In addition, it is always an appropriate probe to test a new catalyst since this reaction does not give any byproduct.

13.1 Co_3O_4 and Pt/Co_3O_4

13.1.1 Gas Composition-dependent Reducibility

Co_3O_4 is active for WGS at a temperature of 180 °C and above. AP-XPS was used in uncovering the authentic surface of Co_3O_4 nanorods under catalytic conditions of WGS. Co 2p peaks during pretreatment in H_2 were collected in Figure 13.1a. As shown in Figure 13.1a2, the surface of Co_3O_4 was reduced to CoO since the characteristic satellite peak of Co $2p_{3/2}$ of CoO at 786.5 eV appeared at 200 °C in pure H_2. This peak is a lack in Co $2p_{3/2}$ of Co_3O_4 in H_2 at 25 °C (Figure 13.1a1). It originates at the Co^{2+} in an octahedral coordination with oxygen atoms of CoO. Notably, since Co^{2+} and Co^{3+} of Co_3O_4 are in a tetrahedral and octahedral coordination sphere, respectively, no Co^{2+} is in octahedral coordination in the lattice of Co_3O_4. Thus, the peak at 786.5 eV is absent in Co 2p of Co_3O_4. There is a weak satellite peak at 798.5 eV rooting from Co^{3+} in octahedral coordination sphere of oxygen atoms in Co_3O_4. Simply speaking, an easy way to identify whether Co_3O_4 is reduced to CoO is to check whether the characteristic satellite peak of CoO at 786.5 eV appears. In addition, the formation of CoO was supported by the O/Co ratio measured with AP-XPS during the pretreatment with H_2. The O/Co ratios of Co_3O_4 at 25 and 200 °C in H_2 are 1.33 and 0.85, respectively. The O/Co ratio upon reduction of Co_3O_4, 0.85 suggests that the surface is not Co_3O_4 anymore. Clearly, the formed CoO is non-stoichiometric since the measured O/C ratio of the catalyst surface is lower than 1.0. Notably, pure Co_3O_4 remains not reduced at a temperature as higher as 260 °C in a mixture of CO_2 and H_2 and only partially reduced at 300 °C in the mixture (Figure 12.1a), but it is reduced to CoO by pure H_2 at a temperature as low as 200 °C (Figure 13.1a2). This difference results from the fact that CO_2 is an oxidizing gas in Figure 12.1a. The obviously different reduction temperature of Co_3O_4 suggests

Application of Ambient Pressure X-ray Photoelectron Spectroscopy to Catalysis, First Edition. Franklin Tao.

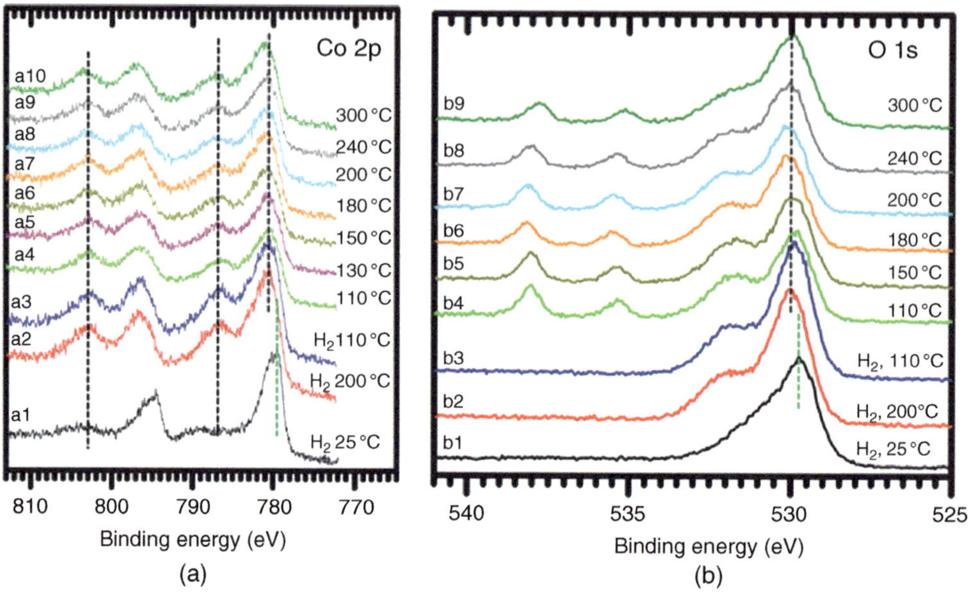

Figure 13.1 AP-XPS studies of pure Co_3O_4 during pretreatment in H_2 (a1–a3 and b1–b3) and under the following catalytic conditions in the temperature range of 110–300 °C in a mixture of CO and H_2O (a4–a10 and b4–b10). The partial pressure ratio of CO to H_2O is 3 : 1. All spectra were collected when the gaseous environment existed around the catalyst. (a) Co 2p. (b) O 1s. *Source:* Reproduced with permission from Zhang et al.[1]/American Chemical Society.

the significant role of composition of a gas (pure H_2 in Figure 13.1a versus mixture of H_2 and CO_2 in Figure 12.1a) in surface chemistry of a catalyst. Having an oxidizing constituting gas such as CO_2 in the mixture of reactants likely requests a higher reduction temperature as demonstrated in Figure 12.1a.

13.1.2 Active Phase of Co_3O_4 during Water-Gas Shift

The authentic surface of Co_3O_4 during water-gas shift (WGS) was investigated with AP-XPS in the mixture of CO and H_2O (Figure 13.1a4–a10 and b4–b9). The active surface of Co_3O_4 during WGS in the temperature range of 110–300 °C is CoO_{1-x} instead of Co_3O_4. Thus, this non-stoichiometric CoO surface is active for WGS. Kinetic studies for WGS on CoO_{1-x} show the apparent activation barrier is about 91.0 kJ/mol in the temperature range of 180–230 °C (Figure 13.2a). Based on the AP-XPS studies of Co_3O_4 during WGS and the confirmed activity for WGS, the authentic catalytic phase of the nominal catalyst, Co_3O_4, is the non-stoichiometric CoO formed during pretreatment and remained during WGS.

13.1.3 Active Phase of 0.5 wt% Pt/Co_3O_4 at 150–200 °C

Co_3O_4 nanorods with singly dispersed Pt atoms are highly active for WGS.[1] The nominal catalyst was designated as 0.5 wt%Pt/Co_3O_4. Before introducing mixture of CO and H_2O, the support, cobalt oxide is Co_3O_4 and there is a lack of CoO due to the absence of the characteristic satellite peak of Co^{2+} of CoO at 786.5 eV. Extended X-ray absorption fine structure

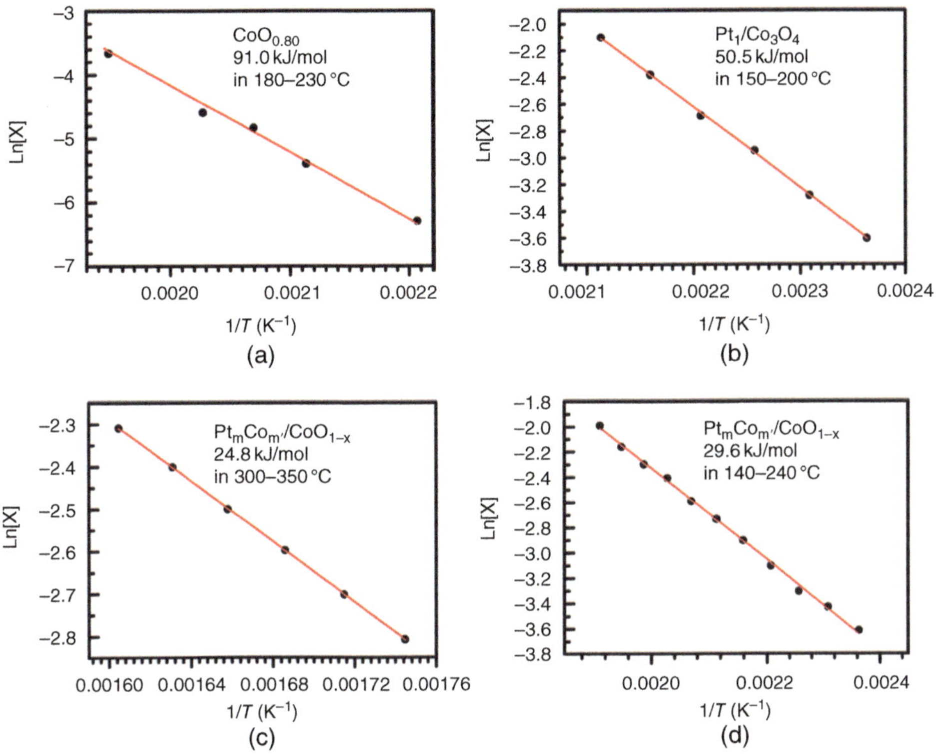

Figure 13.2 Kinetic studies of WGS on (a) pure Co_3O_4 nanorod catalyst in the temperature range of 180–240 °C, (b) Pt_1Co_n/Co_3O_4 catalyst in the temperature range of 180–240 °C, (c) $Pt_mCo_{m'}/CoO_{1-x}$ catalyst in the temperature range of 300–350 °C, and (d) $Pt_mCo_{m'}/CoO_{1-x}$ catalyst in the temperature range of 140–240 °C upon cooling it from 350 to 140 °C. *Source:* Reproduced with permission from Zhang et al.[1]/American Chemical Society.

(EXAFS) spectroscopy studies suggest the supported Pt atoms are singly dispersed, and each Pt atom is coordinated with four oxygen atoms on average and Pt atoms do not directly bond with any Co atoms of Co_3O_4 at 150–200 °C.[1] Without pretreatment in H_2, a mixture of CO and H_2O was flowing through this catalyst. Notably, the singly dispersed Pt atoms were partially reduced at 150–200 °C by the mixture of CO and H_2O, evidenced by the downshift of Pt $4d_{5/2}$ to 316.4 eV (Figures 13.3b2 and b3). Corresponding to this partial reduction of Pt^{x+}, the coordination environment of the Pt_1 atoms changed.[1] At 250 °C in WGS condition, Pt atoms are still singly dispersed but they bond with Co atoms. As no Pt–Pt bonds are observed in this temperature range, Pt atoms of the active phase of 0.5 wt% Pt/Co_3O_4 are still singly dispersed on the surface of Co_3O_4 during WGS below 280 °C. It is called a singly dispersed bimetallic site, Pt_1Co_n. Its structure is schematically shown in the middle of Figure 13.4. Kinetics studies show that 0.5 wt% Pt/Co_3O_4 is more active than CoO without any loaded Pt, evidenced by the obviously lower activation barrier of 50.1 kJ/mol for 0.5 wt% Pt/Co_3O_4 (Figure 13.2b), in contrast to 90.1 kJ/mol for pure Co_3O_4 at a temperature lower than 280 °C (Figure 13.2a). Notably, here the support of 0.5 wt% Pt/Co_3O_4 under WGS condition at 150–200 °C retains its original support Co_3O_4, and the active phase 0.5 wt% Pt/Co_3O_4 in the temperature range of 150–200 °C is a singly dispersed Pt_1Co_n site supported on Co_3O_4.

13.1.4 Active Phase of 0.5 wt% Pt/Co_3O_4 at 280–350 °C

The function of AP-XPS in tracking surface of a catalyst during catalysis under a sequential catalysis temperature was demonstrated in exploring how the active phase of a nominal catalyst morphs under the catalytic condition along increase of catalysis temperature.[1] As shown in Figure 13.3a4 and a5, Co_3O_4 of the active phase Pt_1Co_n/Co_3O_4 was reduced to CoO at a temperature 280 °C or higher. In addition, a shoulder of Pt $4d_{5/2}$ at ~314.4 eV appeared at 280 °C in Figure 13.3b4. It is marked as "3" in Figure 13.3b. This shoulder in Figure 13.3b5 becomes obvious at 350 °C. The Pt $3d_{5/2}$ observed at 350 °C during WGS by AP-XPS can be deconvoluted into Pt atoms of Pt–O at 317.5 eV (blue peak in Figure 13.3b5), Pt atoms of Pt–Co–O at 316.4 eV (green peak without a mark in Figure 13.3b5), and Pt atoms of Pt–Pt bonds at 314.4 eV (green peak marked with "4" in Figure 13.3b5), respectively. Their atomic fractions of Pt represented by peaks 2, 3, and 4 are 36%, 32%, and 32%, respectively. The active phase of 0.5wt%Pt/Co_3O_4 at 280–350 °C is designated as $Pt_mCo_{m'}/CoO$. The apparent activation barrier of this active phase evaluated in the temperature range of 280–350 °C (Figure 13.2c) is much lower than 50.1 kJ/mol of Pt_1Co_n/Co_3O_4 at 150–200 °C (Figure 13.2b). This significantly different catalytic activity in terms of quite different activation barriers suggests that the active phase in 280–350 °C ($Pt_mCo_{m'}/CoO$) must be different from the active phase at 150–200 °C, Pt_1Co_n/Co_3O_4. AP-XPS studies uncovered that the authentic active phase in this temperature range of 280–350 °C is Pt–Co bimetallic nanoclusters supported on CoO ($Pt_mCo_{m'}/CoO$). The formation of Pt–Co bimetallic nanoclusters $Pt_mCo_{m'}$ at 280–350 °C under the catalytic condition was supported by EXAFS studies of Pt L_3 edge of the catalyst.[1]

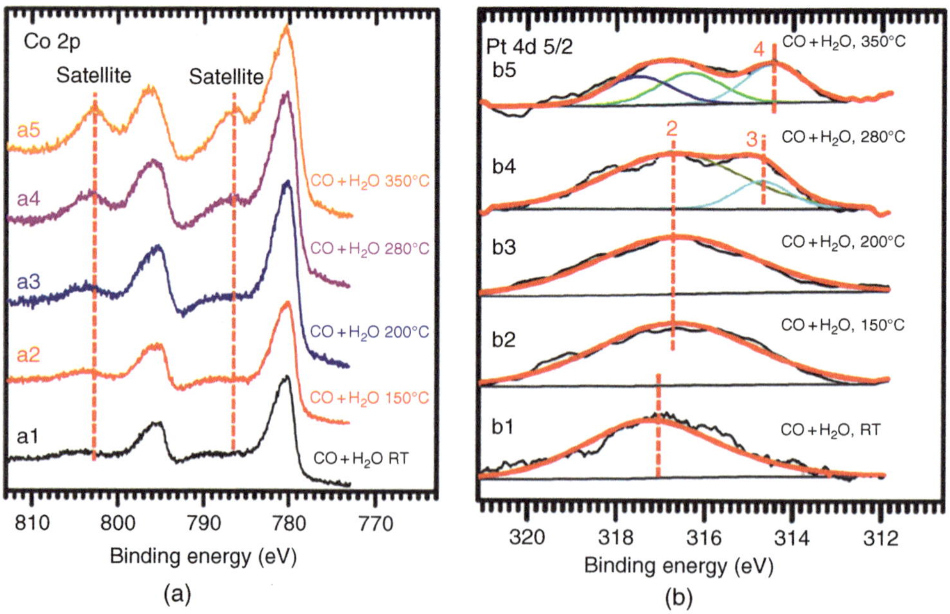

Figure 13.3 Photoemission features of Co 2p and Pt $4d_{5/2}$ of as-synthesized Pt_1/Co_3O_4 catalyst during catalysis at different temperatures. (a) Co 2p and (b) Pt $4d_{5/2}$. As Co 3p overlaps photoemission features of Pt 4f, Pt $4d_{5/2}$ was collected for identifying oxidation state and chemical environment of Pt anchored on Co_3O_4 although the cross section of Pt 4d is much smaller than Pt 4f.[2] *Source:* Reproduced with permission from Zhang et al.[1]/American Chemical Society.

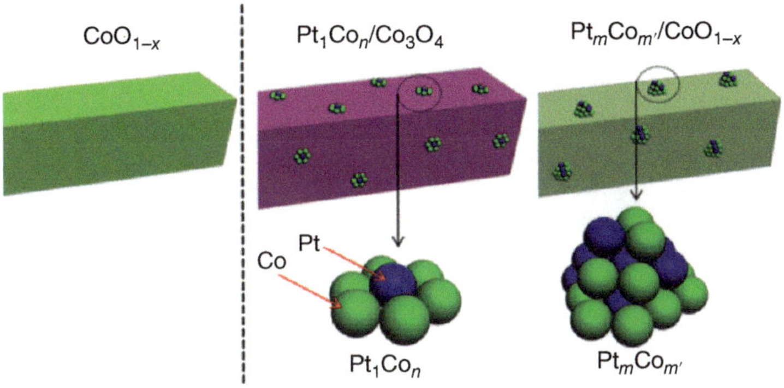

Figure 13.4 Schematics of the three identified catalysts CoO_{1-x}, Pt_1Co_n/Co_3O_4, (catalyst of singly dispersed bimetallic sites) and $Pt_mCo_{m'}/CoO_{1-x}$ (catalyst of bimetallic nanoparticles). Pt_1Co_n/Co_3O_4 is formed from an as-prepared Pt_1/Co_3O_4 at 150–200 °C in the mixture of CO and H_2O; $Pt_mCo_{m'}/CoO_{1-x}$ is formed through the evolution of Pt_1Co_n/Co_3O_4 driven by a higher catalysis temperature, 280–350 °C. *Source:* Reproduced with permission from Zhang et al.[1]/American Chemical Society.

13.1.5 Temperature-dependent Evolution of Active Phase

Tracking surface of Co_3O_4 loaded with 0.5 wt% Pt revealed a significant evolution of surface phase along increase of catalysis temperature under a specific catalytic condition. Two active phases were uncovered in the temperature range of 150–200 °C and 280–350 °C, respectively. One is the low-temperature phase, a singly dispersed bimetallic site, Pt_1Co_n, formed on Co_3O_4. It is active for WGS with an activation barrier of 50 kJ/mol in 150–200 °C. Along the increase of catalysis temperature, it evolves into a high-temperature phase. The high-temperature phase was formed through restructuring the low-temperature phase Pt_1Co_n/Co_3O_4 in the mixture of CO and H_2O. This restructuring formed a new catalyst surface consisting of CoO and its supported Pt–Co bimetallic nanoclusters, $Pt_mCo_{m'}$ where m′ is different from m. This high-temperature phase $Pt_mCo_{m'}/CoO$ exhibits a higher activity than the low-temperature phase Pt_1Co_n/Co_3O_4. Figure 13.4 schematically presents the evolution of active phases of a nominal catalyst 0.5wt%Pt/Co_3O_4 of low temperature to high temperature.

This work clearly demonstrated the unique function of AP-XPS in tracking surface of a catalyst under a reaction or during catalysis condition, particularly the evolution of catalyst surface driven by changes of catalysis temperature. Without in situ studies using AP-XPS, an ex situ study of either the 0.5 wt% Pt/Co_3O_4 before catalysis or a used 0.5 wt% Pt/Co_3O_4 could have provided information that Pt of the active phases would be cationic Pt and the support would be Co_3O_4. Overall, AP-XPS allows for identifying authentic active phases corresponding to a specific catalytic condition.

13.2 Pt, Au, Pd, and Cu Supported on CeO_2 Nanorods

Metal nanoparticles supported on CeO_2 are a type of WGS catalyst studied extensively in literature. The metal nanoparticles can be various transition metals including Au, Cu, Pt, Pd, and many others.[3–5] Here Au and Pt nanoparticles supported on CeO_2 nanorods were

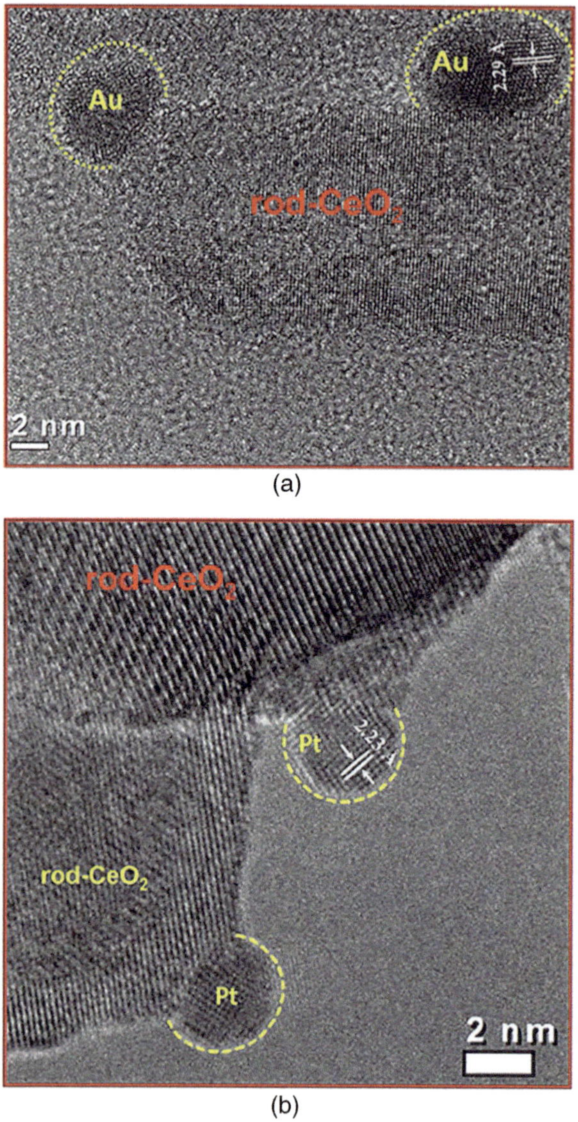

Figure 13.5 High-resolution transmission electron microscope (TEM) images of Au nanoclusters on CeO_2 rods (a) and Pt nanoclusters on CeO_2 rods (b). *Source:* Wen et al.[3]/American Chemical Society.

chosen as examples to demonstrate how AP-XPS was used to study this type of catalyst (Figure 13.5).

Metal nanoparticles such as Au, Pt, Pd, or Cu supported on CeO_2 nanorods were highly active for WGS. These supported catalysts were typically prepared through two steps including preparation of CeO_2 nanorods through a hydrothermal route and a followed deposition precipitation of metal to surface of the CeO_2 nanorods.[3] Figure 13.5 presents representative high-resolution transmission electron microscopy (HRTEM) images of Au and Pt nanoparticles supported on CeO_2 nanorods. These catalysts are highly active for WGS in the temperature range of 100–200 °C.[3] The apparent activation barriers of Pt NPs/rod-CeO_2 and Au NP/rod-CeO_2 are 78 and 51 kJ/mol, respectively.[3]

Surface chemistries of the two catalysts during pretreatment in H_2 at 300 °C and during the followed WGS catalysis at 130, 200, and 270 °C were tracked with AP-XPS. Figure 13.6 presents the XPS peaks of Ce 3d and Au 4f as a function of reaction conditions in the ordering of H_2 at 25 °C, H_2 at 300 °C, $CO+H_2O$ at 130 °C, $CO+H_2O$ at 200 °C, and $CO+H_2O$ at 270 °C. The photoemission features of Ce 3d and Pt 4f were studied as a function of reaction conditions as well (Figure 13.7). The common feature of the evolution of Ce 3d of the two catalysts Au NPs/CeO_2 and Pt NPs/CeO_2 is the formation of oxygen vacancies during H_2 reduction at 300 °C (the spectra of 300 °C in H_2 in Figures 13.6a and 13.7a). Surprisingly, the fraction of Ce^{3+} among all Ce atoms in the surface region of Au/rod-CeO_2 and Pt/rod-CeO_2 is as high as 35% or so (Figure 13.8). Obviously, the exposure to H_2 at 300 °C created oxygen vacancies. The formation of oxygen vacancies results from reduction of CeO_2 surface by atomic H formed by dissociation of molecular H_2 catalyzed by Pt or Au nanoparticle on CeO_2. Notable features at the bottom of Figures 13.6b and 13.7b are (i) Au and Pt atoms of the as-prepared Au/CeO_2 and Pt/CeO_2 are at cationic state before the pretreatment in H_2 at 300 °C and (ii) they appear in metallic state after the pretreatment in H_2.

Another important finding of these AP-XPS studies in Figures 13.6a and 13.7a is the obvious decrease of density of oxygen vacancies of CeO_2 of these catalysts upon the introduction

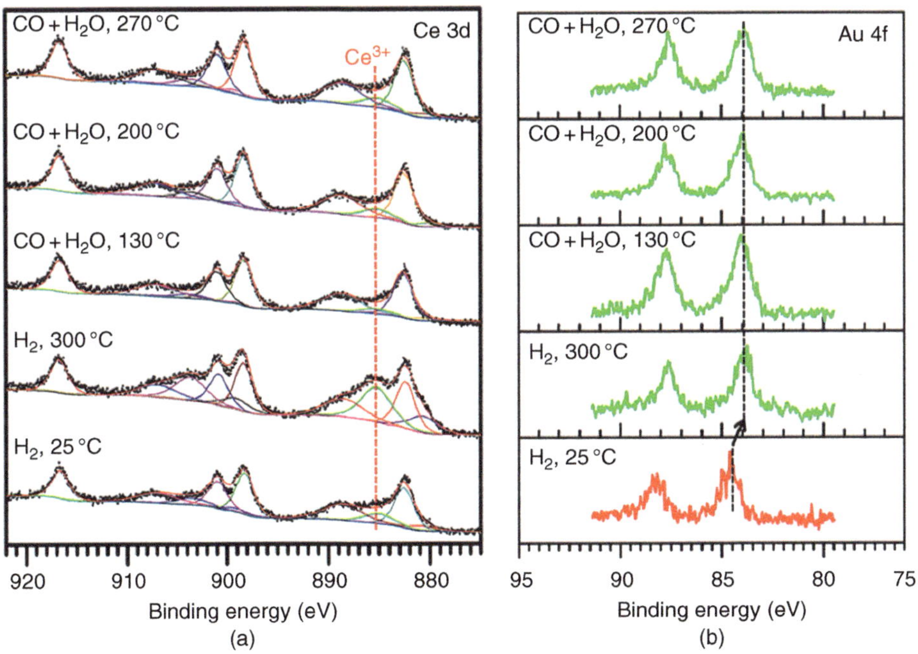

Figure 13.6 AP-XPS studies of (a) Ce 3d and (b) Au 4f of Au/CeO_2 nanorods under reaction conditions studied with AP-XPS. P_{CO} = 0.3 Torr, P_{H_2O} = 1.2 Torr, and P_{H_2} = 1 Torr. Reaction temperatures are given in the figure. *Source:* Reproduced with permission from Wen et al.[3]/American Chemical Society.

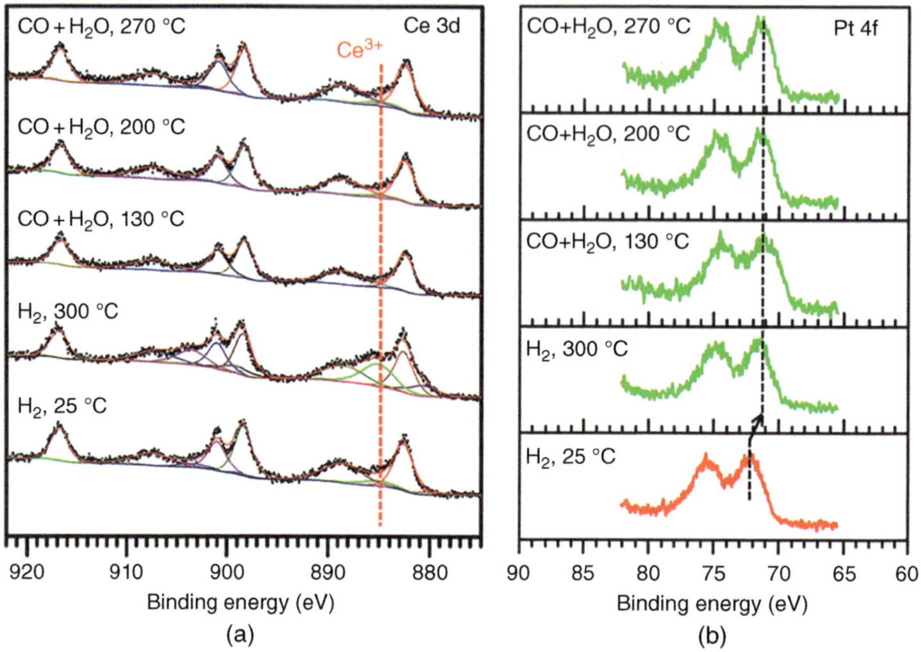

Figure 13.7 AP-XPS studies of (a) Ce 3d and (b) Pt 4f of Pt/CeO_2 under reaction conditions studied with AP-XPS. P_{CO}= 0.3 Torr, P_{CO_2} = 1.2 Torr, and P_{H_2} = 1 Torr. Reaction temperatures are given in the figure. *Source:* Reproduced with permission from Wen et al.[3]/American Chemical Society.

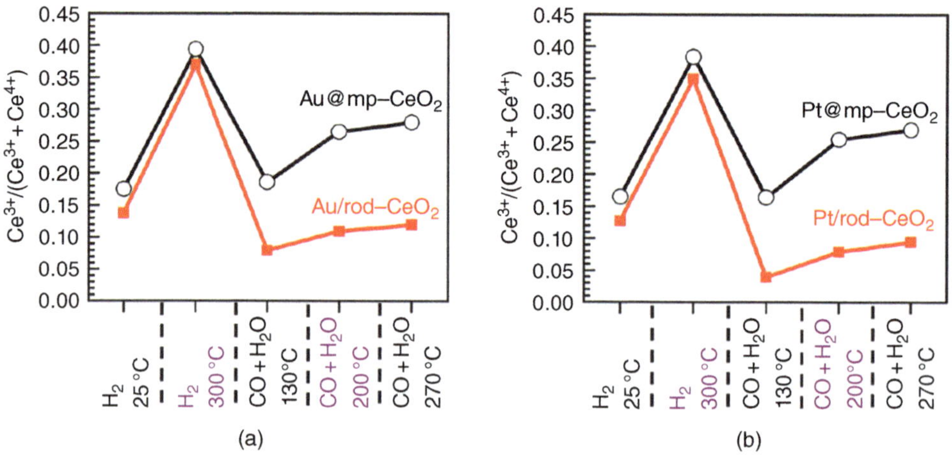

Figure 13.8 Evolution of atomic fractions of Ce^{3+} of Au NPs/CeO_2 (a) and Pt NPs/CeO_2 during WGS. *Source:* Reproduced with permission from Wen et al.[3]/American Chemical Society.

of reactant mixture of H_2O and CO.[3] The filling of some of these vacancies created by H_2 reduction in the pretreatment with H_2 was done by OH groups that were formed through dissociation of H_2O to H and OH at oxygen vacancies. This type of filling to oxygen vacancies can be expressed as $H'O' + O_{vac} - Ce^{3+} - O \rightarrow H'O' - Ce^{4+} - O$. Here " ′ " stands for "external" instead of a native atom of the as-prepared catalysts. The filling by these external HO during WGS decreases the density of oxygen vacancies as shown in Figure 13.8.

Here the coordination environment of Ce cations in surface region of CeO_2 rod of Au NPs/CeO_2 and Pt NPs/CeO_2 is designated as $(O)_n$–Ce or $(O)_{n-x}$–Ce–O_x. AP-XPS studies revealed there are still small portion of Ce^{3+} at the level of 10–15% of Ce atoms during WGS (Figures 13.6a and 13.7a) although it is much lower than 35% in H_2 (Figure 13.8). Some of these oxygen vacancies could be a type of false vacancies. They could be contributed from Ce atoms that directly bond to metal atoms of metal nanoparticles at the interface between metal nanoparticles and CeO_2. The coordination environment of this type of Ce atoms can be written as $(O)_{n-y}$–Ce–M_z; here n is the number of O atoms coordinating a Ce atom on surface of CeO_2 without any oxygen vacancies; $n–y$ is the actual number of oxygen atoms bonded to the Ce; M is the metal element of a metal nanoparticle such as Pt or Au. z is the number of M atoms directly bonded to the Ce atom. Metal atoms M have a lower electronegativity than O atoms, electron density of metal atoms is transferred to the Ce atom upon some O atoms bonding to a Ce atom are replaced by M atoms. Compared to $(O)_{n-x}$–Ce–O_x, the increase of electron density on Ce atom in $(O)_{n-y}$–Ce–M_z due to electron transfer from M to Ce makes Ce^{4+} appear as Ce^{3+} in terms of electron density. Thus, the oxygen vacancies of the catalyst evaluated with the atomic fraction of Ce^{3+} among all Ce atoms in the surface region of the catalyst could not necessarily be the physical vacancies around Ce atoms in surface region of CeO_2. Thus, the high fraction of Ce^{3+} in H_2 at 300 °C is likely partially contributed from these Ce atoms which directly bond to metal atoms of the supported metal or metal oxide nanoparticles.

Transition metal oxides such as ceria are significant supports of a great number of catalysts and often play crucial role in catalysis. Their surface chemistry is complicated as cations of a transition metal with different valences often coexist and they can be even readily interchanged by switching gaseous environment between reducing and oxidizing gas or changing temperature. In a reducing or oxidizinggas, the high-valence cations can be reduced or low-valence cations can be oxidized, respectively. Such reduction or oxidization generates or fills oxygen vacancies. As a transition metal oxide is typically sensitive for gaseous environment of pretreatment such as H_2 and oxidizing or reducing reactant of a reaction, a characterization of gaseous environment- or temperature-sensitive surface by ultrahigh vacuum XPS likely gives inappropriate information as demonstrated in this subsection. AP-XPS is the right technique to track the surface chemistry of a transition metal oxide, measure the authentic atomic ratio of different valance states of a transition metal, and quantify the density of surface vacancies such as density of oxygen vacancies under a catalytic condition.

13.3 $CuO–Cr_2O_3–Fe_2O_3$

Industrial WGS includes two processes, the high-temperature and low-temperature WGS reactions. The high-temperature WGS catalysis is performed at ~350–400 °C; the catalyst is iron oxide with loaded chromium. In terms of low-temperature WGS, it occurs at 200–250 °C; the catalyst of low-temperature WGS is copper-based catalyst. The purpose of having the low-temperature WGS is to gain high conversion of CO since the WGS is slightly exothermic and reversible.[6] $CuO–Cr_2O_3–Fe_2O_3$ is an industrial WGS catalyst. Identification of the active phases of the industrial catalyst has remained nonconclusive and the community has kept debating the catalytic mechanism at a molecular level. Different mechanisms were proposed in literature. For instance, the nature of chromium and its distribution on and in the support were explained with different models in literature.[7–11] The role of copper in this high-temperature WGS was studied but remained debating.[12–14] A conclusive understanding has not reached due to the lack of information on the catalyst surface during catalysis. AP-XPS was used in investigation of surface chemistry of this catalyst during WGS.

$CuO–Cr_2O_3–Fe_2O_3$ was prepared with ammonia-assisted coprecipitation followed by drying at 80 °C for 12 hours and calcination at 400 °C in air for 3 hours.[6] The prepared catalyst consists of 3 wt% CuO, 8 wt% Cr_2O_3, and 89 wt% Fe_2O_3. X-ray diffraction (XRD) of this prepared catalyst is mainly the diffraction pattern of Fe_2O_3. Neither CuO nor Cr_2O_3 phase was identified in XRD pattern, probably because CuO and Cr_2O_3 have become a part of iron oxide-based solid solution. Another explanation for the lack of Cr_2O_3 and CuO diffraction peaks in XRD pattern of the as-prepared catalyst is that sizes of Cr_2O_3 and CuO nanoparticles are smaller than 2 nm even if they could exist in separate phases.

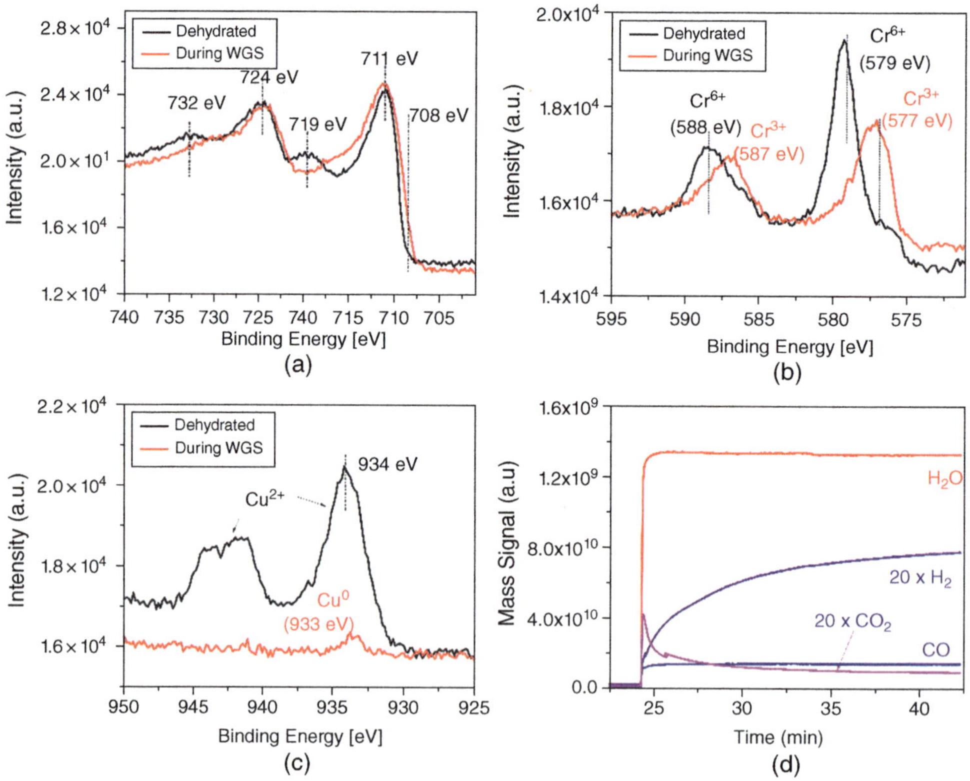

Figure 13.9 AP-XPS studies of $CuO–Cr_2O_3–Fe_2O_3$ surface and mass spectrometry analysis of gas sampled from the reaction cell. (a) Fe 2p, (b) Cr 2p, and (c) Cu 2p regions from the $CuO–Cr_2O_3–Fe_2O_3$ catalyst under dehydrated conditions at 400 °C and during the high-temperature WGS (P = 0.3 mbar, T = 400 °C, and H_2O:CO ratio = 10), and (d) the corresponding mass spectrometer signals as a function of time. *Source:* Reproduced with permission from Zhu et al.[6]/American Chemical Society.

By using AP-XPS, the prepared catalyst was studied by Wachs and Schlögl groups under conditions including (i) dehydrated condition at 400 °C and (ii) high-temperature WGS condition in terms of 400 °C in the mixture of 0.02 Torr H_2O and 0.2 Torr CO.[6] Figure 13.9 presents the Fe 2p, Cr 2p, and Cu 2p XPS spectra and mass spectra collected during characterizations of catalysts with AP-XPS. The data of Fe 2p, Cr 2p, and Cu 2p of the catalyst under the dehydrated condition are shown with black spectra in Figure 13.9a, b, and c.[6] Under the dehydrated condition, Fe exists as Fe_2O_3 evidenced by the main peak of Fe $2p_{3/2}$ at 711 eV and the characteristic satellite peak of Fe $2p_{3/2}$ at 719 eV (black spectrum in Figure 13.9a). The peak position of Cr $2p_{3/2}$ at 579 eV suggests that Cr atoms are +6 (black spectrum in Figure 13.9b). In terms of Cu atoms, its oxidation state is +2 based on the broad satellite peak at 940–945 eV (black spectrum in Figure 13.9c). These observations of Cu and Cr by the surface sensitive analytical technique, AP-XPS suggest that Cu and Cr are distributed on surface region of the catalyst under the dehydrated condition.

Compared to $CuO–Cr_2O_3–Fe_2O_3$ under the dehydrated condition, surface of the $CuO–Cr_2O_3–Fe_2O_3$ during WGS catalysis at 400 °C in the mixture of CO and H_2O was distinctly different.[6] The red lines in Figure 13.9a, b, and c are the Fe 2p, Cr 2p, and Cu 2p spectra collected during the WGS at 400 °C. The ongoing WGS catalysis while AP-XPS data was being collected was confirmed by the observation of products of WGS, H_2 and CO_2 in the online mass spectrum (Figure 13.9d). In the mixture of CO and H_2O at 400 °C, the characteristic satellite peak of Fe_2O_3 at 719 eV disappeared in red spectrum in Figure 13.9a. Detailed analysis of the AP-XPS data suggests that Fe_2O_3 is transformed to Fe_3O_4 that is the active phase for the high-temperature WGS catalysis. This is consistent with the XRD pattern of a used $CuO–Cr_2O_3–Fe_2O_3$ catalyst after reversal WGS. XRD studies show the activated catalyst is a Fe_2O_3 phase while neither CuO nor Cr_2O_3 phase could be identified.[6]

In terms of Cr, AP-XPS studies show that Cr^{6+} observed under the dehydrated condition was significantly reduced to Cr^{3+}under high-temperature WGS condition at 400 °C, evidenced by the obvious downshift of peak position of Cr $2p_{3/2}$ from 579 eV of Cr^{6+} to 577 eV of Cr^{3+} (Figure 13.9b). Compared to the Cu 2p of the catalyst under a dehydrated condition, a surprising change for Cu 2p during WGS at 400 °C is the significant decrease of peak intensity of Cu 2p during WGS (Figure 13.9c). In addition, the peak positions of Cu obviously downshift from 940–945 eV for the catalyst under the dehydrated condition to 933 eV for the catalyst during catalysis, suggesting a partial reduction of Cu^{2+}. These changes show that Cu in surface region was progressively covered by other metal or metal oxide, suggesting that copper cations or copper-containing species were buried in subsurface or deep layers. Figure 13.10 is the Cu 2p spectra collected by AP-XPS during catalysis as a function of time. The intensity of Cu 2p decreased by 90% within four minutes. Meanwhile, the Cu^{2+} was reduced to metallic Cu within two minutes. These time-dependent changes tracked by AP-XPS suggest (1) Cu^{2+} was reduced to metallic Cu and (2) other metal oxides diffuse to surface region to cover the formed metallic Cu nanoparticles. As no XRD diffraction patterns of Cu metal nanoparticle was observed from a used $CuO–Cr_2O_3–Fe_2O_3$ catalyst, it is deduced that the formed metallic Cu nanoparticles are smaller than 2 nm and buried in Fe_2O_3. Figure 13.11 schematically shows the restructuring of the as-prepared $CuO–Cr_2O_3–Fe_2O_3$ under the high-temperature WGS condition at 400 °C.

A transition metal oxide is typically sensitive for gaseous environment such as H_2 or O_2 of pretreatment or the oxidizing or reducing reactant of a catalytic reaction. The extent of

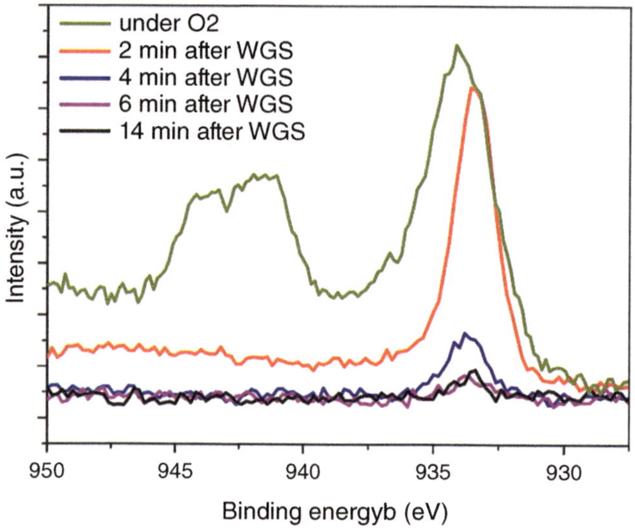

Figure 13.10 Time-resolved AP-XPS studies of Cu 2p region of $CuO-Cr_2O_3-Fe_2O_3$ upon switching from dehydrated oxidizing condition to WGS conditions ($P = 0.3$ mbar, $T = 400\,°C$, and H_2O:CO ratio = 10). *Source:* Reproduced with permission from Zhu et al.[6]/American Chemical Society.

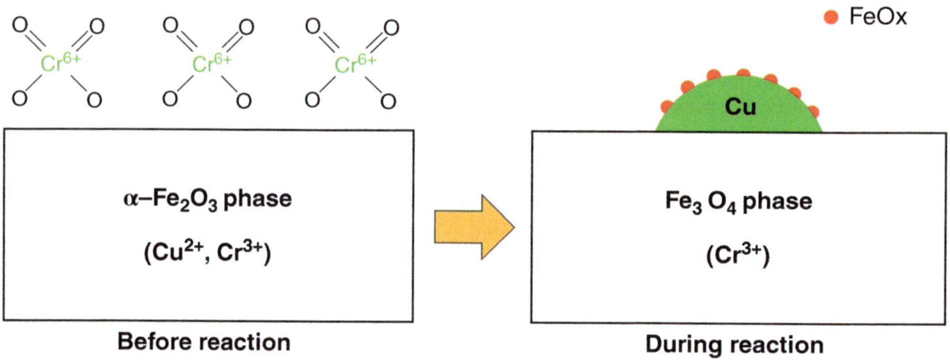

Figure 13.11 Schematics of the copper-chromium-iron oxide catalyst before and during the high-temperature WGS. *Source:* Reproduced with permission from Zhu et al.[6]/American Chemical Society.

being oxidized or reduced depends on the type of constituting gases such as oxidizing gas CO_2 or reducing gas H_2 and even the composition such as the molar ratio of reducing gas to oxidizing gas in a gas mixture. Thus, it is crucial to use AP-XPS to track the surface chemistry of a transition metal oxide during pretreatment and under authentic conditions in order to uncover the crucial surface chemistry. In addition, the time-resolved AP-XPS study demonstrated a time-dependent evolution of concentration and chemical states of the transition metal element during catalysis. It provides first-hand information on how the chemical environment of a constituting element of a catalyst evolves under a catalytic condition.

References

1 Zhang, S., Shan, J., Zhu. Y. *et al.* 2013. "WGS catalysis and in situ studies of CoO(1-x), PtCo(n)/Co3O4, and Pt(m)Co(m′)/CoO(1-x) nanorod catalysts." *J. Am. Chem. Soc.* 135, 8283–8293.

2 Yeh, J. J. and Lindau, I. 1985. "Atomic subshell photoionization cross sections and asymmetry parameters: $1 \leqslant Z \leqslant 103$." *Atomic Data Nucl. Data Tables* 32, 1–155.

3 Wen, C., Zhu, Y., Ye, Y. *et al.* 2012. "Water-gas shift reaction on metal nanoclusters encapsulated in mesoporous ceria studied with ambient-pressure X-ray photoelectron spectroscopy". *ACS Nano* 6, 9305–9313.

4 Rodriguez, J. A., Senanayake, S. D., Stacchiola, D. *et al.* 2014. "The activation of gold and the water–gas shift reaction: insights from studies with model catalysts." *Acc. Chem. Res.* 47, 773–782.

5 Rodriguez, J. A., Grinter, D. C., Liu, Z. *et al.* 2017. "Ceria-based model catalysts: fundamental studies on the importance of the metal–ceria interface in CO oxidation, the water–gas shift, CO_2 hydrogenation, and methane and alcohol reforming." *Chem. Soc. Rev.* 46, 1824–1841.

6 Zhu, M., Rocha, T. C., Lunkenbein, T. *et al.* 2016. "Promotion mechanisms of iron oxide-based high temperature water–gas shift catalysts by chromium and copper." *ACS Catal.* 6, 4455–4464.

7 Chinchen, G. C., Logan, R. H., and Spencer, M. S. 1984. "Water-gas shift reaction over an iron oxide/chromium oxide catalyst.: I: mass transport effects." *Appl. Catal.* 12, 69–88.

8 Chinchen, G. C., Logan, R. H., and Spencer, M. S. 1984. "Water-gas shift reaction over an iron oxide/chromium oxide catalyst.: II: stability of activity." *Appl. Catal.* 12, 89–96.

9 Meshkani, F. and Rezaei, M. 2015. "Dry reforming over CeO_2-promoted Ni/MgO nano-catalyst: effect of Ni loading and CH_4/CO_2 molar ratio." *J. Indust. Eng. Chem.* 21, 717–722.

10 Meshkani, F. and Rezaei, M. 2015. "Preparation of mesoporous nanocrystalline iron based catalysts for high temperature water gas shift reaction: effect of preparation factors." *Chem. Eng. J.* 260, 107–116.

11 Latifi, S. M. and Salehirad, A. 2016. "High temperature water gas shift reaction over Fe-Cr-Cu nanocatalyst fabricated by a novel method." *Korean J. Chem. Eng.* 33, 473–480.

12 Andreev, A., Idakiev, V., Mihajlova, D. *et al.* 1986. "Iron-based catalysts for the water—gas shift reaction promoted by first-row transition metal oxides." *Appl. Catal.* 22, 385–387.

13 Reddy, G. K., Gunasekera, K., Boolchand, P. *et al.* 2011. "Ceria-modified manganese oxide/titania materials for removal of elemental and oxidized mercury from flue gas." *J. Phys. Chem. C* 115, 24300–24309.

14 Puig-Molina, A., Cano, F. M., and Janssens, T. V. W. 2010. "The Cu promoter in an Iron—Chromium—Oxide based water—gas shift catalyst under industrial conditions studied by in-situ XAFS." *J. Phys. Chem. C* 114, 15410–15416.

14

Complete Oxidation of Methane

14.1 Complete Oxidation of Methane on $NiCo_2O_4$

Complete oxidation of methane exhibits significant applications in removal of the greenhouse gas, CH_4. For instance, complete oxidation of CH_4 at high temperature is important for powerplant since it can transform the unburned CH_4 to CO_2 and H_2O before it is released to ambient environment. To remove the CH_4 left in the exit of a gasoline engine, complete oxidation of CH_4 at relatively high temperature up to 900 °C is necessary. Other than these applications at a high or relatively high temperature regimes, the low-temperature application in terms of catalytic complete oxidation of methane at a relatively low temperature range (<500 °C) is important for removal of CH_4 released from exhaust of an automobile engine combusting CH_4 or liquefied petroleum gas because (1) methane and liquified petroleum gas are still widely used in different types of automobiles in some countries and (2) combustion temperature in these engines is performed at a relatively low temperature (<500 °C). Designing a catalyst highly active for complete oxidation of methane at a low-temperature regime (300–500 °C) has been challenging. It is reported that Co_3O_4 doped with Ni cations, $NiCo_2O_4$, is quite active for complete oxidation of methane at a temperature as low as 300 °C. Surface chemistry of $NiCo_2O_4$ during complete oxidation of CH_4 was tracked with AP-XPS. A stable intermediate involved in the catalysis was identified through the temperature-dependent AP-XPS studies. The identification of this stable intermediate aided the understanding of the catalytic mechanism at a molecular level.

$NiCo_2O_4$ was prepared by coprecipitation. X-ray diffraction (XRD) confirmed the preservation of Co_3O_4 phase since there is a lack of any diffraction peaks of NiO in the XRD pattern of $NiCo_2O_4$. For a comparison, Co_3O_4 was prepared with the same method. However, the activity for complete oxidation of CH_4 to CO_2 and H_2O on Co_3O_4 is distinctly different from that on $NiCo_2O_4$. As shown in Figure 14.1, under the same catalytic condition, conversion of CH_4 on 500 mg of $NiCo_2O_4$ reaches 95% at a temperature as low as 300 °C but the conversion on 500 mg of Co_3O_4 nanoparticles or NiO nanocubes at the same temperature is only 3%. Clearly, the doped Ni cations in $NiCo_2O_4$ play a significant role in this low-temperature catalysis on $NiCo_2O_4$.

AP-XPS was used in the identification of surface of the catalyst during catalysis, providing crucial information for screening potential reaction pathways of methane complete oxidation on $NiCo_2O_4$ before an appropriate reaction mechanism is proposed. Figure 14.2

Application of Ambient Pressure X-ray Photoelectron Spectroscopy to Catalysis, First Edition. Franklin Tao.

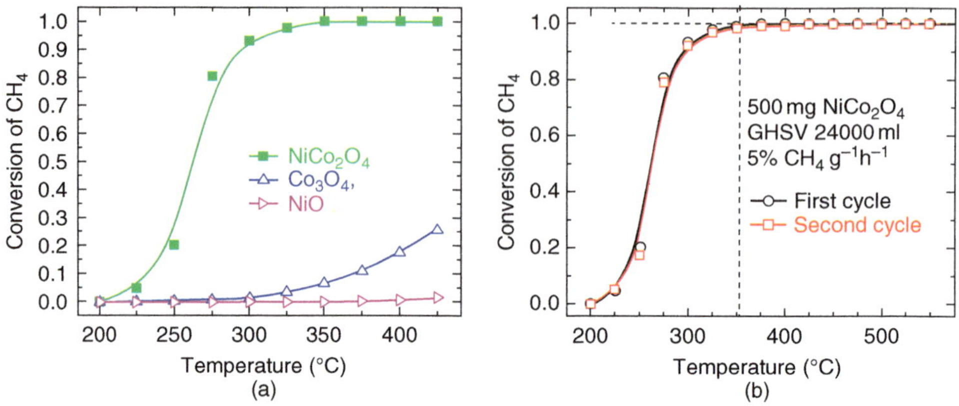

Figure 14.1 Catalytic performance of catalysts for complete oxidation of methane. (a) Conversion of CH_4 in a complete oxidation of CH_4 on $NiCo_2O_4$ in the temperature range of 200–425 °C. The gas composition of the feed gas is 5% CH_4, 25% O_2, and 70% Ar. Molar ratio of CH_4 to O_2 in the gas mixture is always 1 : 5. As 500 mg $NiCo_2O_4$ catalyst was used, the gas hourly space velocity of CH_4 is 24 000 ml 5% CH_4 g^{-1} h^{-1}; at this GHSV the conversion of CH_4 on $NiCo_2O_4$ at 350 °C is 100%. (b) CH_4 can be completely oxidized at a temperature up to 550 °C (black line), and the catalytic performance remains the same after heating to 550 °C and then cooling to 200 °C (red line). *Source:* Reproduced with permission from Tao et al.[1]/Springer Nature.

presents the Ni 2p, Co 2p, O 1s, and C 1s collected with AP-XPS from $NiCo_2O_4$ in the mixture of CH_4 and O_2 during complete oxidation of methane. Both Ni 2p and Co 2p peaks were clearly observed in the temperature range of 60–400 °C, suggesting cations of Ni and Co elements coexist in the surface region of the catalyst during catalysis. Before tracking the evolution of C 1s XPS peak under the catalytic condition, surface of the catalyst was oxidized with O_2 at 300 °C and then cooled to 25 °C. This pretreatment provides a carbon-free surface before the introduction of CH_4 and O_2, confirmed by the bottom spectrum in Figure 14.2d. Then, the mixture of CH_4 and O_2 was introduced to the catalyst. Surprisingly, carbon-containing species was formed even at a temperature as low as 60 °C (Figure 14.2d) when the mixture of CH_4 and O_2 was flowing through the carbon-free $NiCo_2O_4$. Peaks 1 and 2 in Figure 14.2d mark the two carbon-containing species formed on the catalyst surface in the mixture of CH_4 and O_2. In terms of peak 1, it remained in the whole temperature range of 25–400 °C although its binding energy (BE) slightly downshift to 285.1 eV at 400 °C (Figure 14.2d). As intensity of peak 1 in Figure 14.2d remains nearly same, it is considered as a spectator formed at a relatively low temperature at 60 °C but remained or experienced a slight change of molecular structure along the increase of catalysis temperature.

AP-XPS studies uncovered that the intensity of peak 2 exhibits significant change along the increase of catalysis temperature. Peak 2 of C 1s spectra was observed at 60 °C (Figure 14.2d). It has a quite high BE at 288.2 eV, suggesting it is a species probably containing C=O. More interestingly, the intensity of this peak 2 increases at 100 °C in contrast to 60 °C. However, it starts to decrease at 150 °C. It eventually disappears at 250 °C. The evolution pattern of peak 2 in terms of first increase and then decrease as a function of catalysis temperature exactly fits the temperature dependence of intensity of a stable intermediate of a catalytic pathway. Here a stable intermediate is defined as an intermediate

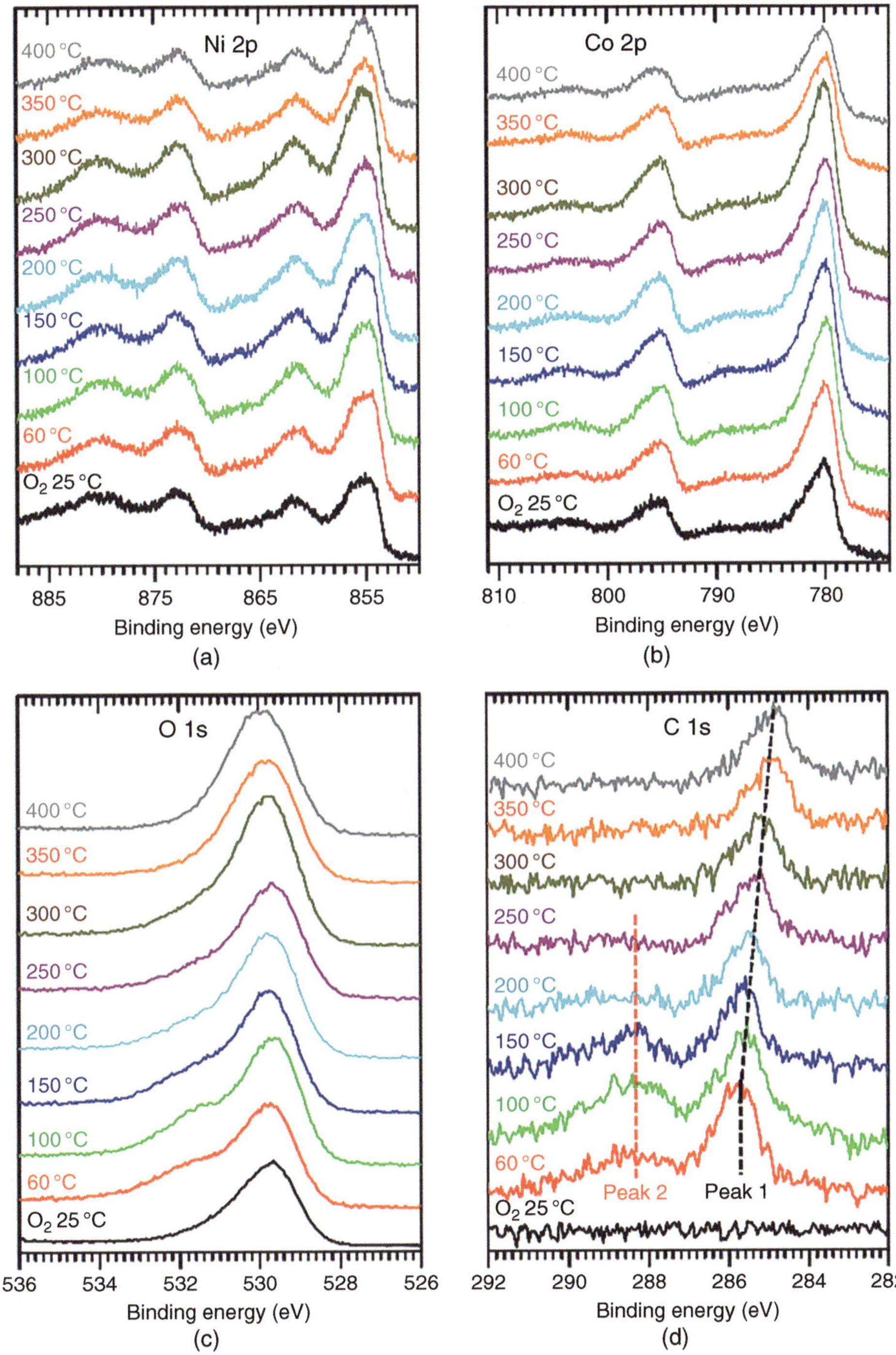

Figure 14.2 Photoemission features of $NiCo_2O_4$ at different temperatures during catalysis on $NiCo_2O_4$ in the mixture of CH_4 and O_2 with a molar ratio of 1 : 5 at Torr pressure range. (a) Ni 2p, (b) Co 2p, (c) O 1s, and (d) C 1s. *Source:* Reproduced with permission from Tao et al.[1]/Springer Nature.

formed at a low temperature in one of the early elementary steps before the rate-determining step. At low temperature, this intermediate does not have enough energy to cross the activation barrier of the transition state of the rate-determining step. Thus, a stable intermediate can be observed at low temperature; its amount increases along the elevation of catalyst temperature. However, when the temperature is high enough, this intermediate has enough energy to cross the activation barrier of the rate-determining step. Then, it is transformed to the next intermediate and forms a product that desorbs from the catalyst surface. The stable intermediate could not be observed at a high temperature. Thus, the temperature-dependent evolution of the intensity of stable intermediate represented as C/(Ni+Co) appears as a volcano-like curve when the intensity of this stable intermediate is plotted as a function of catalysis temperature (red lines in Figure 14.3).

As mentioned above, the high C 1s BE of peak 2 at 288.2 eV suggests that the stable intermediate contains a C=O bond. Diffuse reflectance infrared Fourier transform spectroscopy (DRIFTS) of the stable intermediates shows that the stable intermediate exhibits a O=C(H)–O-like structure. The identification of this stable intermediate facilitates the finding of the pathway of CH_4 complete oxidation on $NiCo_2O_4$ in which OCHO is a stable intermediate formed through a low-barrier step from CHO to OCHO. Based on the density functional theory (DFT) calculation,[1] the intermediate OCHO has a lower energy compared to its competing intermediate (adsorbed CO and O–O) (Figure 14.4b). In other words, OCHO is thermodynamically more stable than the adsorbed CO and O–O. Thus, OCHO can be formed readily at a low temperature. On the other hand, compared to the path of CO + O (Figure 14.4b), the transformation of OCHO to final product CO_2 and H_2O must cross a higher activation barrier as shown in Figure 14.4b. The successful observation of OCHO stable intermediate and its disappearance at a relatively high temperature suggest the path of OCHO must be one of the paths if not the only one, of CH_4 complete oxidation at ≥250 °C although it is not a path at low temperature (<250 °C). When the temperature is high enough, the OCHO can readily cross the high barrier to form final products CO_2 and H_2O, by which OCHO is transformed and the concentration of OCHO is nearly zero and invisible at high temperature. Thus, the successful identification of the stable intermediate with AP-XPS by uncovering the volcano evolution of intensity of the stable intermediates (red line in Figure 14.3a) helped screen numerous potential reaction pathways.

In this work, AP-XPS successfully identified the stable intermediate, OCHO which is formed at low temperature and is then transformed to products at high temperature by a quantitative analysis of its surface concentration along increase of catalysis temperature. The experimental identification of stable intermediate by AP-XPS can largely help narrowing a great number of potential pathways down to a specific pathway, facilitating fundamental studies of catalytic mechanisms at a molecular level.

14.2 Complete Oxidation of Methane on $NiFe_2O_4$

$NiFe_2O_4$ was prepared through coprecipitation of Ni^{2+} and Fe^{3+} with a following calcination at 500 °C in air.[3] It exhibits an inverse spinel structure of Fe_3O_4. It is highly active for complete oxidation at a temperature higher than 300 °C (Figure 14.5). AP-XPS tracked surface chemistry of this catalyst during complete oxidation of CH_4 (Figure 14.6). Before

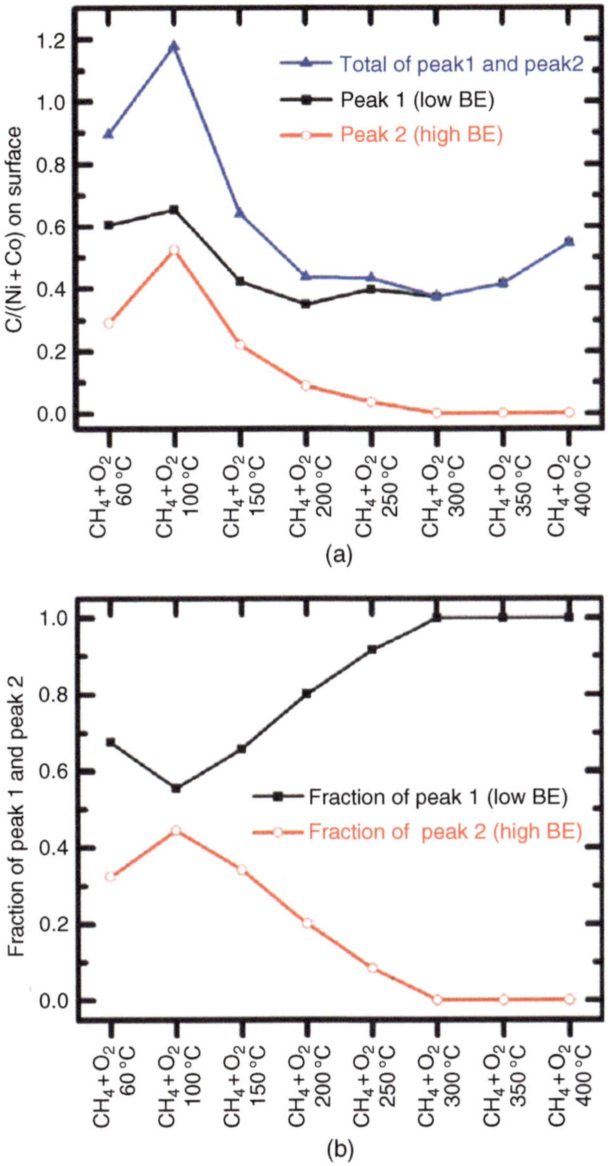

Figure 14.3 Evolution of surface concentration of carbon-containing species during catalysis at different temperature based on in situ studies of $NiCo_2O_4$ surface by AP-XPS. (a) Atomic ratio of $\frac{C_{total}}{Ni+Co}$, $\frac{C_{peak\,1}}{Ni+Co}$, and $\frac{C_{peak\,2}}{Ni+Co}$; (b) atomic fraction of carbon species 1 (peak 1), $\frac{C_{peak\,1}}{Ni+Co}$, and carbon species 2 (peak 2), $\frac{C_{peak\,2}}{Ni+Co}$, at different temperatures. These atomic ratios or atomic fractions are not normalized since what matters is the change of these atomic ratios or atomic fractions. The error of measurement of peak area is ±5% of the corresponding peak area. *Source:* Reproduced with permission from Tao et al.[1]/Springer Nature.

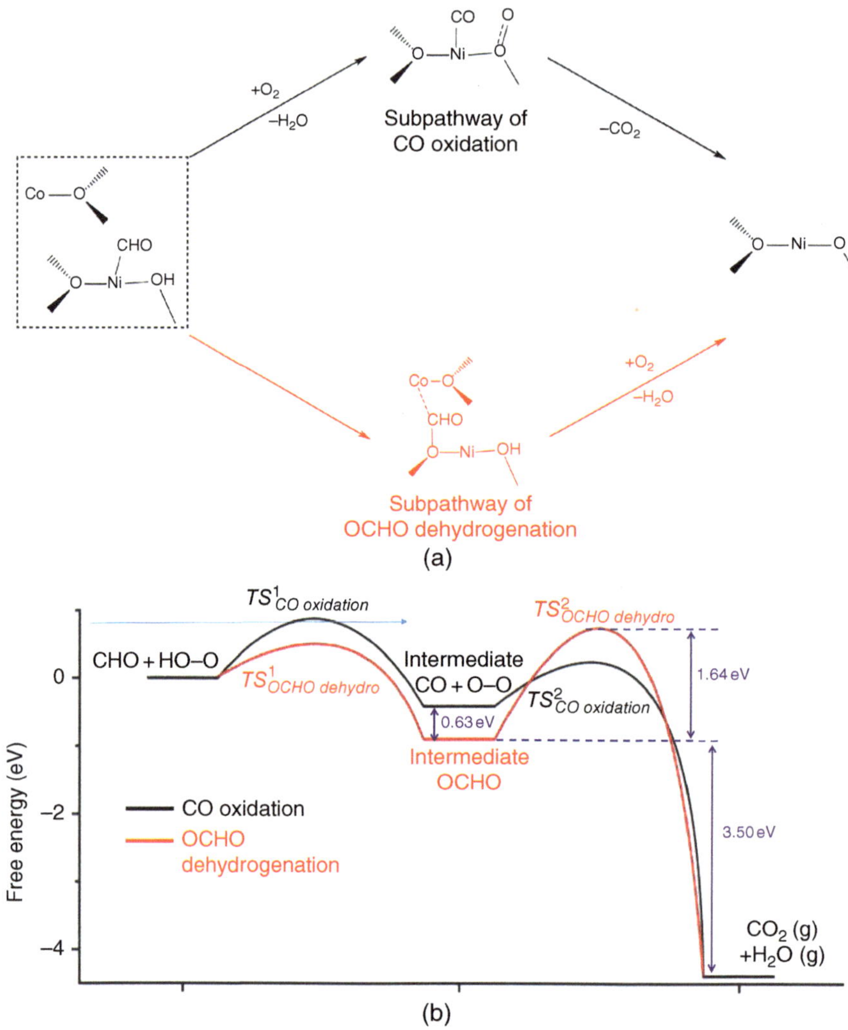

Figure 14.4 DFT calculations of subpathways for transforming CHO to product molecules. Each subpathway includes two elemental steps with two transition states and one intermediate. (a) Two different subpathways for the transformation of CHO to products, including the subpathway of CO oxidation (shown in black) that is the last part of pathway B in Figure 14.12 and the subpathway of OCHO dehydrogenation (shown in red). (b) Calculated energy profiles and optimized intermediates (CO + O–O versus OCHO) and transition states for the transformation of CHO to product CO_2 on $NiCo_2O_4$ of type I crystal after CHO is formed in pathway B. The elemental steps involved in this transformation (from CHO to products) are called a subpathway. Other than this subpathway, subpathway of OCHO dehydrogenation (red line) was proposed and calculated. *Source:* Reproduced with permission from Tao et al.[1]/Springer Nature.

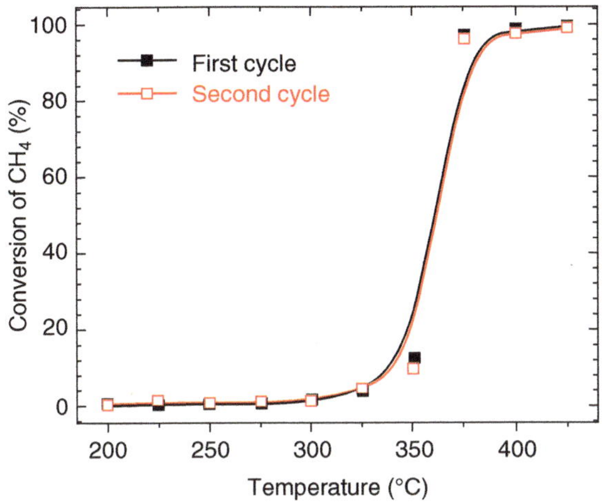

Figure 14.5 Catalytic activity of $NiFe_2O_4$ in terms of conversion of CH_4 as a function of reaction temperature; 500 mg of $NiFe_2O_4$ (40–60 mesh) was measured. The reactant gas mixture consists of 100 ml/min 10% CH_4/Ar and 50 ml/min O_2. The second cycle started after the catalyst was completely cooled down to room temperature in reactant gas mixture. *Source:* Reproduced with permission from Zhang et al.[3]/Elsevier.

flowing the mixture of reactants CH_4 and O_2, pure O_2 was introduced to the reaction cell to flow through the catalyst. At 300 °C, the carbon-based contaminants reacted with O_2 to form CO_2 and H_2O for forming a carbon-free surface. Then, the mixture of CH_4 and O_2 was introduced to the carbon-free surface of catalyst $NiFe_2O_4$. In the mixture of reactants CH_4 and O_2, three C 1s peaks at 284.8 eV (peak 1), 285.6 eV (peak 2), and 288.6 eV (peak 3) were observed at 60 °C in Figure 14.6d. Definitely, these peaks represent different types of carbon-containing species formed from CH_4.

Atomic ratio of carbon atoms represented by C 1s peak area to the total number of Fe and Ni atoms in surface region was calculated with AP-XPS spectra collected at each temperature. The ratios of the three peaks are presented in Figure 14.7. As the atomic ratios of carbon-containing species contributing to peaks 1 and 3 to total of Ni and Fe increase along the evaluation of catalysis temperature and then decrease to zero along further increase of catalysis temperature, the carbon-containing species represented by peaks 1 and 3 are stable intermediates (Figure 14.7a and c). Different from peaks 1 and 3, the atomic ratio of peak 2 in Figure 14.7b does not decrease along the increase of catalysis temperature. Thus, the carbon-containing species represented by peak 2 is obviously a spectator instead of a stable intermediate.

The surface atomic fraction of Ni or Fe to the total of Ni and Fe were analyzed based on AP-XPS studies. As shown in Figure 14.6a and b, the peak positions of Ni 2p and Fe 2p showed no change along the increase of catalysis temperature from 60 to 400 °C. In addition, the atomic fractions of Ni and Fe to the total of Ni and Fe do not change along the increase of the catalysis temperature (Figure 14.8a), suggesting the lack of segregation of Ni or Fe cations during catalysis at 200–400 °C.

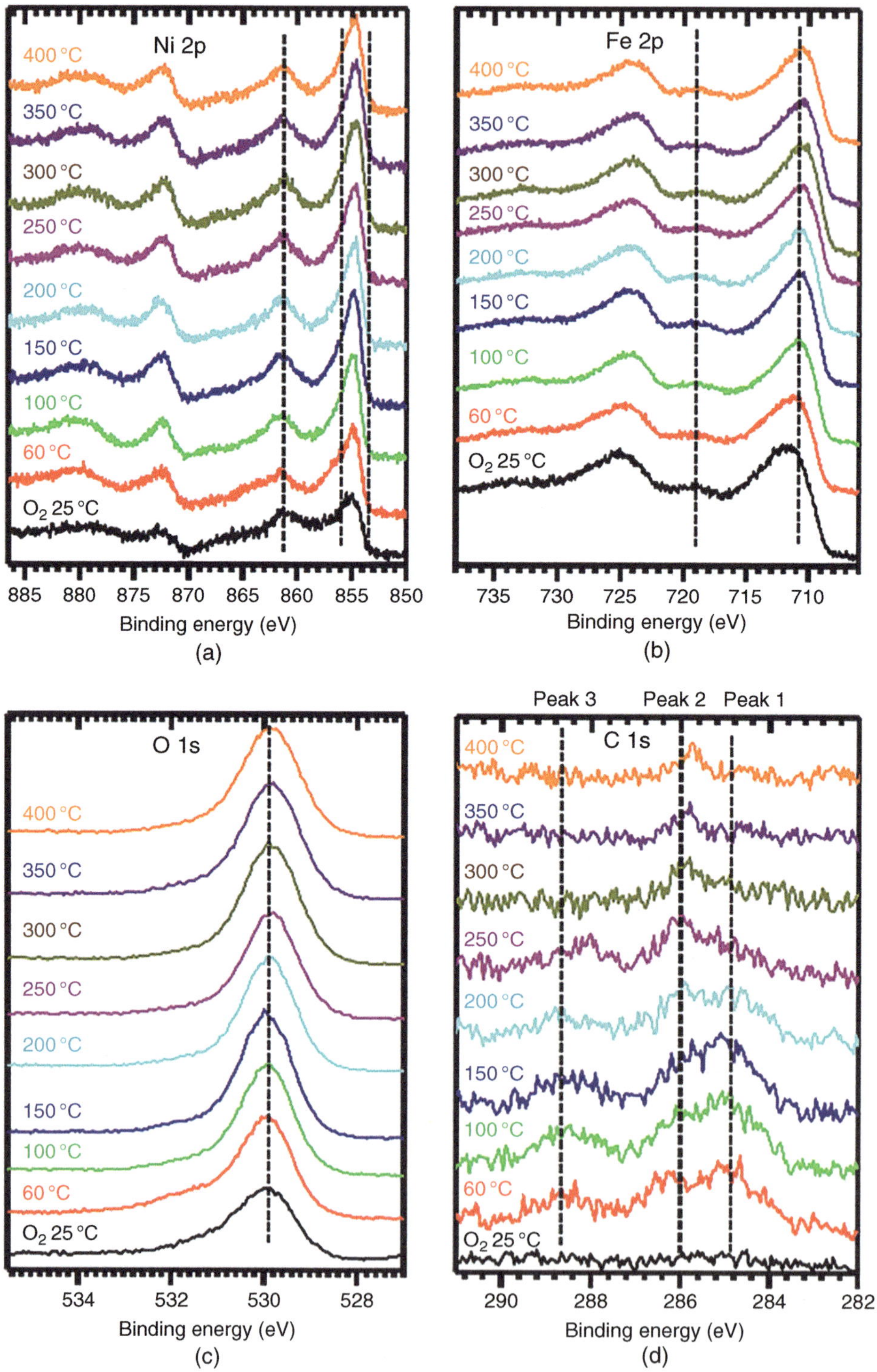

Figure 14.6 XPS peaks of (a) Ni 2p, (b) Fe 2p, (c) O 1s, and (d) C 1s of $NiFe_2O_4$ at different temperatures in 25–400°C in a mixture of 0.1 Torr CH_4 and 0.5 Torr O_2. *Source:* Reproduced with permission from Zhang et al.[3]/Elsevier.

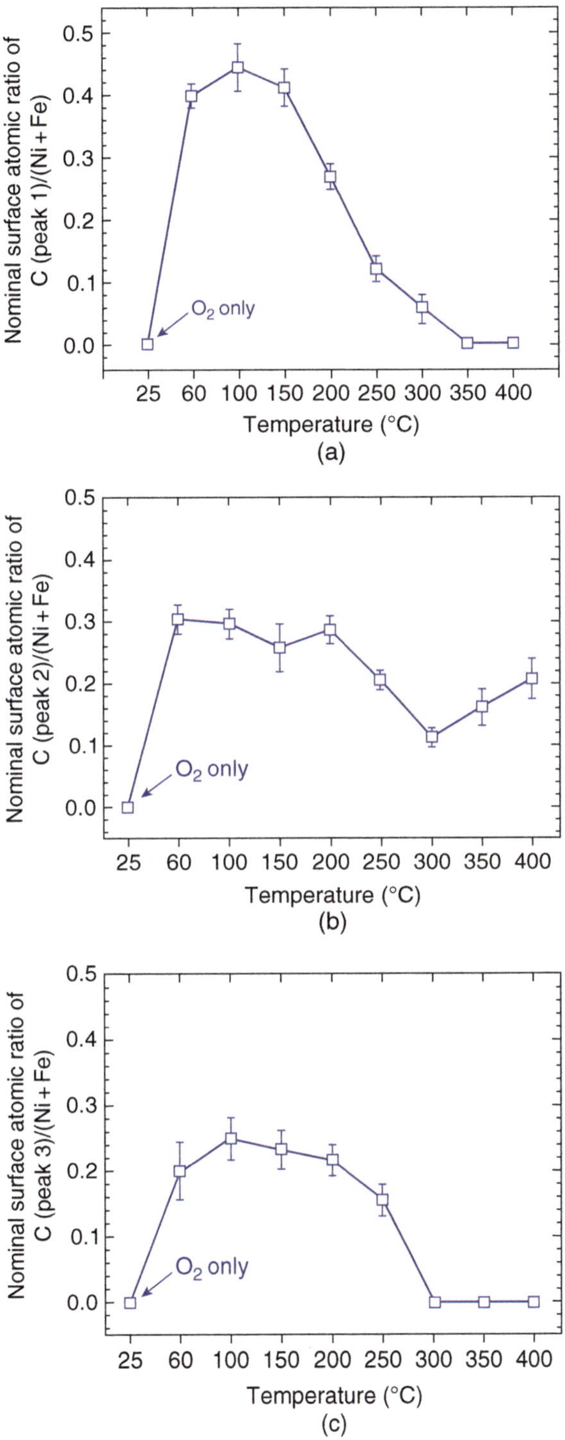

Figure 14.7 Nominal surface atomic ratio, $\frac{\textit{Number of carbon atoms contributing to peak 1}}{\textit{Number of all Ni and Fe atoms on surface}}$ (a), $\frac{\textit{Number of carbon atoms contributing to peak 2}}{\textit{Number of all Ni and Fe atoms on surface}}$ (b), and $\frac{\textit{Number of carbon atoms contributing to peak 3}}{\textit{Number of all Ni and Fe atoms on surface}}$ (c) at different reaction temperatures in the temperature range of 25–400 °C. Photoemission intensities marked with peak 1, peak 2, and peak 3 are contributed from $-CH_n$ ($n = 1-3$), methoxy-like species, and formate-like species, respectively. *Source:* Reproduced with permission from Zhang et al.[3]/Elsevier.

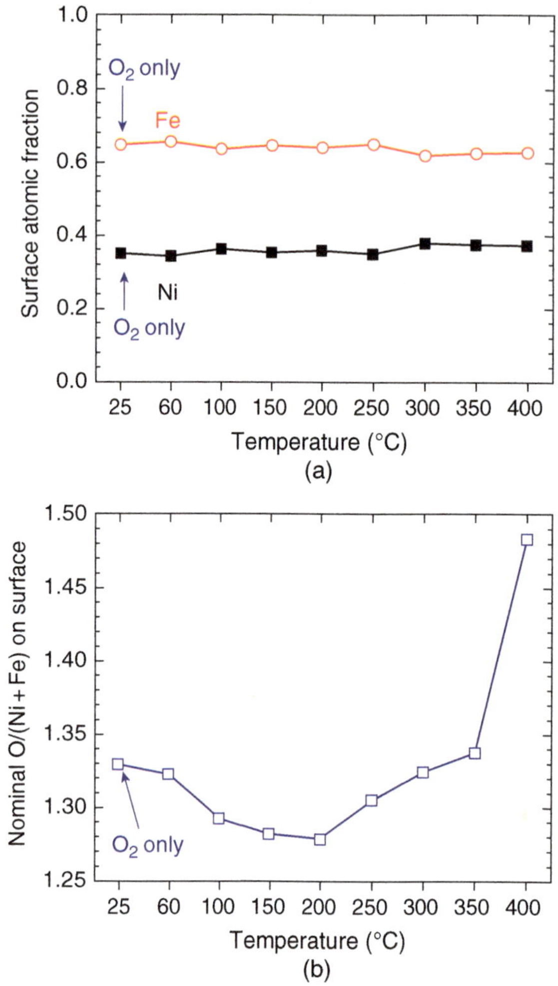

Figure 14.8 Quantitative analysis of surface atomic fraction of Fe or Ni among the total of Ni and Fe element and the ratio of O to the total of Ni and Fe during catalysis at different temperatures. (a) Surface atomic fractions of iron and nickel cations as a function of catalysis temperature. (b) Ratio of number of oxygen atoms to the total number of Ni and Fe atoms on surface as a function of catalysis temperature. *Source:* Reproduced with permission from Zhang et al.[3]/Elsevier.

The atomic ratio of surface oxygen atoms to the total of Ni and Fe atoms exhibits a "u" shape evolution (Figure 14.8b). The atomic ratio of O/(Ni + Fe) first slightly decreases along the increase of catalysis temperature from 60 to 200 °C and then obviously increases from 200 to 400 °C during catalysis. The decrease of the atomic ratio of O/(Ni + Fe) at a relatively low temperature suggests that the surface lattice oxygen participated into the complete oxidation of CH_4. The consumption of surface lattice oxygen atoms implies that a methyl group formed from dissociation of the C—H of CH_4 could couple with a surface lattice oxygen atom to form methoxy. The methoxy could be further transformed to a formate-like intermediate leading to the formation of final products CO_2 and H_2O. Thus, the oxygen vacancies are created by the consumption of surface lattice oxygen in the formation of CO_2 and H_2O. The oxygen vacancies can be filled by atomic oxygen formed through dissociation of molecular O_2. The decrease of O/(Ni + Fe) at 60–200 °C in Figure 14.8b suggests that the generation of atomic oxygen from O_2 is slower than the consumption of oxygen atoms to form product. However, the increase of O/(Fe + Ni) at ≥250 °C suggests that the dissociation of molecular O_2 to atomic oxygen at ≥250 °C is faster than the consumption of surface lattice oxygen for forming products. The "u" shape evolution of atomic ratio of O to the total of Ni and Fe observed by AP-XPS suggests the direct participation of surface lattice oxygen atoms to the catalytic reaction. Obviously, ultrahigh vacuum (UHV) XPS is never able to identify the "u" shape-like evolution of atomic ratio of surface lattice oxygen atoms during catalysis in Figure 14.8b.

In this work, the quantitation function of AP-XPS allows to distinguish a stable intermediate from spectator(s) readily by tracking the evolutions of their quantities as a function of catalysis temperature. The capability of AP-XPS in observing and distinguishing surface species formed under a catalytic condition has largely assisted fundamental understanding of the catalytic mechanism at an atomic scale.

14.3 Complete Oxidation of Methane on NiO with Different Surface Structures

Metal and transition metal oxides are two major categories of heterogeneous catalysts. A great number of works demonstrated the size-dependent and shape-dependent catalytic performances on metal nanoparticle catalysts. Literature has reviewed the size- and shape-dependent catalysis of metal nanoparticles.[4] These dependencies are closely related to coordination environment of a metal atom on surface of a metal catalyst since the local electronic state of a metal atom is essentially determined by its coordination environment.[4] We have recently explored whether the coordination environment of a cation of surface of an metal oxide could be a key factor in modulating catalytic performance for a reaction catalyzed by a metal oxide.[2] NiO was taken as a probe catalyst and CH_4 complete oxidation a probe reaction. It is found that the catalytic activity of NiO for complete oxidation of methane strongly depends on the coordination of Ni cations on a NiO surface.[2]

Three NiO catalysts with relatively uniform, specific but different coordination environments of Ni cations were prepared with three different protocols as described in literature.[2,5,6] They are planar sheet NiO, cubic NiO, and octahedral NiO. As suggested from these high-resolution TEM images in Figure 14.9d–f, the mainly exposed surface of a planar sheet NiO is (110). The optimized surface structure of NiO(110) can be found in Figure 14.10b. On NiO(110), each Ni cation coordinates with four oxygen atoms. In addition, the faces along the edge of the NiO planar sheet are (100) and (111). As the fraction of area of the ultra thin edge of a NiO planar sheet is less than 5%, most of the surface of a planar sheet NiO is (110). Thus, most of the exposed Ni cations of a planar sheet NiO coordinate with four oxygen atoms.

Another type of NiO catalyst is NiO nanocubes (Figure 14.9g and h). Different from the planar sheet NiO, the exposed surface of a NiO cube is (100). Surface structure of NiO(100) is shown in Figure 14.10a. In addition, octahedral NiO nanoparticles were prepared. The surface of an octahedral NiO nanoparticle is (111). Figure 14.10c presents the packing of Ni and O atoms on NiO(111) wherein each Ni cation coordinates with 6 oxygen atoms. Thus, the

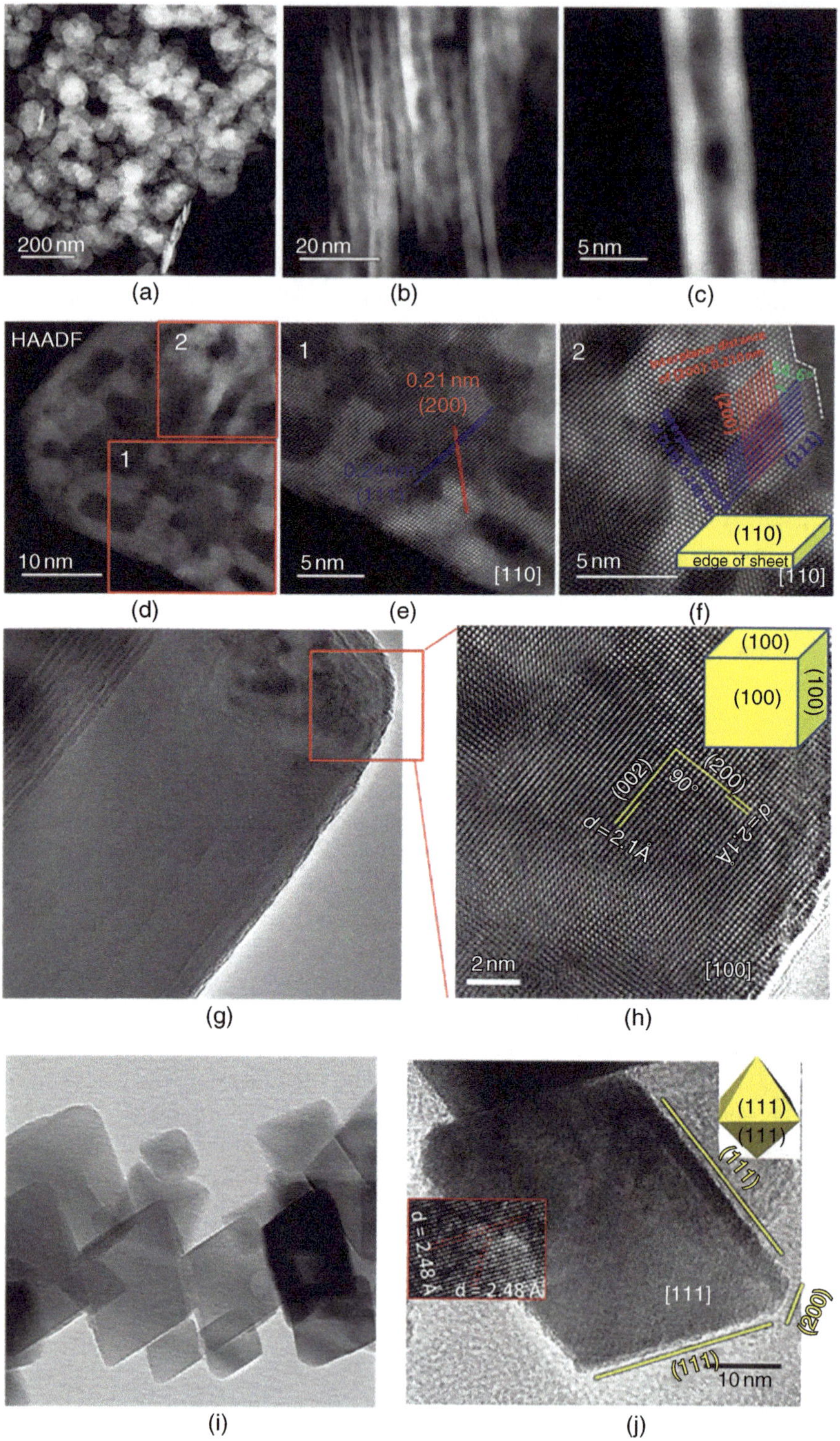

Figure 14.9 High-angle annular dark field scanning transmission electron microscope (HAADF-STEM) studies of morphology and lattice fringes of planar sheet NiO, cubic NiO, and octahedral NiO. (a) HAADF-STEM image of planar sheet NiO. (b) HAADF-STEM image of planar sheet NiO (side view). (c) HAADF-STEM image of a single planar sheet NiO; it is a side view of a planar sheet. (d–f) HAADF-STEM images of a planar sheet of NiO. (g, h) High resolution transmission electron microscope (HRTEM) of cubic NiO; the exposed planes are (100). (i, j) HRTEM images of octahedral NiO; the exposed faces are mainly (111). Insets of (f), (h), and (j) are schematics of planar sheet, cube, and octahedron of NiO, respectively. *Source:* Reproduced with permission from Zhang et al.[2]/American Chemical Society.

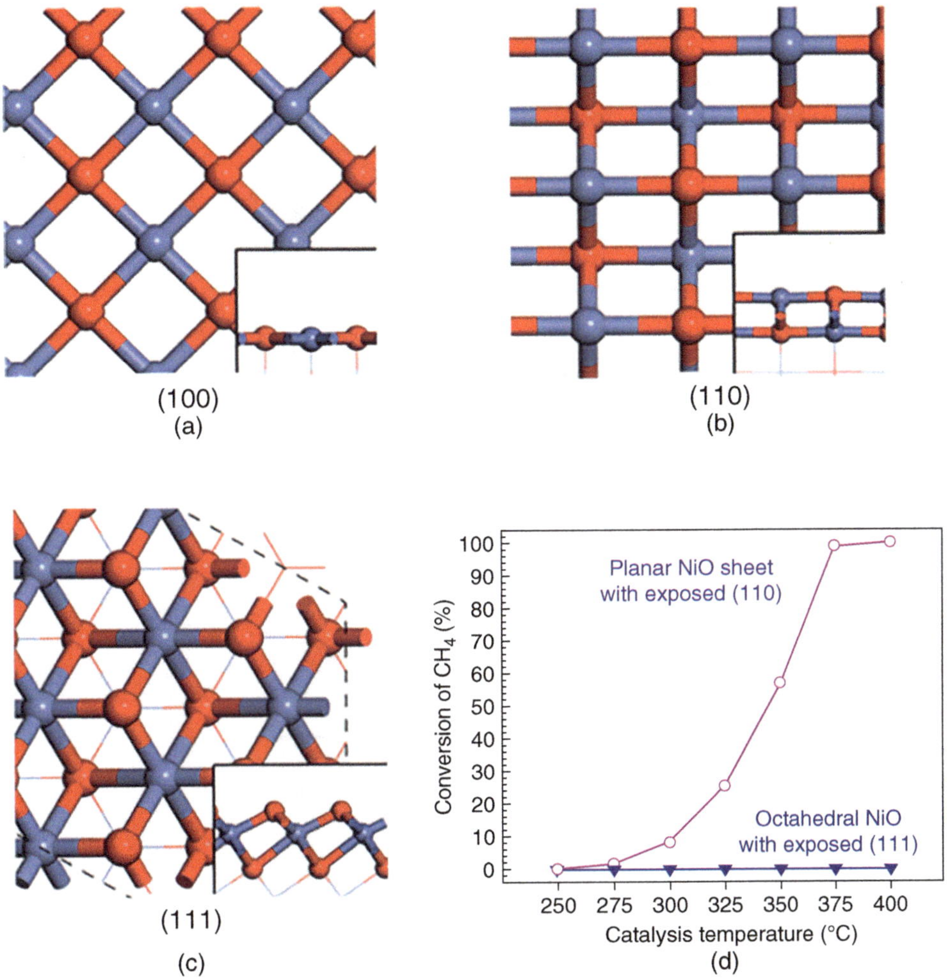

Figure 14.10 Optimized surface structures of NiO(100), NiO(110), and NiO(111) and experimentally measured catalytic performances of NiO. (a) (100) surface of cubic NiO; each Ni cation (in blue) of (100) surface coordinates with given oxygen atoms (in red). (b) (110) surface of planar sheet NiO; each Ni cation (in blue) of (110) surface coordinates with four oxygen atoms (in red). (c) (111) surface of octahedral NiO; each Ni cation of (111) surface coordinates with six oxygen atoms. (d) Conversion of CH_4 by planar sheet NiO and octahedral NiO as a function of temperature in complete oxidation of CH_4 in the temperature range of 250–400 °C; 100 mg of planar sheet NiO and 500 mg of octahedral NiO were used in measurement of catalytic performances in (d). *Source:* Reproduced with permission from Zhang et al.[2]/American Chemical Society.

coordination number of O atoms to a Ni atom on planar sheet, cubic, or octahedral NiO is 4, 5, and 6, respectively. They were chosen as three different catalysts for exploring the dependence of catalytic performance on crystallographic surfaces of a transition metal oxide catalyst. Their catalytic performances and surface chemistries were examined under the same catalytic condition, by which a coordination environment-dependent catalytic performance was uncovered.[2]

Under the same catalytic condition, 100 mg of planar sheet NiO exhibits much high catalytic activity in terms of conversion of CH_4 in the temperature range of 250–400 °C than 500 mg of octahedral NiO (pink line in Figure 14.10d) although 100 mg of planar sheet NiO

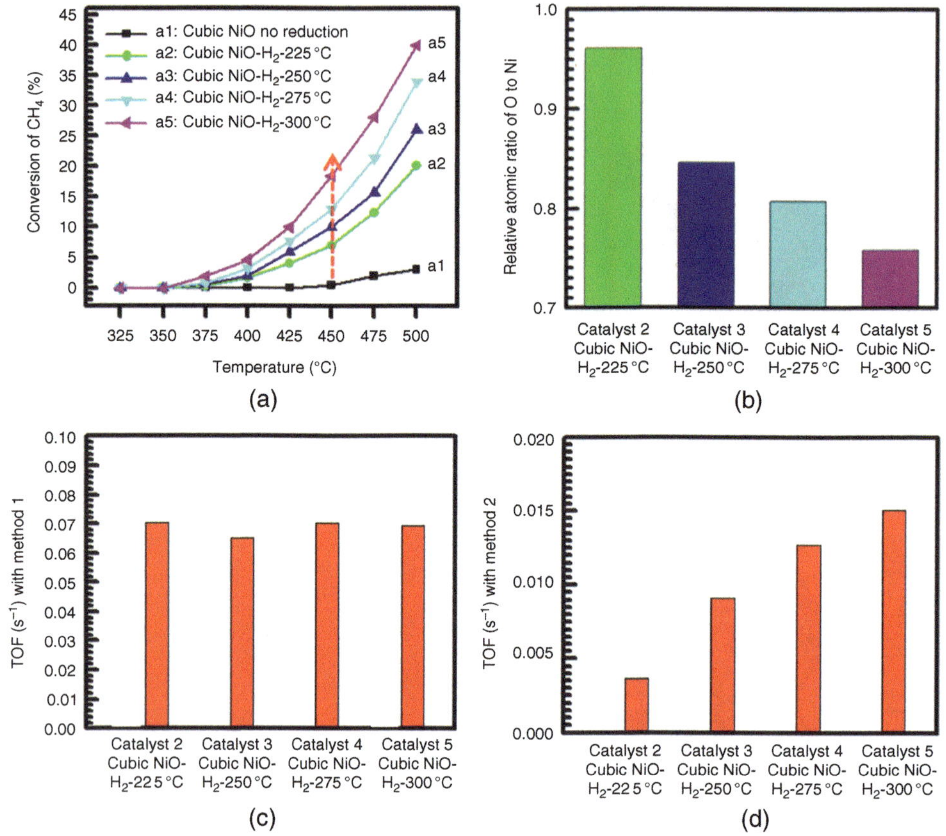

Figure 14.11 Catalysis on partially reduced cubic NiO and their catalytic activity. (a) Catalytic performances of 100 mg of cubic NiO catalyst that was pretreated in 5% H_2 for 1 h at 225, 250, 275, and 300 °C; thus, the four different catalysts are named cubic NiO-H2-225 °C (green), NiO-H2-250 °C (blue), NiO-H2-275 °C (cyan), and NiO-H2-300 °C (magenta). (b) Measured relative surface atomic ratios of O/Ni of the four catalysts. (c) List of TOFs of the four catalysts at catalysis temperature 450 °C; these TOFs were calculated by method 1 by dividing the number of consumed CH_4 molecules at a specific temperature such as 450 °C by the number of Ni atoms in oxygen vacancies measured with XPS. (d) List of TOFs of the four catalysts at catalysis temperature 450 °C; these TOFs were calculated with method 2 by dividing the number of consumed CH_4 molecules at a specific temperature such as 450 °C with the number of all nickel cations on surface measured with XPS. *Source:* Reproduced with permission from Zhang et al.[2]/American Chemical Society.

has a surface area similar to 500 mg octahedral NiO. In fact, octahedral NiO is nearly inert for complete oxidation of CH_4 in the temperature range of 250–400 °C (blue line in Figure 14.10d).

Other than the inertness of (111) surface of octahedral NiO, the low activity of (100) surface of cubic NiO was found in evaluations of catalytic performances of (a) 1.0 mg of planar sheet NiO mixed with purified quartz and (b) 100 mg of cubic NiO that has a surface area roughly similar to 1.0 mg of planar sheet NiO. Catalytic activity on the 100 mg of cubic NiO (total surface area of 1.2 m^2) mixed with quartz was evaluated at 375 °C under a flow of mixture of 20 ml/min 10% CH_4 balanced with Ar and 10 ml/min pure O_2 (99.99%) (Figure 14.11a). The 100 mg of cubic NiO can convert only 4.2% conversion of CH_4 at 375 °C under the same catalytic condition as nanosheet NiO.

The significant difference in catalytic activity in complete oxidation of methane between NiO(110) of nanosheets and (100) of cubic NiO was studied with AP-XPS in parallel.[2] The major finding from AP-XPS studies of the NiO nanosheet and NiO cubes is the formation of both spectator and stable intermediate on (110) of planar sheet NiO but only spectator on (100) of cubic NiO. The lack of a stable intermediate on NiO rationalized the low activity of NiO cubes.

XPS peaks of Ni 2p, O 1s, and C 1s are presented in Figure 14.12a–c. Although there is no significant change of photoemission features of Ni 2p and O 1s along the increase of catalysis temperature in the temperature range of 30–400 °C, significant change of C 1s photoemission feature was found. Before introduction of CH_4 and O_2 to the planar sheet NiO, the surface of the planar sheet at 25 °C was carbon free upon a pretreatment in O_2 at 300 °C. Then, the mixture of CH_4 and O_2 was introduced to the carbon-free catalyst. Interestingly, carbon-containing species formed at 50 °C on the surface once the mixture of CH_4 and O_2 was flowing through the catalyst. It clearly shows high activity of the planar sheet NiO in activation of C–H of CH_4. Evaluation of catalytic performance did not find any products of complete oxidation of CH_4 at 50 °C although carbon-containing species formed on the catalyst surface were observed with AP-XPS. The lack of catalytic activity on planar sheet NiO at 50 °C indicates that the formed species on surface is either a spectator or a stable intermediate that does not have enough energy for crossing the rate-determining step to form product molecules. In fact, the temperature-dependent C 1s feature in Figure 14.12c suggests the peak at 285.0 eV (peak 1) represents a spectator as this peak remained in the whole temperature range of 50–400 °C. Other than peak 1, a relatively weak peak, peak 2 at 288.2 eV, was first formed at 50 °C. As shown in Figure 14.12e, the fraction of the shoulders in terms of peak 2 of Figure 14.12c evolves as a function of catalysis temperature. It presents a volcano evolution. Its fraction increased along the increase of temperature from 50 to 200 °C. Starting at 200 °C, its fraction decreased and eventually peak 2 disappeared at temperature higher than 300 °C. This volcano-like evolution of the peak 2 in Figure 14.12e suggests that the nature of the species represented by the peak 2 of Figure 14.12c is a stable intermediate.

Similar to nanosheet NiO, AP-XPS was used to study cubic NiO under the same condition as nanosheet NiO. The parallel study of cubic NiO uncovered quite different surface species. Figure 14.12f–h is the AP-XPS study of cubic NiO under the same condition as that for the planar sheet NiO. One obvious difference is the lack of the peak 2 at 288.0 eV in the case of cubic NiO (Figure 14.12h). The lack of a stable intermediate in the temperature range of 25–400 °C indicates the inertness of NiO(100) of cubic NiO as shown in Figure 14.11a1.

The difference in surface species on NiO(110) and NiO(100) of NiO cubes under catalytic conditions uncovered by AP-XPS studies rationalized the high activity of NiO(110) of planar sheet NiO but inertness of NiO(100) of cubic NiO. It was hypothesized that the significant difference in catalytic activity could result from the difference in coordination environment of Ni cations in terms of the number of oxygen anions coordinating a Ni cation. As shown in Figure 14.10b and a, the coordination numbers of oxygen atoms of a Ni cation on NiO(110) and NiO(100) are 4 and 5, respectively. This hypothesis was supported

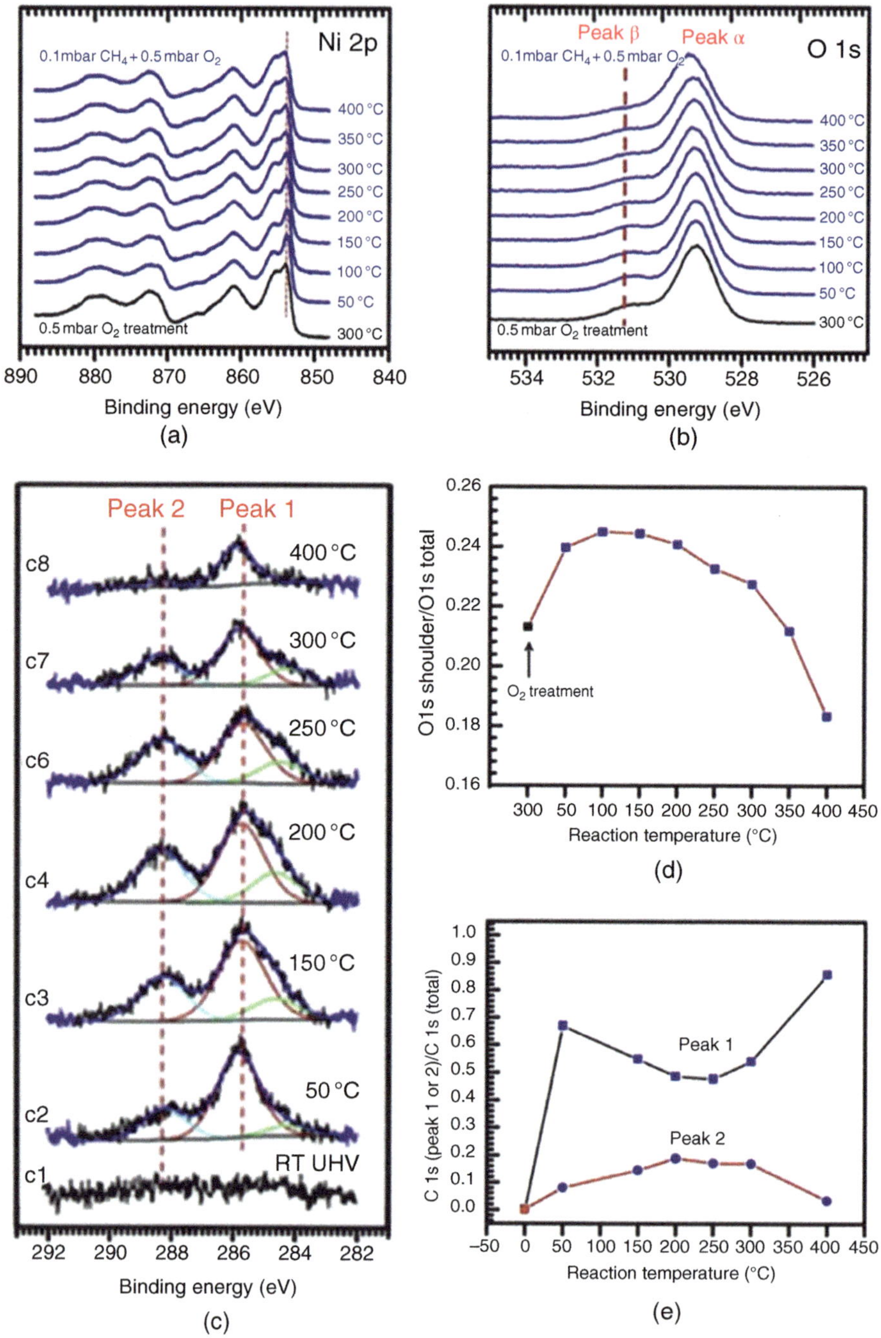

Figure 14.12 Operando studies of (110) surface of planar sheet NiO and (100) surface of cubic NiO during catalysis. (a) Ni 2p of (110) surface of planar sheet NiO in the temperature range of 25–400 °C in the mixture of CH_4 and O_2. (b) O 1s of (110) surface of planar sheet NiO in the temperature range of 25–400 °C in the mixture of CH_4 and O_2 with the same molar ratio as the catalytic condition; the shoulders were marked with a red dashed line. (c) C 1s of (110) surface of planar sheet NiO in the temperature range of 25–400 °C in the mixture of CH_4 and O_2. Spectra were deconvoluted to peaks 1 and 2 and a minor peak at 284.3 eV. (d) Plot of fraction of area of O 1s shoulder to the total area of O 1s photoemission feature on (110) surface of planar sheet NiO during catalysis. (e) Fractions of area of peaks 1 or 2 of C 1s in total area of peaks 1 and 2 as a function of catalysis temperature on (110) surface of planar sheet NiO during catalysis. (f–h) Ni 2p, O 1s, and C 1s of (100) surface of cubic NiO in the same temperature range of 25–400 °C in the same mixture of CH_4 and O_2 as those used for planar NiO sheets. *Source:* Reproduced with permission from Zhang et al.[2]/American Chemical Society.

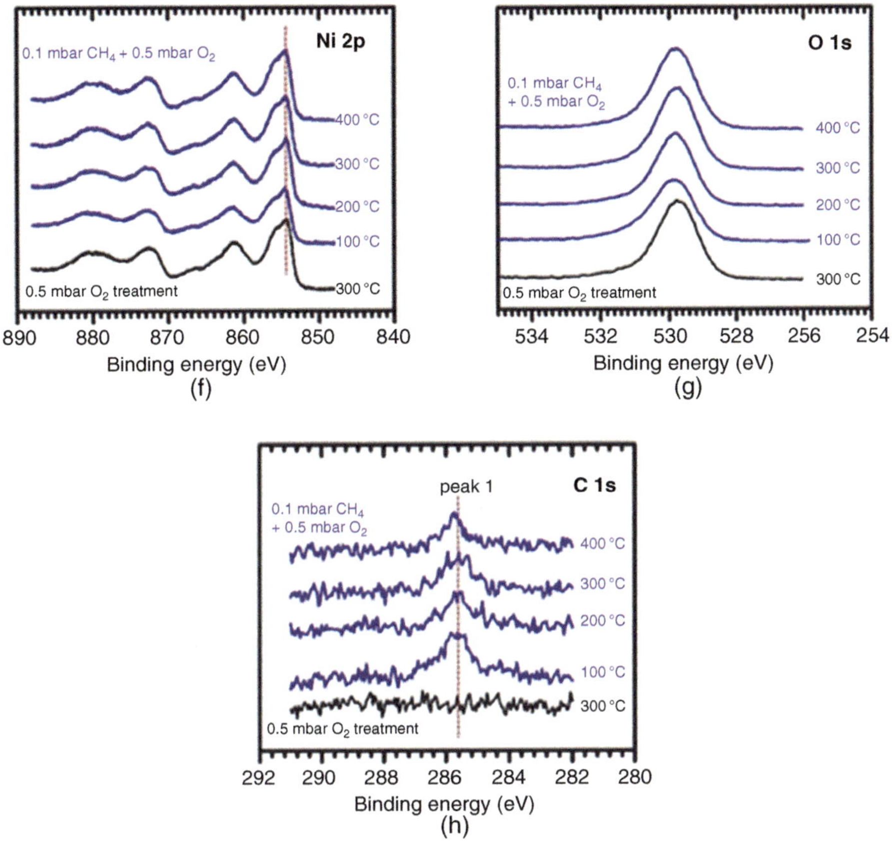

Figure 14.12 (Continued)

by the correlation of catalytic activity of a trimmed NiO(100) surface of cubic NiO and the concentration of Ni atoms with coordination number lower than 5. Modification of the NiO(100) surface was performed through pretreatment of the cubic NiO surface with 5% H_2 in the temperature range of 225–300 °C. As shown in Figure 14.11a, these pretreatments induced activity for complete oxidation of CH_4 on the NiO(100) surface of the cubes. Figures 14.11a2–a5 uncovered a trend that a higher pretreatment temperature generates a surface with a higher activity. Such a trend is marked with a red dash line at 450 °C in Figure 14.11a. AP-XPS played a significant role in finding out the reason driving this trend. Analysis of the surface experienced pretreatments suggests that a surface treated at a higher temperature in 5% H_2 created a surface with a lower O/Ni atomic ratio (Figure 14.11b). In other words, a higher pretreatment temperature in H_2 to cubic NiO generates a surface with a larger concentration of under-coordinated Ni atoms in terms of Ni atoms coordinating with oxygen atoms <5.

Obviously, there are two types of Ni cations on NiO(100) of cubic NiO upon pretreatment in 5% H_2. They are the Ni cations coordinating with four oxygen atoms formed from the H_2 reduction and those Ni cations still coordinating with five oxygen anions. How do we know the Ni cations with CN(Ni-O)<5 instead of those with CN(Ni-O)=5 are active sites. To further check whether these Ni cations with lower coordination number of oxygen atoms [CN(Ni–O) < 5] are active sites on the trimmed cubic NiO, the turnover frequency (TOF) of these four catalysts [cubic NiO-H_2-225 °C, cubic NiO-H_2-250 °C, cubic NiO-H_2-275 °C, and cubic NiO-H_2-300 °C] prepared by pretreatment in 5% H_2 at 225, 250, 275, and 300 °C for the same amount of time were calculated through dividing the amount of consumed CH_4 per second by the number of Ni cations coordinating with four oxygen atoms. The assumption of this calculation is that only Ni cations coordinating with four or fewer oxygen atoms are active sites for complete oxidation of CH_4. TOFs of the four catalysts at the catalysis temperature 450 °C calculated under this assumption are plotted in Figure 14.11c; they are nearly same. The nearly same TOFs of the four catalysts suggest that the assumption that the Ni cations coordinating with four or fewer oxygen atoms are active sites is reasonable. Alternatively, TOFs of the four catalysts were calculated by taking another assumption (only a temporary assumption!) that all Ni cations regardless of the CN(Ni-O) on surfaces of a catalyst are active sites. Figure 14.11d presents the TOFs calculated under this assumption that *both* Ni cations coordinating with five oxygen atoms *and* Ni cations with four or fewer oxygen atoms are active sites for CH_4 complete oxidation. However, the TOFs of the four catalysts calculated under this assumption are quite different (Figure 14.11d). The very different TOFs of the four catalysts suggest the second assumption that all Ni cations on surface are active sites for CH_4 complete oxidation is incorrect since the nature including structure of the active sites of the four catalysts is in fact the same. Thus, we can conclude that these Ni cations coordinating with five oxygen atoms are not active for complete oxidation of methane. This conclusion is supported by the computational studies in literature.[2]

In this work, AP-XPS plays significant roles in fundamentally understanding catalytic mechanism. It identified the stable intermediate for methane complete oxidation on NiO(110) with its quantitative analysis of surface species. By tracking the temperature-dependent evolutions of surface species of two catalysts, NiO(110) versus NiO(100), the fundamental understanding of coordination environment-dependent catalysis on transition metal oxides was demonstrated.

References

1 Tao, F., Shan, J., Nguyen, L. *et al.* 2015. "Understanding complete oxidation of methane on spinel oxides at a molecular level." *Nat. Commun.* 6, 7798. doi:10.1038/ncomms8798.
2 Zhang, S., Li, F., Wang, Z. *et al.* 2021. "Coordination number-dependent complete oxidation of methane on NiO catalysts." *ACS Catal.* 11, 9837–9849.
3 Zhang, S., Shan, J., Nie, L. *et al.* 2016. "In situ studies of surface of $NiFe_2O_4$ catalyst during complete oxidation of methane." *Surf. Sci.* 648, 156–162.
4 Cao, S., Tao, F., Tang, Y. *et al.* 2016. "Size- and shape-dependent catalytic performances of oxidation and reduction reactions on nanocatalysts." *Chem. Soc. Rev.* 45, 4747–4765.
5 Wang, C., Zhao, Y., Su, D. *et al.* 2017. "Synthesis of NiO nano octahedron aggregates as high-performance anode materials for lithium ion batteries." *Electrochim. Acta* 231, 272–278.
6 Liu, B., Wei, A., Zhang, J. *et al.* 2012. "Synthesis and photocatalytic activity of monodisperse single crystalline NiO octahedrons by the selective adsorption of Cl– ions." *J. Alloy. Compd.* 544, 55–61.

15

Partial Oxidation of Methanol

Methanol is available from numerous sources including renewable biomass. Compared to electrooxidation of H_2, direct oxidation of methanol through electrochemical process to CO_2 and H_2O encounters much higher activation barriers in terms of sloppy kinetics. Thus, the chemical transformation of methanol to H_2 is a valuable process before an electrooxidation of H_2. Chemical transformation of methanol to H_2 can be done through partial oxidation of methanol. Singly dispersed bimetallic sites Pd_1Zn_3 supported on ZnO were reported as a highly active and selective catalyst in the production of H_2 through partial oxidation of methanol at a relatively low temperature, 250 °C.[1]

15.1 Partial Oxidation of Methanol on Pd_1Zn_3/ZnO

As shown in Figure 15.1a, ZnO is not active for this reaction at a temperature lower than 330 °C.[2] Upon 0.01 wt% Pd was loaded on the surface of ZnO, high activity was observed at 250 °C (Figure 15.1b).[1] The activity results from the formed Pd_1Zn_3 sub-nanometer clusters, called singly dispersed bimetallic sites schematically shown in Figure 15.2b. As shown in Figure 15.1b, the selectivity for producing CO and H_2 is about 90% in the temperature range of 250–350 °C. Compared to 100 mg of 0.01 wt% Pd/ZnO catalyst, 5.0 wt% Pd/ZnO is active for the conversion of methanol but the selectivity for production of H_2 is obviously lower than 0.01 wt% Pd/ZnO. The win of 0.01 wt% Pd over 5.0 wt% Pd was a surprise definitely. This surprising difference in catalytic performance was confirmed through the measurement of turnover frequencies (TOFs) of the two catalysts through kinetic studies.[1] 10 mg of either 0.01 wt% Pd/ZnO or 5.0 wt% Pd/ZnO mixed with purified quartz was used in kinetic studies in the temperature range of 200–260 °C where the conversions of CH_3OH are lower than 10%. TOF in terms of the number of H_2 molecules produced from each surface Pd atom of the catalyst per second under a kinetics-controlled region was measured (Figure 15.1e and f). Very clearly, the TOF of 0.01 wt% Pd/ZnO (Figure 15.1f) is nearly 400 times of 5.0 wt% Pd/ZnO at the same temperature (Figure 15.1e). The distinctly different TOFs between 0.01 wt% Pd/ZnO and 5.0 wt% Pd/ZnO clearly suggest that the electronic states of Pd atoms of the two catalysts are obviously different.

Application of Ambient Pressure X-ray Photoelectron Spectroscopy to Catalysis, First Edition. Franklin Tao.

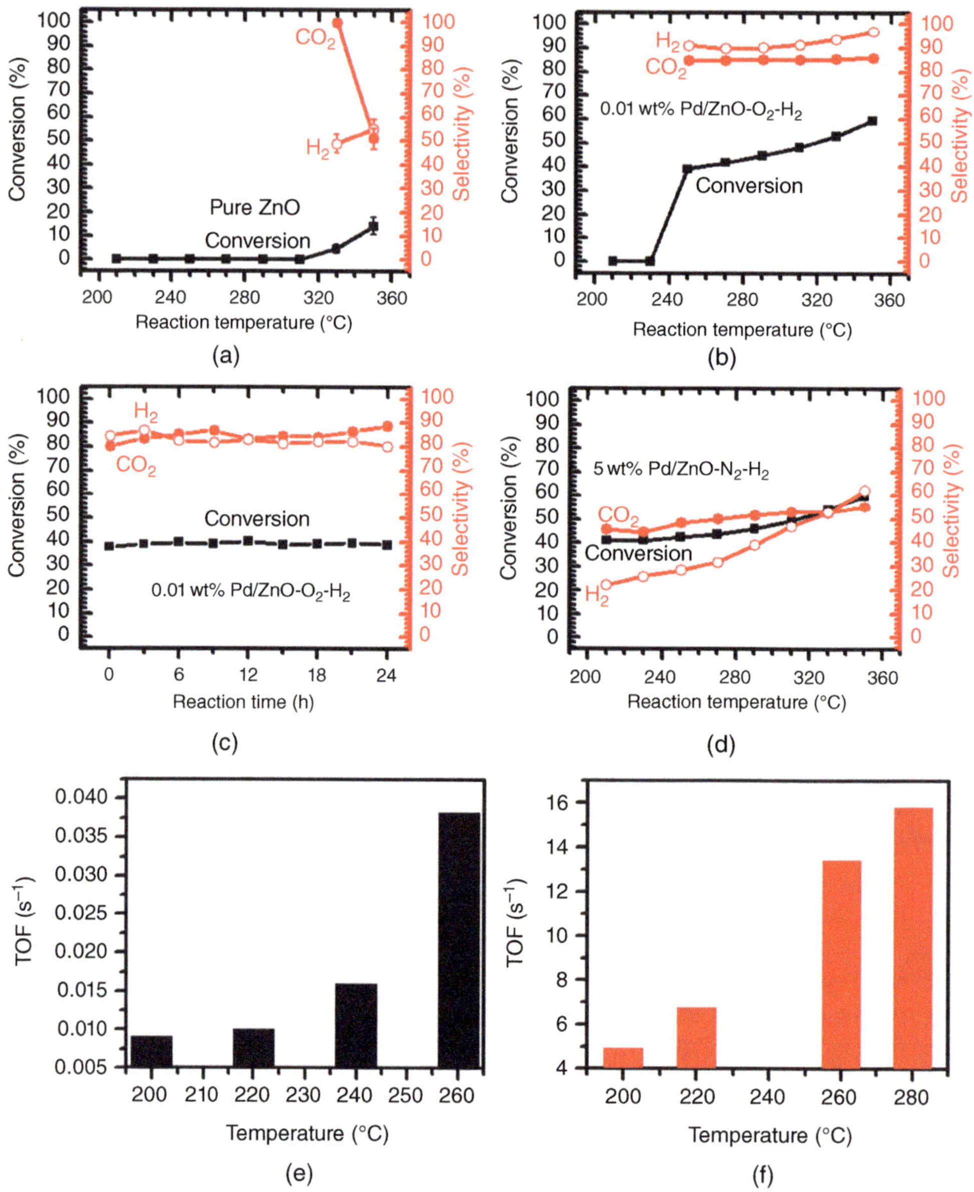

Figure 15.1 Catalytic performances of (a) pure ZnO, (b) 0.01 wt% Pd/ZnO–O_2–H_2 at 250 °C, (c) the stability test of 0.01 wt% Pd/ZnO–O_2–H_2 at 250 °C, (d) 5.0 wt % Pd/ZnO–N_2–H_2, (e) TOF of producing H_2 on Pd-Zn/ZnO, and (f) TOF of producing H_2 on Pd_1Zn_3/ZnO measured under kinetics-controlled regime. 20 mg catalyst was used in each catalytic test of the four catalysts. Liquid methanol was introduced at a flow rate of 10 μl/min through an high performance liquid chromatography (HPLC) pump. The injected liquid methanol was vaporized by keeping the gas line (1/8 in. stainless steel tube) between the source of methanol (HPLC pump) and the reactor at 110 °C. The vapor of methanol was delivered to the reactor by flowing 5% O_2 balanced with Ar (60 ml/min) into the line of methanol vapor. In kinetics studies of (e, f), 10 mg catalyst was used. *Source:* Reproduced with permission from Tang et al.[1]/American Chemical Society.

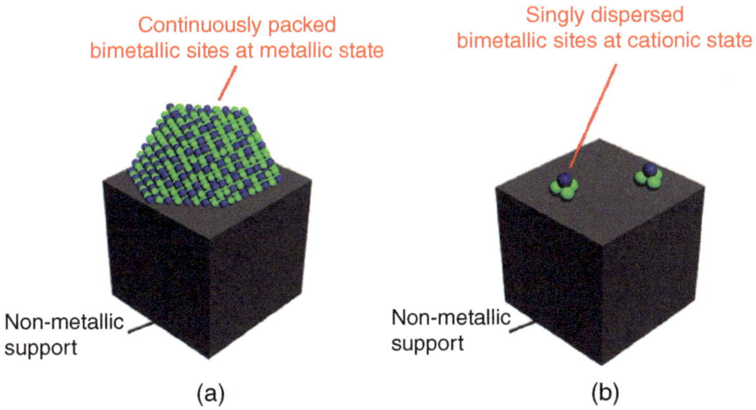

Figure 15.2 Schematics showing (a) nanoscale alloy particle ($M^{\alpha}-M^{\beta}$) consisting of continuously packed bimetallic sites that are at metallic state and (b) singly dispersed, bimetallic site $M_1^{\alpha}-M_n^{\beta}$ (which is at a cationic state) immobilized on a nonmetallic support. M^{α} (blue) and M^{β} (green) are two different metal elements. *Source:* Reproduced with permission from Tang et al.[1]/American Chemical Society.

In site/operando studies of AP-XPS and EXAFS uncovered this difference in chemical and coordination environments of Pd atoms.[1] During the catalysis, Pd atoms of 0.01 wt% Pd/ZnO are at a *cationic* state and they are singly dispersed and bonded to three Zn atoms and oxygen atoms on average. However, the Pd atoms of 5.0 wt% Pd/ZnO are at a metallic state and coordinate with Zn atoms, forming Pd-Zn alloy nanoparticles. This distinct difference in chemical and coordination environments of Pd atoms is responsible for this huge difference in catalytic activity and selectivity. The surprisingly high catalytic activity and selectivity on a singly dispersed bimetallic site suggests a new approach to improve catalytic performance of a bimetallic catalyst. Formation of singly dispersed bimetallic sites is a particularly valuable new approach for developing new catalysts containing a minimal amount of noble metal since noble metals are typically prohibitively expensive.

The singly dispersed Pd_1Zn_3 on ZnO was formed through a controlled reduction by 5% H_2 for ZnO anchored with singly dispersed Pd atoms upon annealing catalyst precursor (0.01 wt% Pd/ZnO) at 250 °C in 5% O_2. The formation process of singly dispersed bimetallic sites was tracked with AP-XPS and its online mass spectrometry. As shown in Figure 15.3a, Pd 3d photoemission feature experienced obvious change in the steps of preparation of the Pd_1Zn_3 anchored on ZnO. Figure 15.3a1 is the Pd 3d XPS peak at 337.3 eV for the 0.01 wt% Pd/ZnO at 250 °C in 5% O_2. Obviously, Pd atoms during pretreatment at 250 °C in O_2 is at cationic state. Under this oxidizing condition, Pd atoms are anchored on ZnO surface through directly bonding to surface lattice oxygen atoms of ZnO though Pd-O bonds. However, in 5% H_2 at 300 °C (Figure 15.3a2), the Pd 3d peaks are obviously downshifted to 335.4 eV. This downshift of 1.9 eV clearly shows that the cationic Pd atoms were reduced to some extent. Notably, potential formation of Pd nanoparticles or Pd–Zn alloy nanoparticles was excluded since the observed Pd $3d_{5/2}$ peak position at 300 °C in 5% H_2 is 335.4 eV that is higher than the 334.6 eV of metallic Pd. Systemic studies of the surface of 0.01 wt% Pd/ZnO in 5% H_2 at various temperatures suggest that 300 °C is a reasonably high temperature for remaining Pd atoms at a nonmetallic state.[1] Upon switching the gas environment from

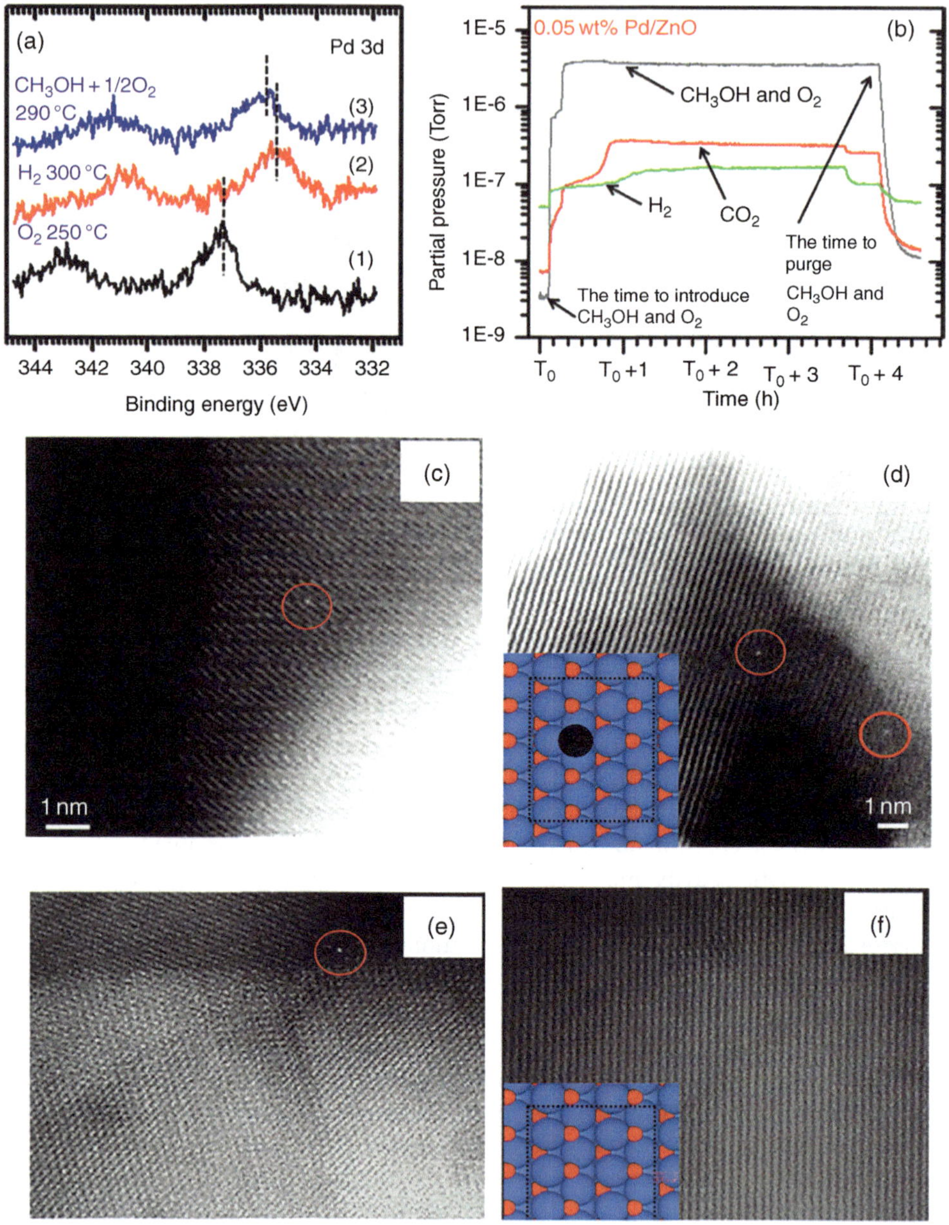

Figure 15.3 Operando studies of the surface of 0.05 wt% Pd/ZnO during pretreatment and during catalysis with AP-XPS and simultaneous tracking of partial pressure of gas components around the catalyst surface during catalysis with mass spectrometer and HAADF-STEM studies of used catalysts. (a) Pd 3d of Pd atoms of 0.05 wt% Pd/ZnO studied with AP-XPS during pretreatment or during catalysis. Spectrum (1) Pd 3d in O_2 at 250 °C; spectrum (2) Pd 3d in H_2 at 300 °C after annealing in O_2 at 250 °C; spectrum (3) Pd 3d during catalysis at 290 °C in the mixture of methanol and O_2 (after pretreatment in O_2 at 250 °C and then H_2 at 300 °C). (b) Analysis of composition of gas near to the catalyst surface during catalysis at 290 °C with mass spectrometer on online mode while AP-XPS was analyzing the catalyst surface during methanol partial oxidation. (c–e) HAADF-STEM images of 0.01 wt% Pd/ZnO–O_2–H_2-catalysis. The red circles in (c–e) mark single Pd atoms anchored on ZnO. Inset in (d) schematically shows that Pd_1 atom bonds to ZnO surface. (f) HAADF-STEM image of ZnO; inset in (f) is the surface structure of pure ZnO. *Source:* Reproduced with permission from Tang et al.[1]/American Chemical Society.

5% H_2 at 300 °C to the catalytic condition in terms of a mixture of CH_3OH and O_2 at 290 °C, the peak position of Pd $3d_{5/2}$ even upshifts 0.3 eV. Thus, the AP-XPS studies of the catalyst surface during catalysis excluded the formation of Pd–Zn alloy nanoparticles. EXAFS studies show that Pd atoms of 0.01 wt% Pd/ZnO at 290 °C in the mixture of CH_3OH and O_2 are still singly dispersed and each Pd atom directly bonds with three Zn atoms on average.[1]

This work demonstrated the benefit of AP-XPS studies to find out the optimum pretreatment temperature. AP-XPS clearly found out the optimum reduction temperature for preparation of Pd1Zn3 nanoclusters. AP-XPS characterized the electronic state of Pd atoms of the 0.01 wt% Pd/ZnO catalyst during catalysis at 290 °C. An online mass spectrometer installed on the AP-XPS system can investigate catalytic activity and selectivity, by which a direct correction between surface chemistry of a catalyst and its corresponding catalytic performance can be readily established.

The unprecedented high activity for production of H_2 through partial oxidation of methanol on Pd_1Zn_3 anchored on ZnO was rationalized through theoretical simulation of methanol partial oxidation on Pd_1Zn_3/ZnO and Pd–Zn alloy on ZnO.[1,3] Based on transmission electron spectroscopy (TEM) studies, ZnO(10–10) was mainly the surface anchoring Pd_1Zn_3 nanoclusters and Pd–Zn alloy nanoparticles. Here Pd_1Zn_3 is a singly dispersed bimetallic site consisting of a Pd atom and three Zn atoms bonding to the Pd atom. It is not an alloy with molar ratio of 1 : 3.[1] A notable difference is that the adsorption energy of CH_3OH on Pd and Zn of Pd_1Zn_3/ZnO (–1.02 eV) is higher than that on Pd–Zn of Pd–Zn alloy surface (–0.40 eV), suggesting that one benefit of cationic Pd atom of Pd_1Zn_3/ZnO is its stronger binding to CH_3OH than Pd–Zn alloy. The higher binding energy to CH_3OH on Pd_1Zn_3/ZnO suggests a low activation barrier for dehydrogenation of CH_3OH in contrast to Pd–Zn alloy supported on ZnO. Computational studies simulated the necessary elementary steps including adsorption of methanol and O_2 molecules, stepwise dehydrogenation of CH_3OH to CO and atomic H, dissociation of O_2 to atomic O, coupling between CO and atomic O to form CO_2, and coupling of H atoms to form H_2.[1] These simulations found that the activation barrier of the rate-determining step of methanol partial oxidation on ZnO, Pd–Zn alloy/ZnO, and Pd_1Zn_3/ZnO are 2.78, 1.33, and 0.80 eV, respectively. The lower activation barrier on Pd_1Zn_3/ZnO in contrast to Pd-Zn/ZnO is consistent with the high adsorption energy of CH_3OH and intermediates on Pd_1Zn_3/ZnO compared to Pd-Zn/ZnO. The lower activation barrier of rate-determining step on Pd_1Zn_3/ZnO compared to Pd–Zn alloy/ZnO results from the positively charged Pd atoms in Pd_1Zn_3/ZnO uncovered by AP-XPS studies during catalysis (blue spectrum in Figure 15.3a). The positive charge of Pd_1 atoms decreased the activation barriers for methanol dehydrogenation steps due to its stronger binding to CH_3OH and the subsequently formed intermediates.

15.2 Partial Oxidation of Methanol on Ir_1Zn_3/ZnO

AP-XPS studied the methanol partial oxidation on another important catalyst, Ir_1Zn_3/ZnO which is designated as 0.01wt%Ir/ZnO as well.[2] The active catalytic site is a singly dispersed bimetallic site consisting of one Ir and three Zn atoms. The single dispersion of Ir atoms on ZnO was well confirmed with high-angle annular dark-field scanning transmission electron microscopy (HAADF-STEM) (Figure 15.4). This catalyst is quite active for methanol partial oxidation at a temperature higher than 250°C (Figure 15.5b). The selectivities for

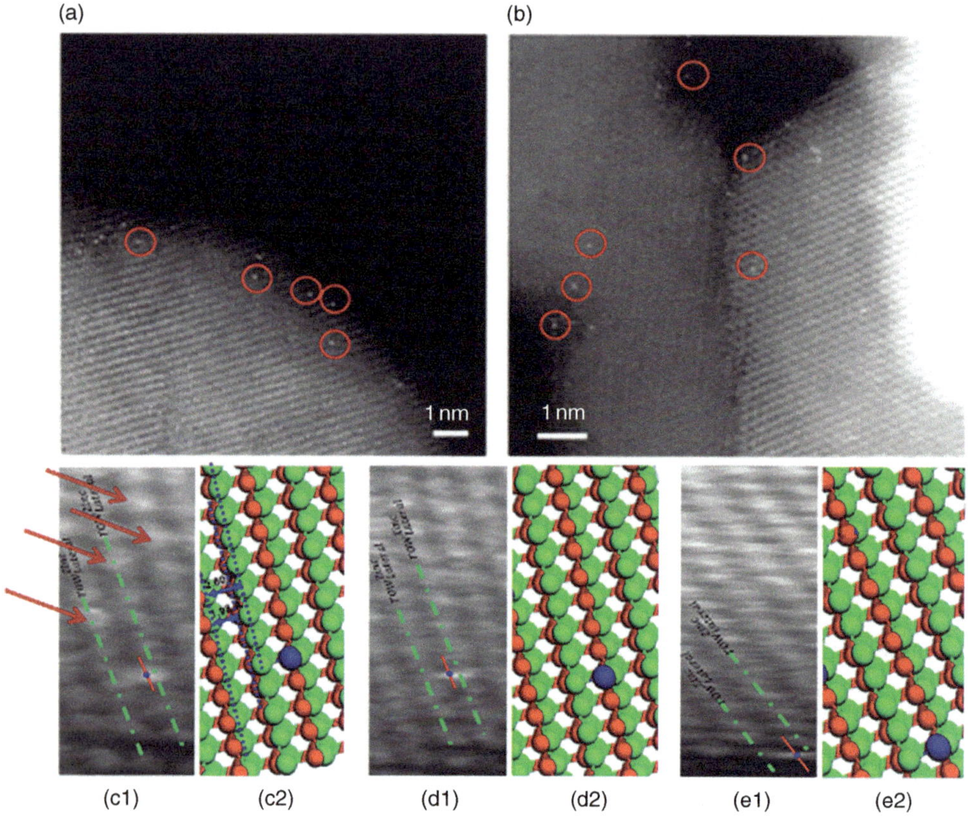

Figure 15.4 HAADF-STEM images of the used 0.01 wt% Ir/ZnO catalyst. (a) and (b) HAADF-STEM micrographs of used 0.01 wt% Ir/ZnO after being subjected to a evaluation of catalytic performance for methanol partial oxidation at 350 °C. (c1, d1, e1) Enlarged HAADF-STEM images of used 0.01 wt% Ir/ZnO. (c2, d2, e2) Surface structure models corresponding to (c1), (d1), and (e1). In (c1), (d1), and (e1), the brighter spots marked with red lines are assigned to Ir atoms; two parallel green dashed lines mark two neighboring zinc rows. Red arrows in (c1) marked these areas with relatively high contrast; however, their contrasts are lower than those marked with red lines in (c1). *Source:* Nguyen et al.[2]/American Chemical Society.

producing H_2 and CO_2 are nearly 95–98% at a temperature >280 °C. Kinetic studies of this catalysis suggest a low apparent activation barrier on Ir_1Zn_3/ZnO.[2] Based on the activity and selectivity for production of H_2 under kinetics-controlled region, turnover rate for producing H_2 was calculated. It produces about 13 H_2 molecules per Ir atom-based site per second.[2] This is a catalyst with catalytic performance comparable or even better than Pd_1Zn_3 anchored on ZnO.[1] The high catalytic activity results from the quite different chemical and coordination environments of Ir atoms of 0.01 wt% Ir/ZnO in contrast to Ir atoms of 5.0 wt% Ir/ZnO. The singly dispersed Ir_1 atom bonding with Zn and O atoms in 0.01 wt% Ir/ZnO exhibit a quite different chemical and coordination environment in contrast to a Ir atom of Ir–Zn alloy nanoparticles of 5.0 wt% Ir/ZnO.[2] The distinctly different chemical environments of Ir atoms on the two catalysts (0.01wt%Ir/ZnO versus 5wt%Ir/ZnO) were confirmed by AP-XPS that identified electronic states of Ir atoms of the two catalysts. As shown in Figure 15.6a, the Ir $4f_{7/2}$ of 0.01 wt% Ir/ZnO during catalysis at 290 °C is 61.7 eV

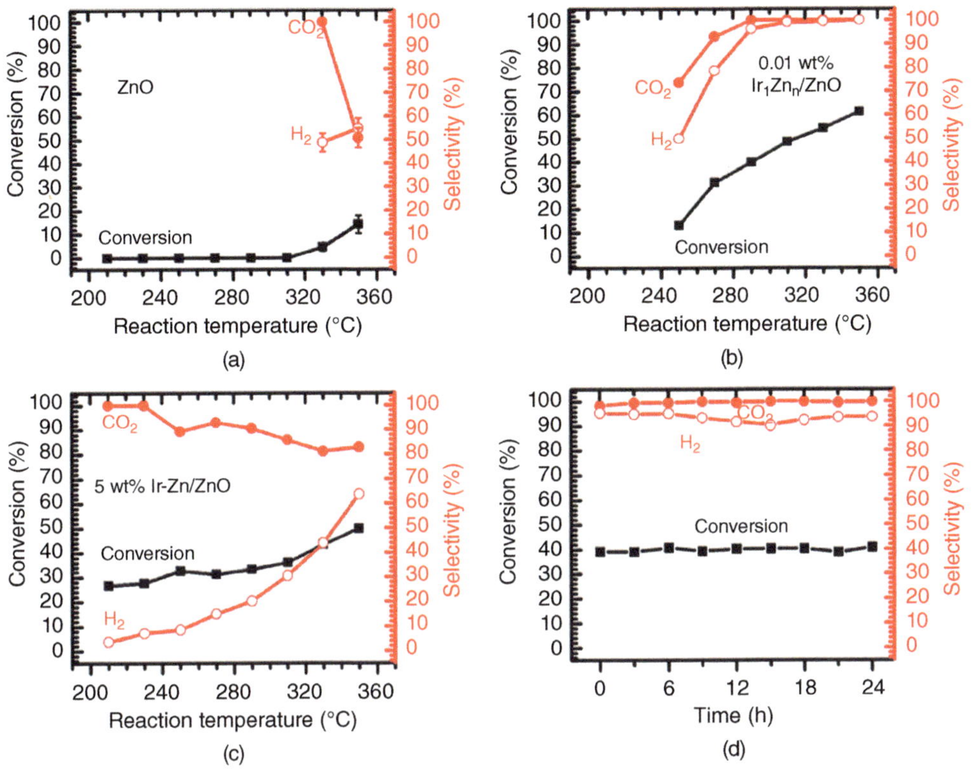

Figure 15.5 Catalytic performances of (a) 20 mg of ZnO, (b) 20 mg of 0.01 wt% Ir/ZnO, (c) 20 mg of 5.0 wt% Ir/ZnO as a function of reaction temperature, and (d) the stability test of 0.01 wt% Ir/ZnO at 290 °C. Methanol (10 μl/min) and O_2 (60 ml/min, 5% O_2 balanced with Ar) with a molar ratio of about 2 : 1 were mixed and then delivered to the reactor. *Source:* Reproduced with permission from Nguyen et al.[2]/American Chemical Society.

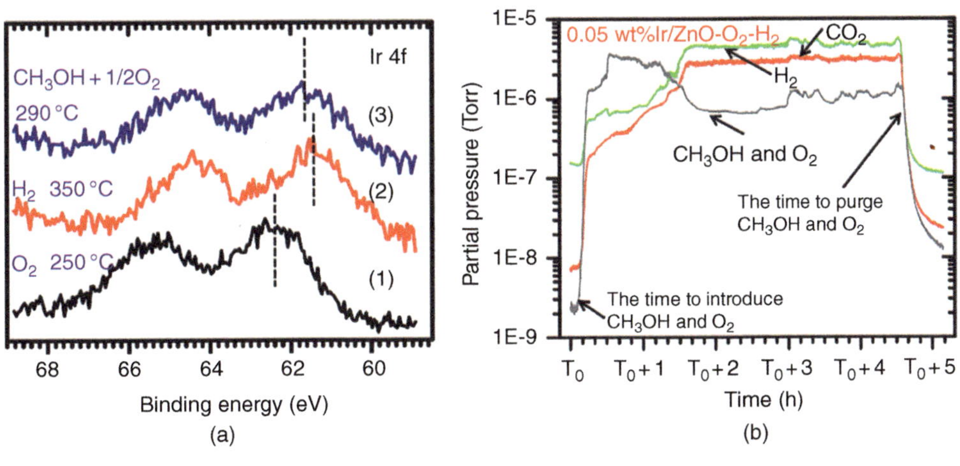

Figure 15.6 AP-XPS studies of Ir 4f of used 0.05 wt% Ir/ZnO and during observation of products formed from 0.05 wt% Ir/ZnO. In (a), Curve 1 was collected during pretreatment in 5% O_2 at 250 °C. Curve 2 was collected during pretreatment in 5% H_2 at 350 °C after the pretreatment at 250 °C in 5% O_2. Curve 3 was collected during methanol partial oxidation at 290 °C after consecutive pretreatments first in 5% O_2 at 250 °C and then in 5% H_2 at 350 °C. In (b), the detection of gas composition using an on-line mass spectrometer during catalysis while AP-XPS was characterizing the catalyst under the catalytic condition. *Source:* Reproduced with permission from Nguyen et al.[2]/American Chemical Society.

that is higher than 60.9 eV of metallic Ir atoms, suggesting the electronic state of the Ir atoms of Ir_1Zn_3/ZnO is cationic or nonmetallic during partial oxidation of methanol.

AP-XPS tracked the evolution of electronic state of Ir atoms of Ir_1Zn_3/ZnO during preparation process (Figure 15.6a). Upon the drying process of the precipitate collection upon centrifugation, the precursor was annealed to 250 °C in 5% O_2 for anchoring the Ir atoms on surface of ZnO. During this annealing, the Ir atoms are definitely at cationic state with a high binding energy of 62.4 eV (black spectrum in Figure 15.6a). A followed reduction to the catalyst precursor at 350 °C in H_2 partially reduced Ir atoms, evidenced by the down-shift of Ir $4f_{7/2}$ to 61.4 eV (red spectrum in Figure 15.6a). The Ir atoms at 350 °C in 5% H_2 are not at metallic state since it is higher than 60.9 eV of metallic Ir. The lack of formation of Ir nanoparticles or Ir–Zn alloy nanoparticles was further confirmed by the HAADF-STEM of the used catalyst.[2] Interestingly, in the mixture of methanol vapor and O_2 at 290 °C in terms of catalytic condition, Ir $4f_{7/2}$ is at 61.7 eV that is 0.8 eV higher than 60.9 eV of metallic Ir. Obviously, Ir atoms of 0.01 wt% Ir/ZnO in terms of Ir_1Zn_3/ZnO during catalysis are at cationic state. Thus, the AP-XPS studies suggest that the extremely high activity and high selectivity of Ir_1Zn_3/ZnO for production of H_2 from partial oxidation of methanol result from the singly dispersed bimetallic sites and identified the appropriate pretreatment temperature for forming singly dispersed bimetallic site.

References

1 Tang, Y., Zhang, S., Rawal, T. B. *et al.* 2020. "Atomic-scale structure and catalysis on positively charged bimetallic sites for generation of H_2." *Nano Lett.* 20, 6255–6262.

2 Nguyen, L., Zhang, S., Tao, L. *et al.* 2019. "Ir1Znn bimetallic site for efficient production of hydrogen from methanol." *ACS Sustain. Chem. Eng.* 7, 18793–18800.

3 Rawal, T. B., Acharya, S. R., Hong, S. *et al.* 2018. "High catalytic activity of Pd1/ZnO($10\bar{1}0$) toward methanol partial oxidation: a DFT+KMC Study." *ACS Catal.* 8, 5553–5569.

16

Partial Oxidation of Methane

The major component of natural gas or shale gas is CH_4. Transformation of CH_4 to syngas is a process before any value-added chemicals including alcohols and liquid hydrocarbons can be produced through the Fischer-Tropsch synthesis. Partial oxidation of CH_4 to CO and H_2 is a route to utilize the inexpensive source, CH_4, to produce syngas, a mixture of CO and H_2, to produce value-added chemicals. Partial oxidation of methane can be performed on a great number of catalysts including CeO_2 doped with 5.0wt% Pd, 5.0wt% Pt, or 5.0wt% Rh. Catalytic partial oxidation of methane on the three catalysts was studied with AP-XPS.[1] These catalysts were prepared with a solution-based hydrothermal procedure.[1] After a hydrothermal reaction at 100 °C for 24 hours and cooling to room temperature, a catalyst precursor was prepared. Then, the sample was dried at 100 °C in air. Calcination at 500 °C in air was performed, offering three catalysts including Pd-CeO_2–air, Pt-CeO_2–air, and Rh–CeO_2–air. Similarly, calcination at 500 °C in 5% H_2 formed three different catalysts including Pd–CeO_2–H_2, Pt–CeO_2–H_2, and Rh–CeO_2–H_2. In terms of nomenclature of these catalysts in this chapter, the first part of the name of a catalyst is the element of the supported metal (either in cationic or metallic state); the term in the middle is CeO_2, the support; the last part is the gas used for pretreatment for this catalyst before partial oxidation of methane is performed.

16.1 Partial Oxidation of Methane on Pd/CeO_2

As shown in Figure 16.1, Pd–CeO_2–air and Pd–CeO_2–H_2 exhibit quite different catalytic performances.[1] Before partial oxidation of methane, Pd of Pd–CeO_2–air is at cationic state at 25 °C (Figure 16.2a1) but the Pd–CeO_2–H_2 before catalysis is at a metallic state at room temperature (Figure 16.2b1). As shown in Figure 16.1a, the conversion of CH_4 increases as a function of temperature and it reaches 84% at 700 °C. The selectivity for producing H_2 increases quickly, and it is 83% at 500 °C and 92.5% at 700 °C. Compared to Pd–CeO_2–air, the Pd–CeO_2–H_2 exhibits a quite different catalytic performance (blue lines in Figure 16.1). Pd–CeO_2–H_2 is not active at a temperature lower than 500 °C. However, its activity rapidly increases in the temperature range of 550–700 °C; in addition, the increase of catalytic selectivity for producing CO on Pd–CeO_2–H_2 (blue in Figure 16.1c) is very similar to that for producing H_2 in the temperature range of 550–700 °C (blue in Figure 16.1b).

Surface chemistries of Pd–CeO_2–air and Pd–CeO_2–H_2 were investigated with AP-XPS under pretreatments and catalytic conditions.[1] As shown in Figure 16.2, Pd atoms of

Application of Ambient Pressure X-ray Photoelectron Spectroscopy to Catalysis, First Edition. Franklin Tao.

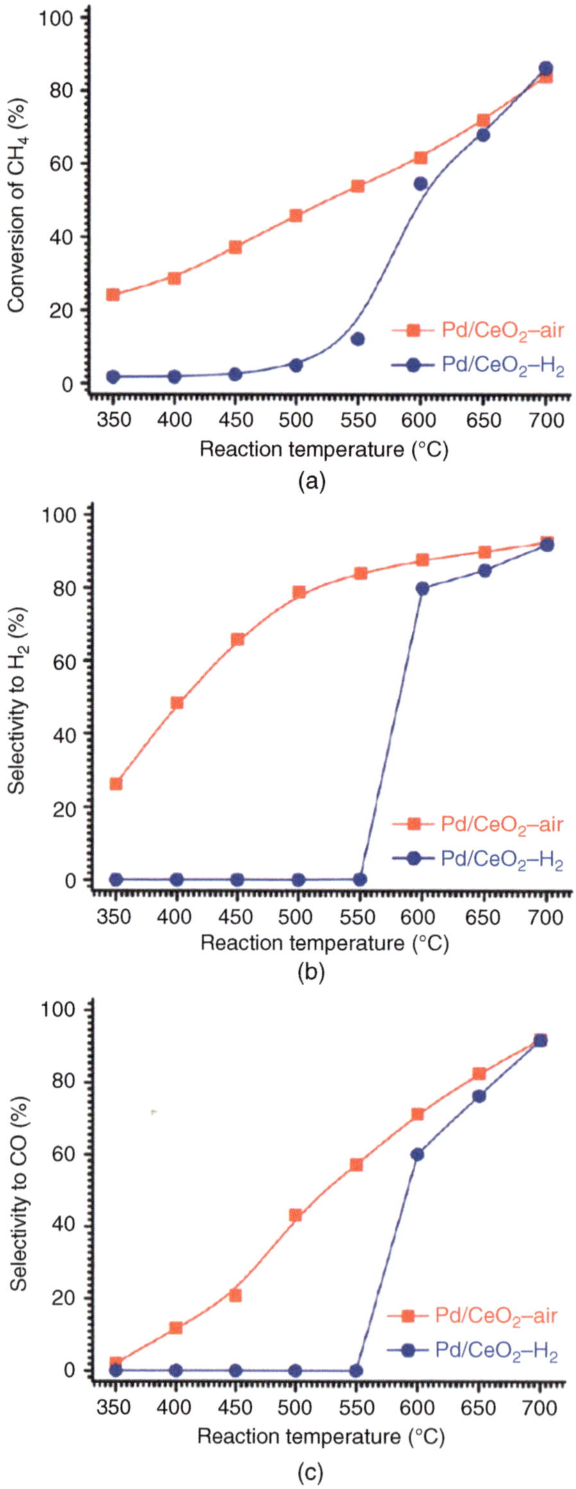

Figure 16.1 Catalytic performance of Pd–CeO_2–H_2 and Pd–CeO_2–air for partial oxidation of methane. Conversion of methane (a), selectivity to H_2 (b), and selectivity to CO (c). For the convenience of comparison, the catalytic performance of Pd–CeO_2–air (red line) under the same catalytic condition as Pd–CeO_2–H_2 (blue line) is listed in a parallel manner. *Source:* Reproduced with permission from Zhu et al.[1]/American Chemical Society.

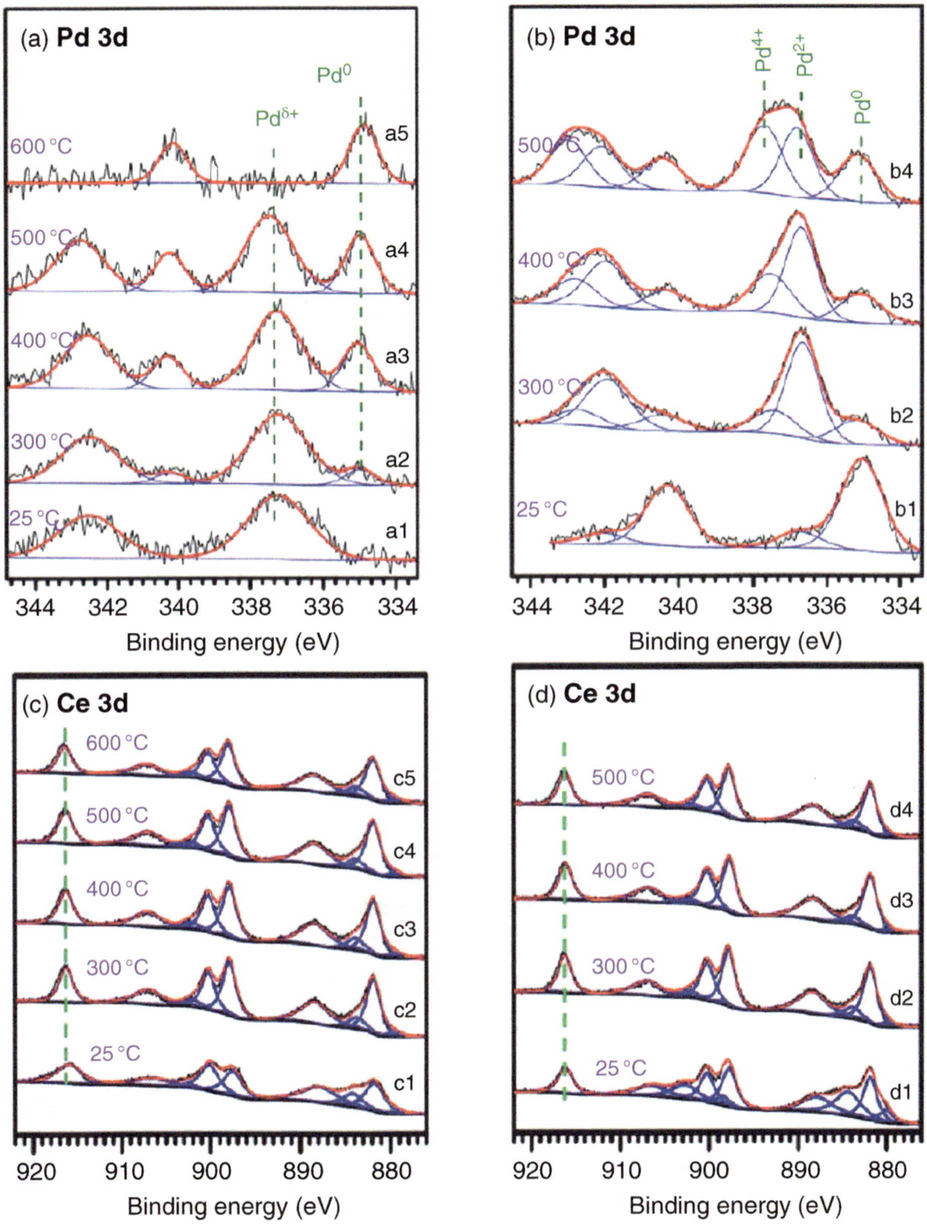

Figure 16.2 XPS spectra of the in situ studies of Pd 3d of Pd−CeO_2−air (a), Pd 3d of Pd−CeO_2−H_2 (b), Ce 3d of Pd−CeO_2−air (c), and Ce 3d of Pd−CeO_2−H_2 (d) during catalysis. The FWHM of the fitted peaks of Pd 3d in panels (a) and (b) were fixed at 1.4 and 1.2 eV, respectively. The ratio of the partial pressures of CH_4 to O_2 in the mixture reactant is 2 : 1, with a total pressure of 3 Torr. The flow rate is 6 ml/min. All AP-XPS data were collected at the same pressure and flow rate of CH_4 and O_2. The only variable is the reaction temperature. The spectra collected at 25 °C is used to represent the surface chemistry of the catalysts before catalysis because none of these catalysts is active for partial oxidation of methane at 25 °C. *Source:* Reproduced with permission from Zhu et al.[1]/American Chemical Society.

Pd–CeO_2–H_2 were at a metallic state at 25 °C before catalysis since it was just pretreated in H_2 at 500 °C (Figure 16.2b1). Notably, Pd–CeO_2–H_2 was oxidized to cationic Pd at a temperature ≤350 °C in the mixture of CH_4 and O_2 (Figure 16.2b2). The preservation of the cationic state of Pd atoms of Pd–CeO_2–H_2 at a temperature of 500 °C suggests that neither CO nor H_2 was produced from partial oxidation of methane to reduce the cationic Pd to metallic. In other words, the lack of the reducing gas, H_2 in the gaseous environment of Pd-CeO_2-H_2 due to the lack of activity of the as-prepared Pd–CeO_2–H_2 at <550 °C in producing H_2 through partial oxidation of CH_4 makes the surface of Pd–CeO_2–H_2 remain oxidized at ≤ 550 °C.

The evolution of the chemical environment of Pd atoms of Pd–CeO_2–H_2 uncovered by AP-XPS in Figure 16.2b allows us proposing a reasonable structural mode. During the pretreatment at 500 °C in H_2, the Pd atoms were reduced and sintered to form metallic Pd nanoparticles (Figure 16.2b1). Then, the exposure of the pretreated Pd–CeO_2–H_2 to the mixture of O_2 and CH_4 at 300 °C made the surface of metallic Pd nanoparticles oxidized; this oxidization formed a Pd metal core @PdO_x shell since (1) peaks of both metallic Pd and cationic Pd were observed at 300 °C in the mixture of CH_4 and O_2 in Figure 16.2b2 and (2) any oxidation of a metal nanoparticle must start at its surface. As the peak intensity of PdO_x in Figures 16.2b2 and 16.2b3 is much larger than metallic Pd, the PdO_x shell is thick. The formed Pd metal core @PdO_x shell is basically stable in the temperature range of 300–500 °C in the mixture of CH_4 and O_2. This existence of PdO_x shell on Pd–CeO_2–H_2 in the mixture of CH_4 and O_2 in 300–500 °C is consistent with the evolution of catalytic performance as shown by the blue lines in Figure 16.1. At a temperature 500 °C or lower, this catalyst is not active. The correlation between surface structure of Pd metal core @PdO_x shell and the lack of catalytic activity (blue lines in Figure 16.1a) suggest that PdO_x shell of Pd@PdO_x formed on Pd–CeO_2–H_2 is not an active phase for partial oxidation of CH_4. Since catalytic activity of Pd–CeO_2–H_2 abruptly increased at 600 °C (blue line in Figure 16.1a), it is expected the PdO_x shell was largely reduced at this temperature although no Pd 3d spectrum of Pd 3d at 600 °C under the catalytic condition was collected.

Compared to Pd–CeO_2–H_2, annealing Pd–CeO_2 at 500 °C in air formed a different catalyst, Pd–CeO_2–air. The surface chemistry of Pd–CeO_2–air under catalytic conditions at different temperatures was tracked with AP-XPS. Pd 3d and Ce 3d XPS peaks of Pd-CeO_2-air are presented in Figure 16.2a and c, respectively. As shown in Figure 16.2a1, Pd atoms of Pd–CeO_2–air are mainly at a cationic state upon the pretreatment at 500 °C in air. Thus, Pd atoms of the Pd–CeO_2–air exist in the format of PdO_x nanoparticles supported on CeO_2. A small portion of cationic palladium was reduced to metallic at 300 °C in the mixture of CH_4 and O_2 (Figure 16.2a2). At 400 °C, a large part of PdO_x was obviously reduced to metallic Pd (Figure 16.2a3). Based on the measurement of catalytic performance of Pd–CeO_2–air (red lines in Figure 16.1), it starts to be active for partial oxidation of methane at 350 °C. The correlation between the observed partial reduction of oxidized Pd atoms of Pd–CeO_2–air with AP-XPS and the appearance of catalytic activity at 350–400 °C suggest the metallic Pd is the active component of the catalytic phase.

These studies in Figures 16.1 and 16.2 in a correlative manner have significantly helped interpreting the catalysis performances of 5.0 wt% Pd/CeO_2. The comparative studies of Pd–CeO_2–H_2 and Pd–CeO_2–air demonstrated that (1) a different pretreatment could result in a profound consequential difference in terms of surface chemistry and catalytic performance for the same reaction, and (2) the catalytic phase of Pd-CeO_2-H_2 and Pd-CeO_2-O_2 is metallic Pd supported on CeO_2. This work demonstrated that, by correlating with temperature-dependent catalytic performance, observation of surface chemistries of a catalyst during pretreatment and catalysis with AP-XPS allows identifying the active phase responsible for a specific catalytic performance, establishing a base for interpreting catalytic mechanisms at a molecular level.

16.2 Partial Oxidation of Methane on Pt/CeO_2

Pt–CeO_2–air and Pt–CeO_2–H_2 were prepared with the same protocols as Pd–CeO_2–air and Pd–CeO_2–H_2, respectively. Both Pt–CeO_2–air and Pt–CeO_2–H_2 are active for methane partial oxidation (Figure 16.3). In terms of Pt–CeO_2–air, conversion of CH_4 increases as a function of catalysis temperature. It is 76% at 700 °C. The selectivity for production of H_2 on Pt–CeO_2–air increases and reaches 72% at 500 °C and 88% at 700 °C. The selectivity for producing CO increases along with the elevation of catalytic temperature. However, it is always lower than H_2 by 10–20%.

In terms of Pt–CeO_2–H_2, the evolution of conversion of CH_4 and selectivities for H_2 and CO on Pt–CeO_2–H_2 as a function of temperature are similar to those of Pt–CeO_2–air. However, a notable difference is that the conversion of CH_4 on Pt–CeO_2–H_2 is always lower than that of on Pt–CeO_2–air by 15% in the temperature range of 400–700 °C (Figure 16.3a).

AP-XPS studies revealed the surface chemistry of Pt–CeO_2–H_2 and Pt–CeO_2–air during catalysis in the temperature range of 25–600 °C (Figure 16.4). Deconvolution of Ce 3d suggests the existence of oxygen vacancies (Figure 16.4c1 and d1). In terms of Pt–CeO_2–air at room temperature (Figure 16.4a1), Pt atoms exist in three different chemical states including Pt(0), Pt(+2), and Pt(+4), associated with different bondings to oxygen atoms of CeO_2 lattice (Figure 16.4a1); the three components of Pt 4f have comparable atomic fractions; about 30% of Pt atoms on Pt–CeO_2–air at room temperature exist in the form of Pt(0) (Figure 16.4a1). The Pt 4f spectrum of Pt–CeO_2–H_2 collected at room temperature was deconvoluted into three components in Figure 16.4b1. However, the majority of Pt atoms in the surface region of Pt–CeO_2–H_2 exist in the format of Pt(0) (Figure 16.4b1). Although XPS studies confirmed the existence of certain fraction of Pt(0) in Pt–CeO_2–air and Pt–CeO_2–H_2 at 25 °C, the TEM studies could not successfully identify Pt nanoparticles. The lack of observable Pt nanoparticles in TEM images likely suggests (1) the formed Pt clusters or Pt–Ce alloy clusters are 2 nm or smaller, or (2) the formed metal or alloy nanoparticles are larger than 2 nm but the thick support such as CeO_2 nanoparticles supporting the metal are the major contributor to the contrast of a TEM image.

As shown in Figure 16.3a, conversions of CH_4 on both Pt–CeO_2–air and Pt–CeO_2–H_2 increase along with the elevation of catalytic temperature. The decrease of atomic fraction of oxidizing Pt atoms in Figure 16.4a and b along the increase of catalytic temperature suggests the oxidized platinum is continuously reduced along the increase of catalysis temperature. Accordingly, the increase of atomic fraction of Pt(0) indicates the metallic Pt(0) is active for partial oxidation of methane.

Although TEM cannot image the Pt or Pt-Ce metal nanoparticles with a size of < 2 nm supported on CeO_2 particles, AP-XPS studies of Pt–CeO_2–H_2 and Pt–CeO_2–air during catalysis show that majority of Pt atoms of Pt–CeO_2–H_2 and a portion of Pt atoms of Pt–CeO_2–air exist at a metallic state. From this point of view, AP-XPS can provide important information while TEM is limited in distinguishing small clusters (< 2.0 nm) from its support due to (1) the small size (< 2nm) or (2) similar image contrasts of the metal element (Pt) of the supported nanoparticles and metal element (Ce) of the support (CeO_2).

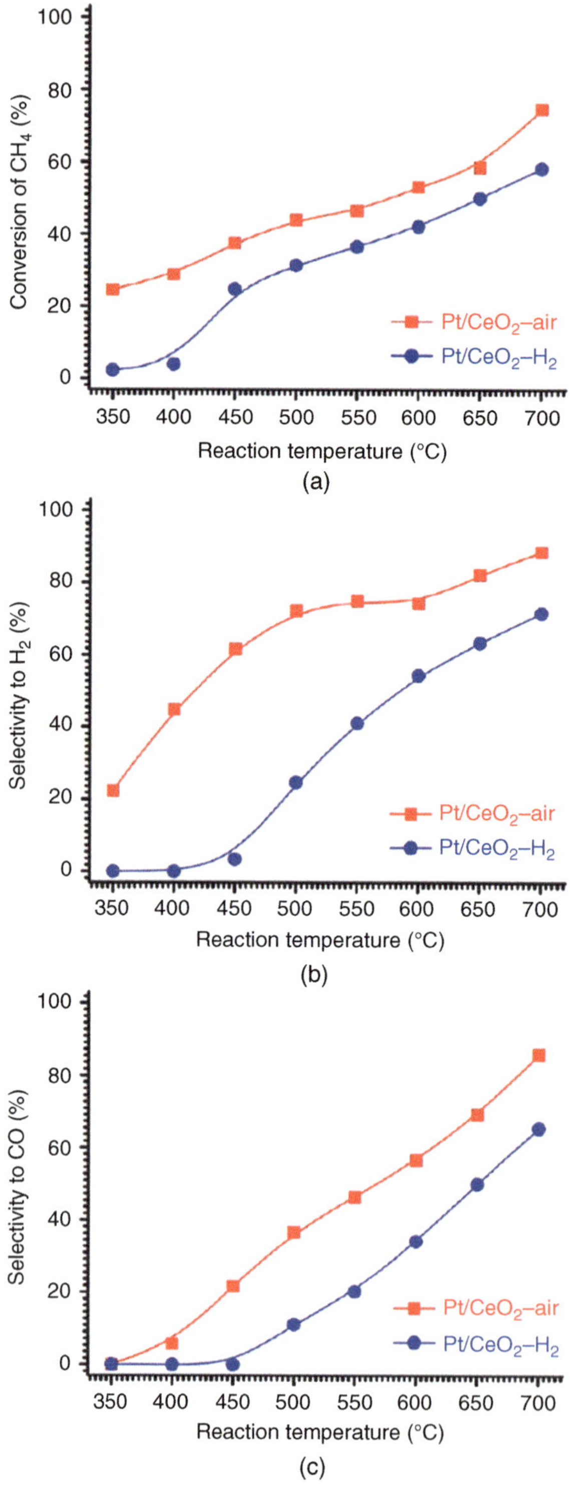

Figure 16.3 Catalytic performances of Pt–CeO_2–H_2 and Pt–CeO_2–air. Conversion of methane (a), selectivity to H_2 (b), and selectivity to CO (c). For the convenience of comparison, the catalytic performance of Pt–CeO_2–air (red line) under the same catalytic condition as Pt–CeO_2–H_2 (blue line) is listed in a parallel manner. *Source:* Reproduced with permission from Zhu et al.[1]/American Chemical Society.

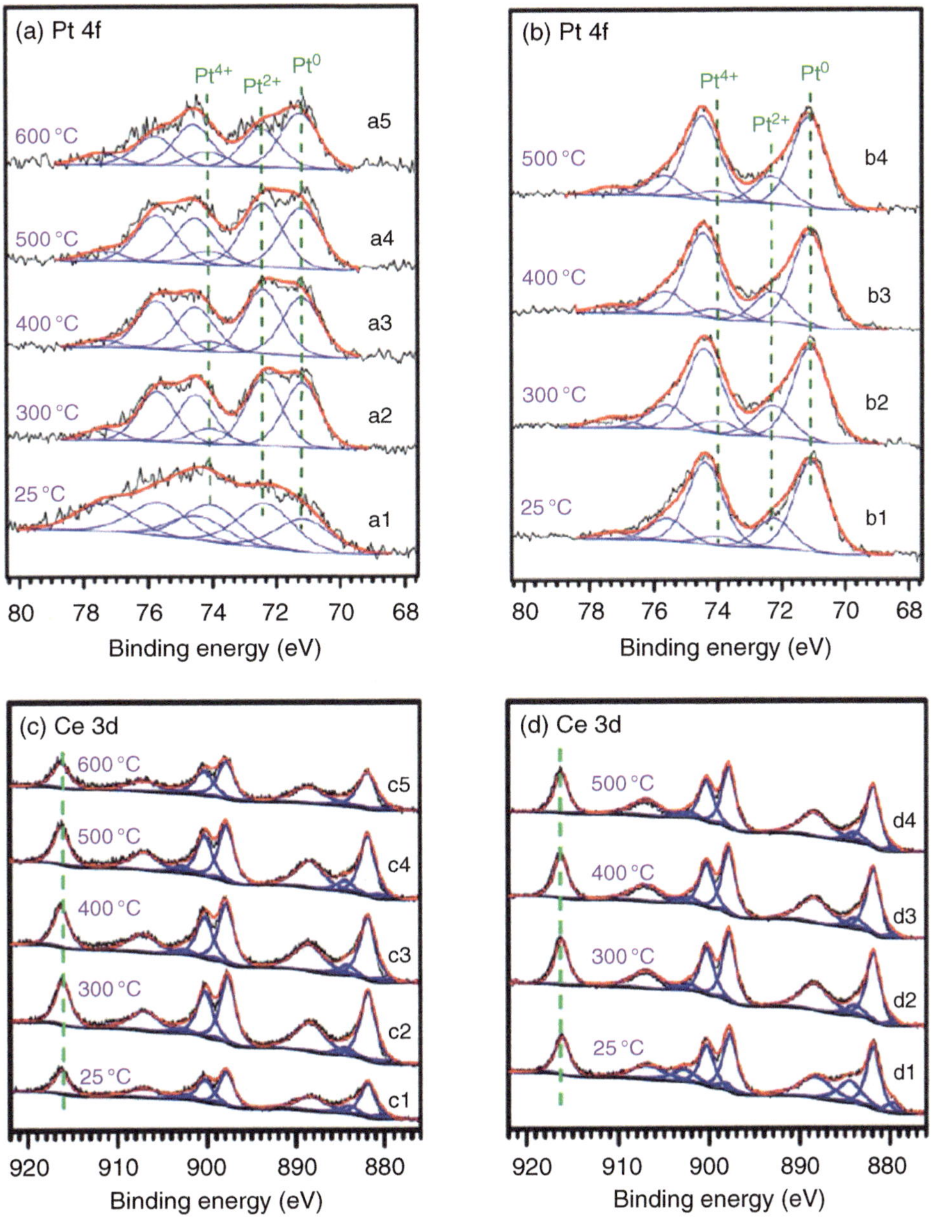

Figure 16.4 AP-XPS studies of Pt 4f of Pt–CeO_2–air (a), Pt 4f of Pt–CeO_2–H_2 (b), Ce 3d of Pt–CeO_2–air (c), and Ce 3d of Pt–CeO_2–H_2 (d) during catalysis. The FWHM of the fitted peaks of Pt 4f in panels (a) and (b) were fixed at 1.2 eV, respectively. The ratio of the partial pressures of CH_4 to O_2 in the mixture reactant is 2 : 1. The flow rate is about 6 ml/min. All data were collected at the same pressure and flow rate of CH_4 and O_2. The only variable is the reaction temperature. *Source:* Reproduced with permission from Zhu et al.[1]/American Chemical Society.

16.3 Partial Oxidation of Methane on Rh/CeO_2

Other than the two pairs of catalysts discusssed above (Pd–CeO_2–air versus Pd–CeO_2–H_2, and Pt–CeO_2–air versus Pt–CeO_2–H_2), Rh–CeO_2–air and Rh–CeO_2–H_2 were prepared with the same method as those discussed above. TEM studies of Rh–CeO_2–air and Rh–CeO_2–H_2 before catalysis suggest the lack of visible Rh nanoparticles. Similarly, no visible Rh nanoparticles were observed on the used Rh–CeO_2–air and used Rh–CeO_2–H_2 after partial oxidation of CH_4 at 700 °C. AP-XPS studies provided significant information on Rh–CeO_2–air and Rh–CeO_2–H_2 during catalysis for excluding the formation of Rh nanoparticles or Rh–Ce bimetallic nanoparticles during catalysis (Figure 16.6a and b).

Although the pretreatments used in Rh–CeO_2–air and Rh–CeO_2–H_2 are quite different, their catalytic performances are nearly the same in the whole temperature range of 350–700 °C (Figure 16.5). Notably, the high similarity of catalytic performances of Rh–CeO_2–air and Rh–CeO_2–H_2 was not found in either the pair of Pd–CeO_2–air and Pd–CeO_2–H_2 or the pair of Pt–CeO_2–air and Pt–CeO_2–H_2.[1] The identical catalytic performance of Rh–CeO_2–air and Rh–CeO_2–H_2 was well rationalized with the results of AP-XPS studies. As shown in Figure 16.6, the two catalysts exhibit nearly the same chemical environments of Rh atoms at each temperature.

Another important feature of the pair of Rh–CeO_2–air and Rh–CeO_2–H_2 during catalysis is the preservation of the cationic state of Rh atoms along with the increase of catalysis temperature even up to 700 °C. This preservation of the cationic state of Rh atoms in the catalysis temperature range (35–700 °C) (Figure 16.6) is quite different from the progressive reductions of Pd of Pd–CeO_2–air and Pd–CeO_2–H_2 during catalysis (Figure 16.2) and Pt atoms of Pt–CeO_2–air and Pt–CeO_2–H_2 during catalysis (Figure 16.4). This distinct difference in catalysis temperature-dependent evolutions of chemical states of metal atoms among the three precious metals probably results from the difference in their affinities to surface lattice oxygen and their reducibility in a gas mixture containing products CO and H_2. Based on the AP-XPS studies, the chemical state of the active Rh atoms of the two catalysts during partial oxidation of CH_4 is at a cationic state. This is different from the metallic state of Pt or Pd of active phase on Pt-CeO_2-O_2, Pt-CeO_2-H_2, Pd-CeO_2-O_2, and Pd-CeO_2-H_2.

It is ready to conclude that cationic Rh atoms are responsible for the activation of CH_4. The lack of visible Rh nanoparticles on CeO_2 suggests that Rh atoms are likely singly dispersed on the surface regions of Rh–CeO_2–air and Rh–CeO_2–H_2 although there is a lack of information on the coordination environment of Rh atoms of the two catalysts during catalysis. The preservation of the chemical state of Rh atoms on both Rh–CeO_2–air and Rh–CeO_2–H_2 inspired exploration of single-atom catalysis on a catalyst consisting of CeO_2 support and singly dispersed Rh atoms on CeO_2. This inspiration has made us successful in the development of Rh single-atom catalyst anchored on CeO_2 for steam reforming of methane.[2] In fact, the high affinity of Rh to surface lattice oxygen atoms of supports has offered itself the best fit for many oxides supports. This insight we gained from this work has inspired us in developing several Rh single-atom catalysts. For instance, Rh atoms were able to anchor on inert support silica, forming a single-atom catalyst active Rh_1/SiO_2 for reduction of NO with CO.[3] Rh_1 was anchored to Co_3O_4 and the formed single-atom catalyst Rh_1/Co_3O_4 exhibits high activity and selectivity for producing N_2 through low-temperature reduction of NO with CO at 110 °C.[4] In addition, Rh_1 was anchored in the microporous aluminosilicate that is highly active for oxidization of CH_4 and CO with O_2 to form acetic

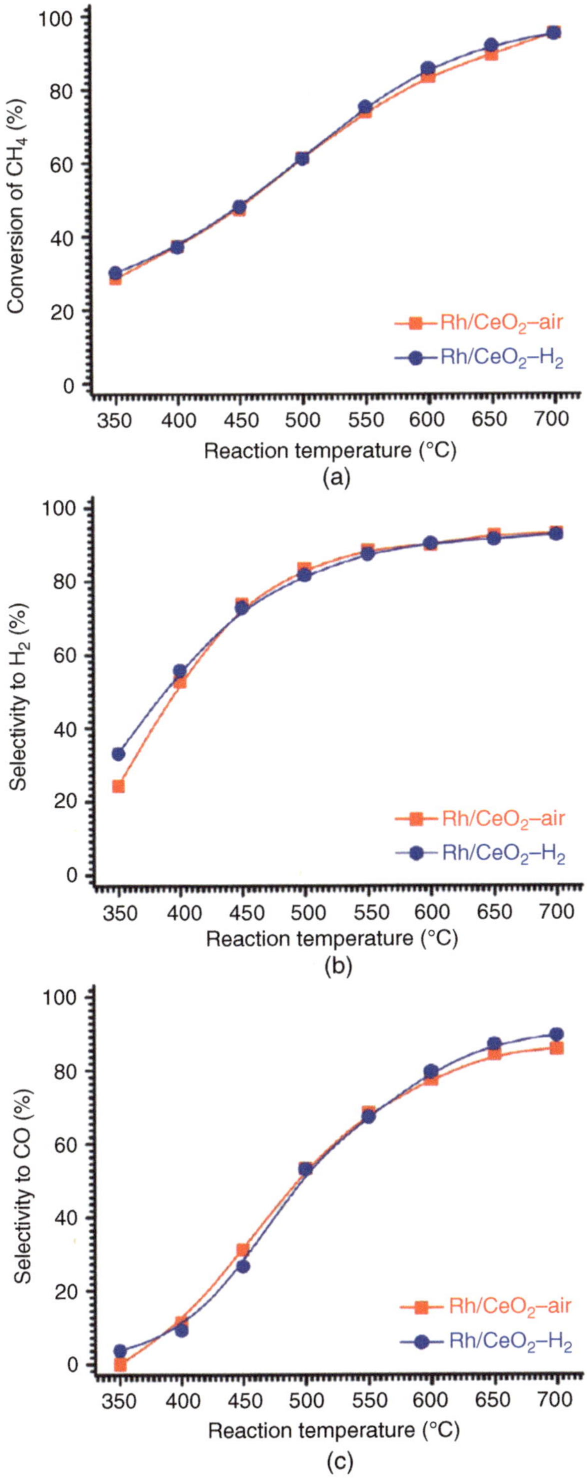

Figure 16.5 Catalytic performance of Rh–CeO_2–H_2 and Rh–CeO_2–air. Conversion of methane (a), selectivity to H_2 (b), and selectivity to CO (c). For the convenience of comparison, the catalytic performance of Rh–CeO_2–air (red line) under the same catalytic condition as Rh–CeO_2–H_2 (blue line) is listed in a parallel manner. *Source:* Reproduced with permission from Zhu et al.[1]/ American Chemical Society.

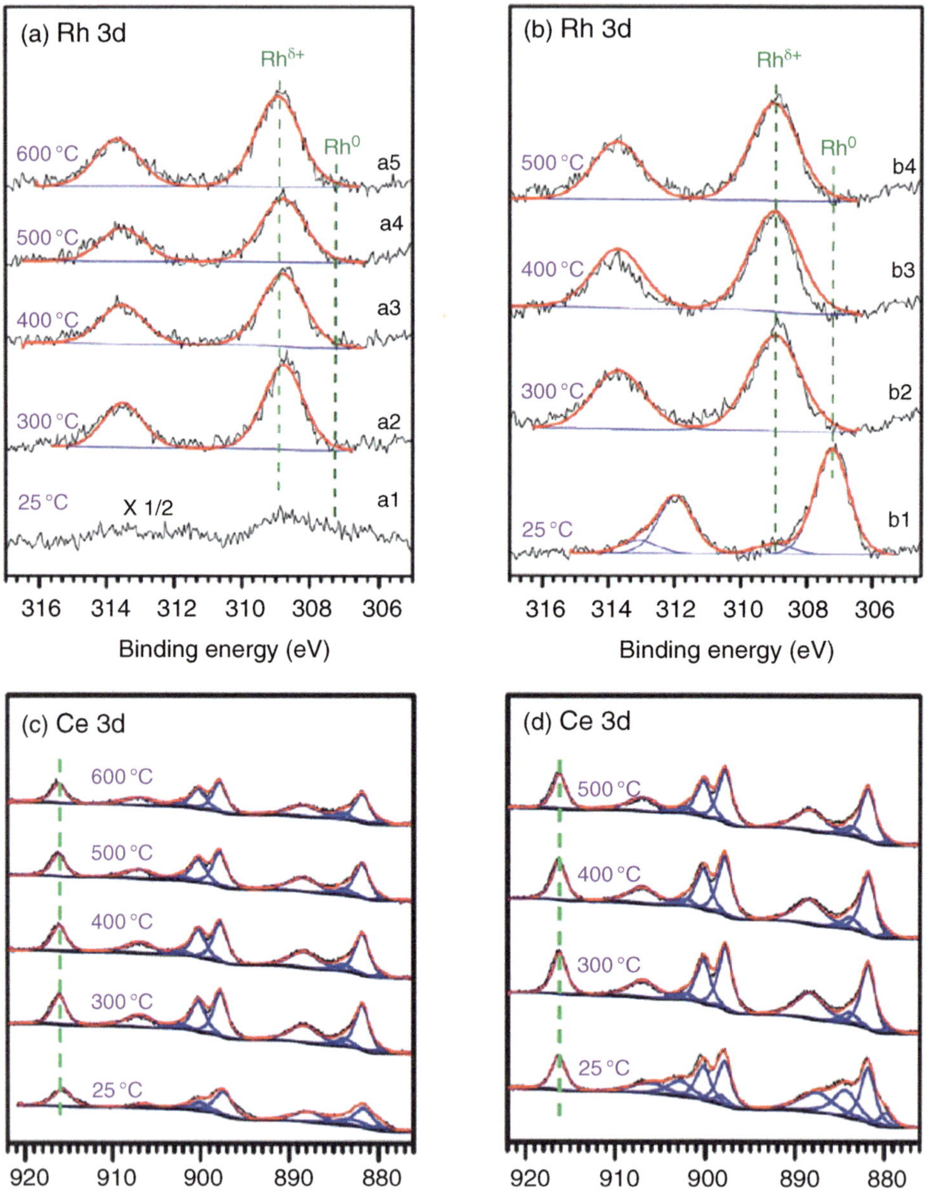

Figure 16.6 XPS spectra of the in situ studies of Rh 3d of Rh–CeO_2–air (a), Rh 3d of Rh–CeO_2–H_2 (b), Ce 3d of Rh–CeO_2–air (c), and Ce 3d of Rh–CeO_2–H_2 (d) during catalysis. The FWHM of the fitted peaks of Rh 3d in panels (a) and (b) were constrained between 1.2 and 1.4 eV, respectively. The ratio of the partial pressures of CH_4 to O_2 in the mixture reactant is 2 : 1, with a total pressure of 3 Torr. The flow rate is 6 ml/min. All AP-XPS data were collected at the same pressure and flow rate of CH_4 and O_2. The only variable is the reaction temperature. *Source:* Reproduced with permission from Zhu et al.[1]/American Chemical Society.

acid at a temperature as low as 110 °C.[5] Another work inspired by the partial oxidation of CH_4 on Rh_1/TiO_2 is that Rh_1 atoms were immobilized on TiO_2 and thus a Rh_1O_5 single-atom structure is formed on TiO_2; The Rh_1O_5 sites are highly active for partial oxidation of CH_4 at high temperature.[6]

References

1 Zhu, Y., Zhang, S., Shan, J. *et al.* 2013. "In situ surface chemistries and catalytic performances of ceria doped with palladium, platinum, and rhodium in methane partial oxidation for the production of syngas." *ACS Catal.* 3, 2627–2639.

2 Yan, G., Tang, Y., Li, Y. *et al.* 2022. "Reaction product-driven restructuring and assisted stabilization of a highly dispersed Rh-on-ceria catalyst." *Nat. Catal.* 5, 119–127.

3 Zhang, S., Tang, Y., Nguyen, L. *et al.* 2018. "Catalysis on singly dispersed Rh atoms anchored on an inert support." *ACS Catal.* 8, 110–121.

4 Zhang, S., Nguyen, L., Liang, J. *et al.* 2015. "Catalysis on singly dispersed bimetallic sites." *Nat. Commun.* 6, 7938. doi:10.1038/ncomms8938.

5 Tang, Y., Li, Y., Fung, V. *et al.* 2018. "Single rhodium atoms anchored in micropores for efficient transformation of methane under mild conditions." *Nat. Commun.* 9, 1231. doi:10.1038/s41467-018-03235-7.

6 Tang, Y., Fung, V., Zhang, X. *et al.* 2021. "Single-atom high-temperature catalysis on a Rh1O5 cluster for production of syngas from methane." *J. Am. Chem. Soc.* 143, 16566–16579. doi:10.1021/jacs.1c06432.

17

Oxidative Coupling of Methane

Oxidative coupling of methane (OCM) is an important reaction directly transforming methane to ethylene or ethane. It is a process of producing value-added intermediates including ethane and ethylene for further chemical transformation although the self-limited activity and selectivity make yield of ethane and ethylene (C_2) limited. The advantage of this reaction is that it bypasses the reforming of CH_4 and a followed Fischer-Tropsch synthesis (FTS) since it directly transforms CH_4 to ethane and ethylene. As OCM is performed at 800 °C or higher, in situ or operando studies of OCM have been extremely challenging. Other than the challenge to instrument brought from the catalysis temperature, the oxidative environment at high temperature gives another challenge in instrumentation for performing in situ/operando studies of OCM. However, from fundamental understanding point of view, it is crucial to perform in situ/operando studies for OCM catalysts before we could gain a fundamental understanding of the catalytic mechanism of OCM since the surface of OCM catalyst performed at such a high temperature is expected to be quite different from that observed under ambient conditions at room temperature.

17.1 OCM on Supported Na_2WO_4 and Hypothesized Active Phase Na_2O_2

Supported Na_2WO_4 is the most studied catalyst for OCM in literatures. However, there has been a lack of fundamental understanding for OCM on most OCM catalysts including supported Na_2WO_4. In early 1980's, scientists hypothesized sodium peroxide is an active phase for OCM performed on Na_2WO_4.[1] The recent kinetics studies suggest that H_2O molecules participate into activation of CH_4 through quasi-equilibrated formation of OH radicals.[2–5] However, there had been a lack of data to confirm the hypothesis in terms of the formation of Na_2O_2 and OH radicals during catalysis although the hypothesis was made four decades ago.[1] A part of the reasons is the lack of in situ or operando studies of surface of Na_2WO_4-based catalysts during OCM catalysis at 800 °C or higher.

A detailed kinetics study of OCM on sodium supported on silica was performed for addressing the effects of H_2O on OCM recently.[2,6] At low conversion rate of CH_4 in OCM on $Mn/Na_2WO_4/SiO_2$, the kinetic expression for the OCM is $r = r_{O^*} + r_{OH*} = k'P_{CH_4}P_{O_2}^{0.5} + k''P_{CH_4}P_{O_2}^{0.25}P_{H_2O}^{0.5}$.[2] In this kinetic expression, the $k'P_{CH_4}P_{O_2}^{0.5}$

Application of Ambient Pressure X-ray Photoelectron Spectroscopy to Catalysis, First Edition. Franklin Tao.

corresponds to CH_4 activation by surface O atom, which is quasi-equilibrated with O_2, one of the reactants of OCM.[2] The second term, $k''P_{CH_4}P_{O_2}^{0.25}P_{H_2O}^{0.5}$ is attributed to CH_4 activation through an OH radical formed from coupling of molecular O_2 and H_2O in a quasi-equilibrated step. Based on literatures, [2, 5, 6] OH radical pathway makes most contribution to OCM on Na_2WO_4/SiO_2.

As Na_2O_2 is highly sensitive to H_2O, CO_2, or O_2 of air, ex situ characterization of a Na_2WO_4-based catalyst before or after catalysis is expected to give inappropriate information. In other words, it is impossible to observe Na_2O_2 from a used catalyst which had been exposed to air after OCM even though Na_2O_2 surface phase is formed under a catalytic condition. As OCM is performed at 800 °C or higher, in situ or operando studies of OCM have been extremely challenging. However, from a fundamental understanding point of view, it is crucial to perform in situ/operando studies of OCM catalysis before we could gain a fundamental understanding of the catalytic mechanism of OCM performed on the support Na_2WO_4. To examine whether the hypothesized Na_2O_2 could be observed, the first AP-XPS study of OCM catalytic reaction was performed on a lab-based AP-XPS system in Tao group.[7] To reduce surface charge of the nonconductive surface of support, silica in AP-XPS studies, TiO_2 instead of silica was used as a support to load Na_2WO_4. The catalyst precursor was prepared through first loading Na_2WO_4 to surface of TiO_2 with impregnation and then annealing Na_2WO_4/TiO_2 at 900 °C in air for 15 hours. Surface area is $3\,m^2/g$. The prepared $NaWO_4/TiO_2$ at 800 °C exhibits high selectivity of nearly 80% for producing C_{2+} molecules through OCM in the presence of O_2, H_2O and CH_4 (Figure 17.1).

We hypothesized OH radicals were formed through the following three steps: (i) $2\ Na_2O + O_2 \leftrightarrow 2\ Na_2O_2$, (ii) $Na_2O_2 + H_2O \leftrightarrow Na_2O + H_2O_2$, and (iii) $H_2O_2 \leftrightarrow 2\ OH\cdot$. The overall reaction between O_2 and H_2O is $O_2 + 2\ H_2O \leftrightarrow 4\ OH\cdot$, a highly endothermic reaction with $\Delta H = 650\,kJ/mol$. The capability of producing OH· was confirmed by a laser-induced fluorescent (LIF) spectrometer.[7] OH· radicals were detected after the mixture of O_2 and H_2O was delivered to the reactor containing Na_2WO_4/TiO_2 at 900 °C.[7] This experiment clearly confirmed that OH· radicals can be clearly formed during catalysis on Na-based catalyst in the mixture of O_2 and H_2O. In terms of OCM, H_2O is one of the products while C_{2+} is formed. Thus, O_2 and H_2O naturally coexist in the gaseous environment near to a catalyst. Based on the equation, $O_2+2H_2O=4OH\cdot$, OH· can be readily formed when Na_2O_2 is a surface phase.

17.2 First Observation of Na_2O_2 through AP-XPS Studies at 800 °C

Similar to the pretreatment for detection of OH radicals, the catalyst precursor was prepared through annealing Na_2WO_4/TiO_2 at 900 °C in air for 15 hours and then transferred to an AP-XPS system to perform OCM at 800 °C in the mixture of CH_4 and O_2 to check whether Na_2O_2 could form during OCM catalysis. Surface chemistry of Na_2WO_4/TiO_2 at 800 °C in either the mixture of O_2 and H_2O (red spectrum in Figure 17.2) or the mixture of O_2 and CH_4 (pink spectrum in Figure 17.2) was investigated with the AP-XPS system by placing the catalyst sample in a reaction cell where the mixture of gases was flowing through the catalyst surface. XPS peaks of Na 1s, O 1s, and W 4f were collected while the catalyst remained at 800 °C in the flowing gas (Figure 17.2). Notably, the binding energies of all spectra shown in Figure 17.2 were calibrated to Au $4f_{7/2}$ of Au foil that is the foil

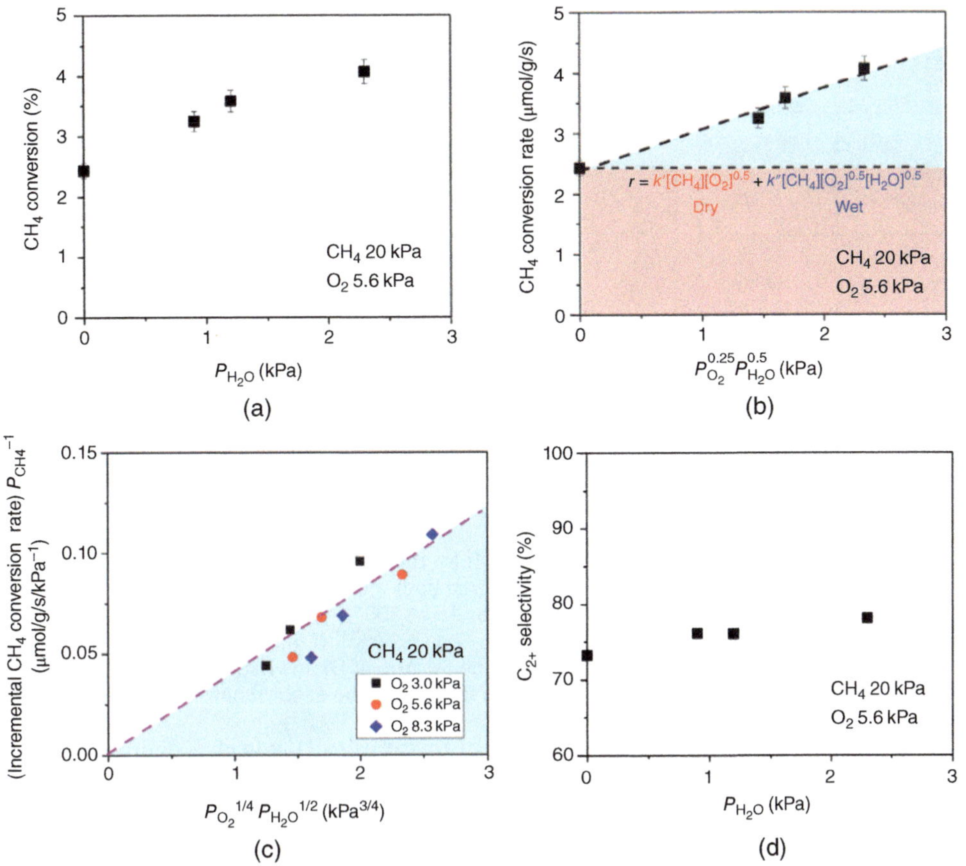

Figure 17.1 Catalytic results for the $CH_4/O_2/H_2O$ reaction using the Na_2WO_4/TiO_2 catalyst (30 mg, 800 °C, 20 kPa CH_4, 3.0–8.3 kPa O_2, 0–2.3 kPa H_2O, total pressure 101 kPa balanced by He, total flow rate 25 cm^3/min). (a) CH_4 conversion as a function of the H_2O pressure. (b) CH_4 conversion rate as a function of the $P_{O_2}^{1/4}P_{H_2O}^{1/2}$. (c) Incremental CH_4 conversion rate $P_{CH_4}^{-1}$ (rate difference between the presence and absence of H_2O) as a function of the $P_{O_2}^{1/4}P_{H_2O}^{1/2}$. (d) C_{2+} selectivity as a function of the H_2O pressure. *Source:* Adapted from Takanabe et al.[7]

loading the Na_2WO_4/TiO_2 catalyst. These Au 4f spectra were collected under the same experimental conditions (UHV 20 °C, or $O_2 + H_2O$ at 800 °C, or $O_2 + CH_4$ at 800 °C) for calibration of binding energies of XPS peaks for a reasonable comparison.

To uncover the authentic chemical state of Na_2WO_4/TiO_2 during OCM at 800 °C, O_2 and CH_4 were mixed and then flew through the catalyst placed in the reaction cell of AP-XPS. XPS peaks of Na 1s, O 1s, and W 4f at 800 °C during OCM were collected (pink spectra in Figure 17.2). Before AP-XPS studies at 800 °C, XPS peaks of Na 1s, O 1s, and W 4f at room temperature in the mixture of CH_4 and O_2 were collected (blue spectra in Figure 17.2). They are the same as those at room temperature in ultrahigh vacuum (UHV) (black spectra in Figure 17.2). Upon heating the catalyst in the mixture of CH_4 and O_2 to 800 °C, the catalyst remained at 800 °C in the flowing mixture of CH_4 and O_2. XPS peaks of Na 1s, O1s, and W 4f of the catalyst at 800 °C in the mixture of CH_4 and O_2 were collected as shown as the pink spectra in Figure 17.2. In terms of these pink spectra in Figure 17.2, the peak position of Na

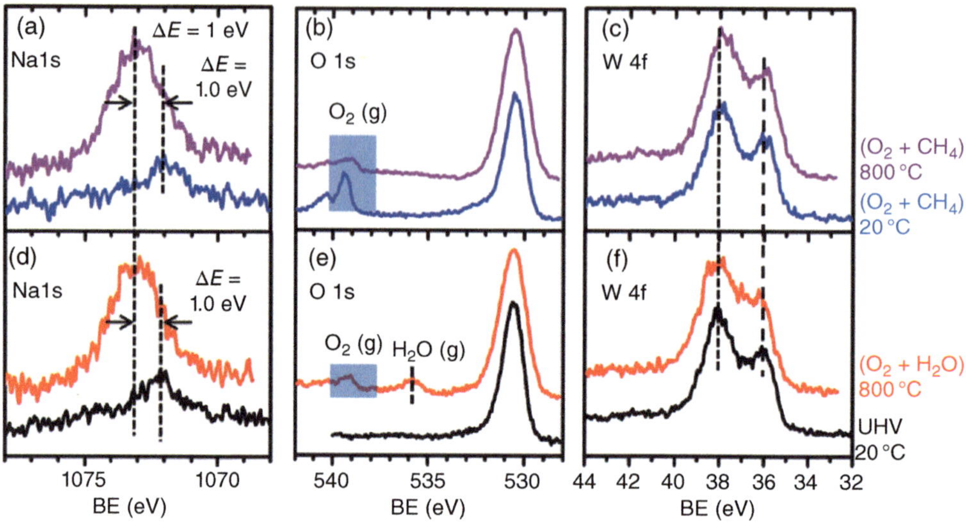

Figure 17.2 AP-XPS studies of Na 1s, O 1s, and W 4f for the catalyst collected at UHV, 20 °C (black line), mixture of 66 Pa O_2 and 66 Pa H_2O at 800 °C (red line), mixture of 66 Pa O_2 and 66 Pa CH_4 at 20 °C (blue line), and mixture of 66 Pa O_2 and 66 Pa CH_4 at 800 °C (pink line). (a) Na 1s in the mixture of O_2 and CH_4. (b) O 1s in the mixture of O_2 and CH_4. (c) W 4f in the mixture of O_2 and CH_4. (d) Na 1s in the mixture of O_2 and H_2O. (e) O 1s in the mixture of O_2 and H_2O. (f) W 4f in the mixture of O_2 and H_2O. *Source:* Reproduced with permission from Takanabe et al.[7]/John Wiley & Sons.

1s centered at 1073.1 eV observed at 800 °C in the mixture of CH_4 and O_2 is definitely higher than 1072.0 eV of Na_2WO_4.[8] The peak position of Na 1s in the pink spectra observed at 800 °C in the mixture of CH_4 and O_2 (Figure 17.2a) is very similar to that observed at 800 °C in the mixture of O_2 and H_2O (red spectrum in Figure 17.2d). Thus, the observed high binding energy at 1073.1 eV for Na 1s of the supported Na_2WO_4 during catalysis is assigned to Na 1s of Na_2O_2 based on the value of binding energy of Na 1s of Na_2O_2 reported in literature.[9] Obviously, the AP-XPS studies of the surface of Na_2WO_4/TiO_2 catalyst at 800 °C during OCM show the Na atoms on surface of this catalyst during catalysis exist in the form of Na_2O_2 instead of Na_2WO_4 observed before catalysis or NaOH likely formed after exposing a used catalyst to atmosphere. To the best of our knowledge, this is the first report of data confirming that Na_2O_2 forms on a Na_2WO_4-based OCM catalyst after the hypothesis was made nearly four decades ago.[1] Definitely, without AP-XPS studies, it is impossible to observe the Na_2O_2 surface phase that is formed during OCM catalysis at 800 °C or higher temperatures. The Na_2O_2 surface phase plays a dominant role in the pathway of OCM involving the formation of OH· radicals and the subsequent activation of CH_4 by OH· radicals.

To track the surface chemistry of Na_2WO_4 in the condition of forming OH· radicals a mixture of O_2 and H_2O was introduced to the reaction cell of the AP-XPS system where the Na_2WO_4/TiO_2 was heated to 800 °C in the flowing mixture of O_2 and H_2O and then remained at 800 °C in the mixture of O_2 and H_2O. The existence of gaseous O_2 and H_2O around the catalyst at 800 °C was confirmed by the observed XPS peaks of O 1s at 539.1 and 535.8 eV of the gas phase O_2 and H_2O (see the marked peaks of O_2 and H_2O in the red spectrum in Figure 17.2e). Surprisingly, the peak position of Na 1s of the catalyst upshift by 1.0 eV at 800 °C in the mixture of O_2 and H_2O (red spectrum in Figure 17.2d) if compared to the Na 1s peak position of the catalyst at 20 °C in UHV (black spectrum in Figure 17.2d). The upshift of Na 1s by 1.0 eV does mean an obvious change of the chemical environment of Na atoms on the surface of the catalyst after the condition of the catalyst was switched from room temperature UHV (black spectrum in Figure 17.2d) to the condition of 800 °C in the mixture of O_2 and H_2O (red spectrum in Figure 17.2d). The assignment of this upshift to an intrinsic change of chemical environment of Na atoms on the catalyst surface is supported by the following two arguments. First, the upshift of Na 1s binding energy does not result from potential surface charging; in fact, having a gaseous environment around a sample always makes the nominal binding energy of subshell electrons downshift instead of upshifting because X-ray can ionize the gas molecules to generate photoelectrons in the gas environment and these photoelectrons can behave as an "electronic clouds" around the catalyst particles to neutralize the positively charged surface. In other words, there is no issue of surface charging for either the red or black spectrum of Figure 17.2. Thus is further supported by the calibration of binding energies of Na 1s and others by Au foil under the same condition. Another argument supporting the assignment of the upshift of Na 1s peak position to the change of chemical environment of Na atoms is based on the fact that the O 1s peak position of the catalyst at room temperature in UHV (black spectrum in Figure 17.2e) is the same as that at 800 °C in the mixture of O_2 and H_2O (red spectrum in Figure 17.2e). The preservation of peak position of O 1s clearly shows there is no surface charging for the catalyst in either at 25 °C in UHV or at 800 °C in the mixture of O_2 and H_2O. In summary, upon its condition was changed from room temperature in UHV to 800 °C in the mixture of O_2 and H_2O, the upshift of Na 1s binding energy and preservation of O 1s binding energy of the catalyst suggest the formation of Na_2O_2 on surface.

17.3 Formation of a Thin Layer of Na_2O_2 Supported on Na_2WO_4

Another finding is the preservation of the peak position of W 4f of the supported Na_2WO_4 upon the experimental condition was switched from room temperature UHV to 800 °C in the mixture of O_2 and H_2O (black spectrum versus red spectrum in Figure 17.2f). It seems this preservation of W 4f binding energy contradicts the upshift of Na 1s. In fact, the preservation of W 4f is reasonable and consistent with the upshift of Na 1s. As the binding energy of W 4f is 34–38 eV, its kinetic energy is as high as >1440 eV. However, the kinetic energy of Na 1s photoelectrons is only about 400 eV. Obviously, the inelastic mean free path of W 4f photoelectrons is larger than that of Na 1s based on the universal curve (Figure 6.1). Thus, the intensity of W 4f peak is mainly contributed from subsurface region but that of Na 1s is mainly from surface region. It is understandable that W atoms of the subsurface of the catalyst during catalysis still retains its original chemical environment of Na_2WO_4. Thus, the peak position of W 4f of the catalyst during catalysis nearly shows no change upon the switching of condition.

Quantitative analysis of AP-XPS using the peak intensity of Na 1s and O 1s of AP-XPS studies was presented in Figure 17.3a. Figure 17.3a shows the relative ratios of Na 1s to O 1s at 800 °C in the mixture of O_2 and H_2O (red spectra in Figure 17.2) or 800 °C in the mixture of O_2 and CH_4 (pink spectra in Figure 17.2) were obviously higher than that at 20 °C in the mixture of O_2 and CH_4 (blue spectra in Figure 17.2) or 20 °C in UHV (black spectra in Figure 17.2), respectively. The observed increase in the Na to O atomic ratio upon

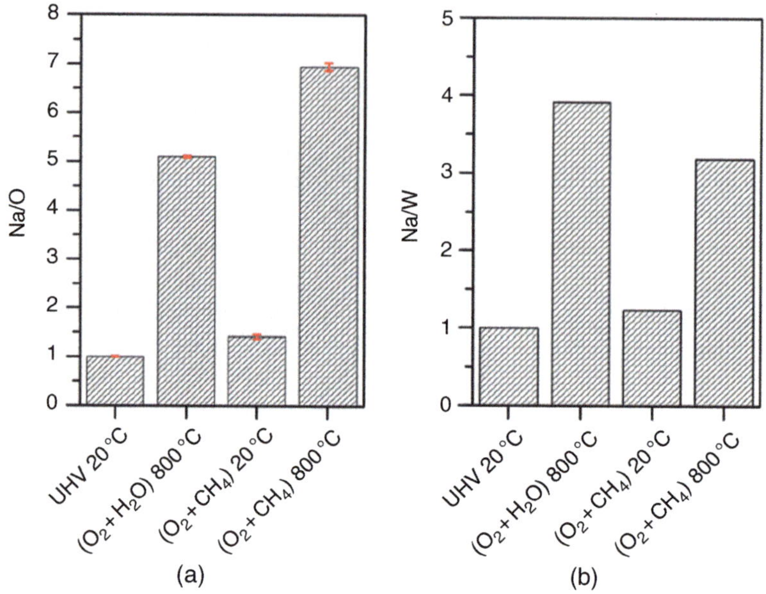

Figure 17.3 Area ratios of (a) Na 1s/O 1s and (b) Na 1s/W $4f_{7/2}$ of the catalyst surface at 800 °C in a mixture of 66 Pa O_2 and 66 Pa H_2O and at 20 and 800 °C in a mixture of 66 Pa CH_4 and 66 Pa O_2, relative to the ratios obtained at 20 °C in UHV. *Source:* Reproduced with permission from Takanabe et al.[7]/John Wiley & Sons.

switching the experimental condition from room temperature in UHV or room temperature in the mixture of CH_4 and O_2 to the catalytic condition (800 °C in the mixture of O_2 and CH_4) or high-temperature condition (800 °C in the mixture of O_2 and H_2O) supports the transformation of Na_2WO_4 to Na_2O_2 since Na/O ratio in Na_2WO_4 is ½ but 1/1 in Na_2O_2.

To reasonably calculate the area ratio of Na 1s to W 4f7/2, the W 4f spectra must be deconvoluted. Each W 4f spectrum was deconvoluted into a peak center at 36.3 eV contributed solely from W 4f7/2 and a peak at 38.0 eV contributed from both W 4f5/2 and Ti 3p. Through such a deconvolution, the accurate intensity of W 4f7/2 at 36.3 eV was known, by which the Na/W ratios under the four conditions including UHV 20 °C, (O_2+H_2O) 800 °C, (O_2+CH_4) 20 °C, and (O_2+CH_4) 800 °C were calculated (Figure 17.3b). As seen in Figure 17.3b the analysis of AP-XPS spectra also found that the Na/W ratio under high temperature condition (800 °C in the mixture of O_2 and H_2O) or catalysis condition (800 °C in the mixture of O_2 and CH_4) is higher than those at 20 °C UHV or at 20 °C in the mixture of CH_4 and O_2. It suggests the Na_2WO_4 of surface region is changed to a Na_2O_2-based phase. Since this Na_2O_2-based phase is covering on the underneath Na_2WO_4, the Na/W ratio increased after surface region of Na_2WO_4 was restructured to Na_2O_2. The Na/W ratio, about 3-4 as listed in Figure 17.3b shows the thickness of the formed Na_2O_2 is in the range of a couple to a few nanometers since the Na/W ratio is not large and the W 4f peaks were readily observed in the pink spectrum in Figure 17.2c and red spectrum in Figure 17.2f. Clearly, such a transformation of Na_2WO_4 of surface region to Na_2O_2 was driven by the high-temperature reaction condition (800 °C in the mixture of O_2 and H_2O) or high-temperature catalytic condition (800 °C in the mixture of O_2 and CH_4).

Without AP-XPS characterization of the Na_2WO_4 based OCM catalysts during catalysis at 800 °C, an ex situ study of the used catalyst could have provided inappropriate information on what is the active phase of Na_2WO_4 formed during OCM because the formed ultrathin Na_2O_2 layers on a used Na_2WO_4 catalyst must have reacted with the by-products CO_2 and H_2O of OCM, or H_2O and CO_2 in air to form NaOH, Na_2CO_3, or/and $NaHCO_3$ in the process of cooling the Na_2O_2 atomic layers to room temperature in air. An ex situ study of a used catalyst with UHV XPS is done typically with a protocol: (i) OCM is performed on Na_2WO_4-based catalyst at high temperature, (ii) the used catalyst is cooled to room temperature, (iii) the used catalyst is removed from the reactor to expose to air, and (iv) the used catalyst is placed in the chamber of a high vacuum XPS to perform ex situ characterization. Then, the Na_2O_2 formed during OCM at high temperature must have been contaminated by H_2O and CO_2 in the process of cooling or exposure to air. Thus, it makes sense that the Na_2O_2 was not observed in the past four decades although a great number of ex situ XPS studies were done since the hypothesis of Otsuka was put forward in 1987.

References

1 Otsuka, K., Said, A. A., Jinno, K. *et al.* 1987. "Peroxide anions as possible active species in oxidative coupling of methane." *Chem. Lett.*, 16, 77–80.

2 Takanabe, K. and Iglesia, E. 2008. "Rate and selectivity enhancements mediated by OH radicals in the oxidative coupling of methane catalyzed by Mn/Na2WO4/SiO2." *Angew. Chem. Int. Ed.* 120, 7803–7807.

3 Takanabe, K. and Iglesia, E. 2009. "Mechanistic aspects and reaction pathways for oxidative coupling of methane on Mn/Na2WO4/SiO2 catalysts." *J. Phys. Chem. C* 113, 10131–10145.

4 Takanabe, K. 2012. "Catalytic conversion of methane: carbon dioxide reforming and oxidative coupling." *J. Jpn. Pet. Inst.* 55, 1–12.

5 Liang, Y., Li, Z., Nourdine, M. *et al.* 2014. "Methane coupling reaction in an oxy-steam stream through an OH radical pathway by using supported alkali metal catalysts." *ChemCatChem* 6, 1245–1251.

6 Li, D., Yoshida, S., Siritanaratkul, B. *et al.* 2021. "Transient potassium peroxide species in highly selective oxidative coupling of methane over an unmolten K2WO4/SiO2 catalyst revealed by in situ characterization." *ACS Catal.* 11, 14237–14248.

7 Takanabe, K., Khan, A. M., Tang, Y. *et al.* 2017. "Integrated in situ characterization of a molten salt catalyst surface: evidence of sodium peroxide and hydroxyl radical formation." *Angew. Chem. Int. Ed. Engl.* 56, 10403–10407. doi:10.1002/anie.201704758.

8 Jiang, Z. C., Yu, C. J., Fang, C. P. *et al.* 1993. "Oxide/support interaction and surface reconstruction in the NaiWO4/SiO2 system." *J. Phys. Chem.* 97, 12870–12875.

9 Wu, Q. H., Thiben, A., and Jaegermann, W. 2005. "XPS and UPS study of Na deposition on thin film V2O5." *Appl. Surf. Sci.* 252, 1801–1805.

18

Dry and Steam Reforming of Methane

Methane, the second largest resource of carbon-based material, has been one of the major sources of valuable chemicals and energy. However, CH_4 must be transformed to intermediate compounds before carbon-based and hydrogen-based commodities can be produced. Methane can be transformed to one of the most important intermediates of chemical industries, syngas consisting of CO and H_2 since a significant number of important chemicals including alcohols, liquid fuels and diesels, alkenes, and even waxes can be produced from syngas through Fischer-Tropsch synthesis and many other processes. Methane reforming includes dry reforming and steam reforming. Industries use Ni nanoparticles supported on inert oxides as catalysts for methane steam reforming. These supported Ni nanoparticle catalysts are not ideal. They can be readily sintered at high temperatures. On the other hand, the Ni nanoparticle catalysts catalyze pyrolysis of methane. The pyrolysis of methane results in the formation of coke on the Ni nanoparticle catalysts, deactivating them readily. On the other hand, the high catalysis temperature requests large consumption of heat to retain the reaction at a high temperature for achieving a high conversion of CH_4. Driven by its low cost, they are still the main catalysts used for steam reforming to produce syngas in industries. Scientists have made significant effort in developing methane reforming catalysts with high catalytic activity and selectivity and long-lasting life. For instance, the recently reported single atom site-based catalysts, $Ce_{1-x-y}Ni_xRu_yO_2$ exhibit high activity and selectivity for steam reforming and dry reforming of methane at intermediate temperatures. AP-XPS played significant roles in these developments.[1,7]

18.1 Dry Reforming of CH_4 on CeO_2 Anchored with Ni_1 and Ru_1 Sites

A CeO_2-based catalyst anchored with two different types of single-atom sites Ni_1 and Ru_1 was prepared through a modified deposition-precipitation method by Tao group.[1] Figure 18.1 schematically presents the surface structure of this catalyst on which both Ni and Rh atoms are singly dispersed on CeO_2. The single dispersion of Ni and Ru atoms was confirmed with in situ/operando studies using X-ray absorption spectroscopy.[1] Figure 18.2 is the aberration-corrected STEM images of $Ce_{0.95}Ni_{0.025}Ru_{0.025}O_2$ catalyst after it was used for reforming CH_4 with CO_2 at 600 °C. Unfortunately, the Ni and Ru atoms are invisible in

Application of Ambient Pressure X-ray Photoelectron Spectroscopy to Catalysis, First Edition. Franklin Tao.

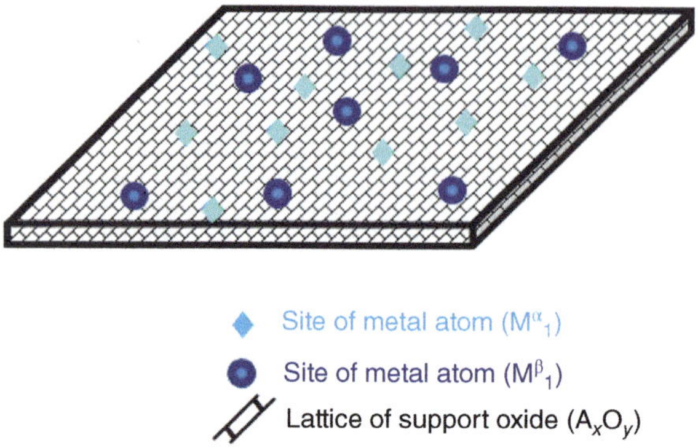

Figure 18.1 The schematic of a catalyst surface consists of two sets of single-atom sites including single-atom M_1^{α} sites (green diamonds) and single-atom M_1^{β} sites (blue balls) anchored on the surface lattice of support (black mesh). This schematic represents the surface structure of $Ce_{0.95}Ni_{0.025}Ru_{0.025}O_2$ on which Ni_1 and Ru_1 cations are single-atom sites. CeO_2 is the support. Both Ni_1 and Ru_1 atoms are singly dispersed on CeO_2. *Source:* Reproduced with permission from Tang et al.[1]/American Chemical Society.

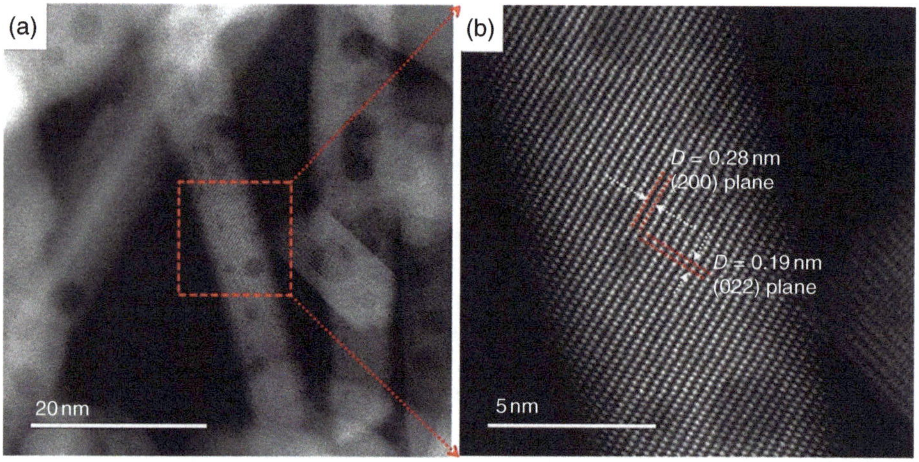

Figure 18.2 Aberration-corrected STEM images of $Ce_{0.95}Ni_{0.025}Ru_{0.025}O_2$ catalyst after it was used for reforming CH_4 with CO_2 at 600 °C performed in a fixed-bed flow reactor. (a) Large scale STEM image. (b) Single nanorod STEM image. *Source:* Reproduced with permission from Tang et al.[1]/American Chemical Society.

scanning transmission electron microscopy images since the contrast of an atom is exponentially proportional to its atomic mass and Ce is heavier than Ni and Ru.[1,2] Figure 18.2 suggests the lattice of CeO_2 remained entirely after steam reforming of methane at 600 °C. Similar to $Ce_{0.95}Ni_{0.025}Ru_{0.025}O_2$, single dispersions of Ni atoms in $Ce_{0.95}Ni_{0.05}O_2$ and Ru atoms in $Ce_{0.95}Ru_{0.05}O_2$ were confirmed by EXAF studies.[1]

Catalytic performance for steam reforming of CH_4 in the temperature range of 500–650 °C was evaluated with a fixed-bed flow reactor under the catalytic condition as listed in the caption of Figure 18.3. Although the total amount of elements Ni or/and Ru is the same on the three catalysts including $Ce_{0.95}Ni_{0.025}Ru_{0.025}O_2$, $Ce_{0.95}Ni_{0.05}O_2$, and $Ce_{0.95}Ru_{0.05}O_2$, they exhibit obviously different catalytic activities (Figure 18.3a). For instance, the conversion of CH_4 on $Ce_{0.95}Ni_{0.025}Ru_{0.025}O_2$ is obviously higher than that of $Ce_{0.95}Ni_{0.05}O_2$ or $Ce_{0.95}Ru_{0.05}O_2$

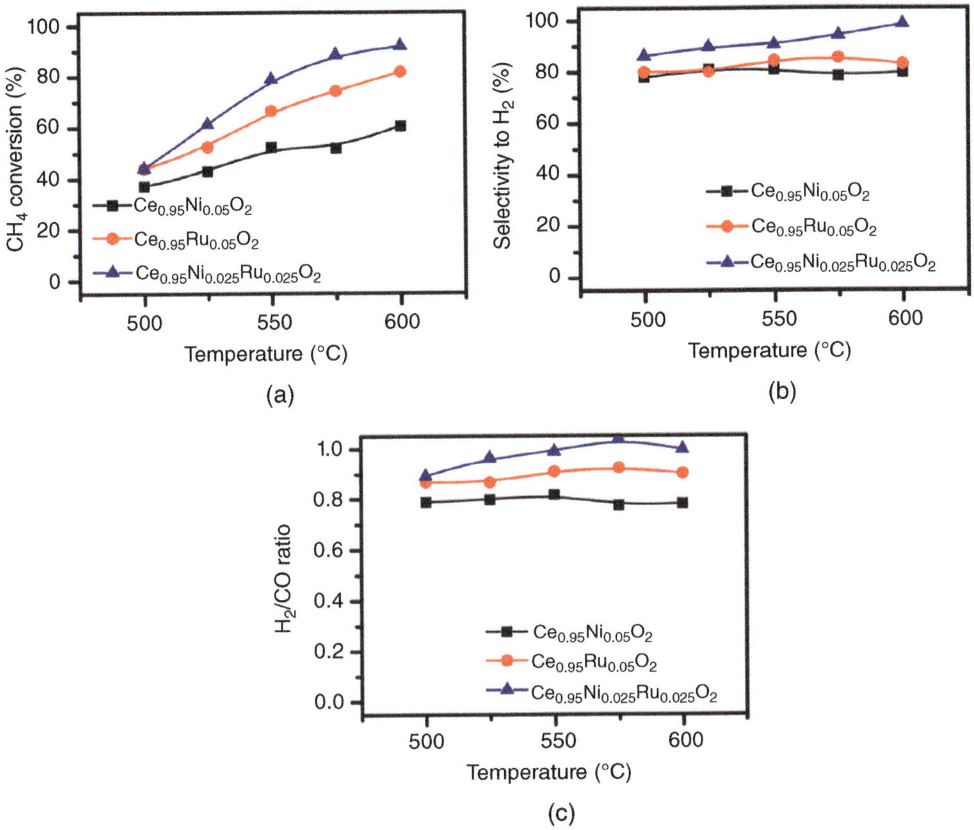

Figure 18.3 Catalytic performances of $Ce_{0.95}Ni_{0.05}O_2$, $Ce_{0.95}Ru_{0.05}O_2$, and $Ce_{0.95}Ni_{0.025}Ru_{0.025}O_2$ for reforming CH_4 with CO_2. (a) Conversion of CH_4, (b) selectivity for producing H_2, and (c) H_2/CO ratio. $Ce_{1-x-y}Ni_xRu_yO_2$ (50 mg) diluted with 300 mg purified quartz was loaded to the fixed-bed flow reactor. CH_4 (20 ml/min) (1.0%) and 20 ml/min CO_2 (1.0%) were mixed and then flown into the reactor. *Source:* Reproduced with permission from Tang et al.[1]/American Chemical Society.

(Figure 18.3a). In addition, the selectivity for production of H_2 or CO on $Ce_{0.95}Ni_{0.025}Ru_{0.025}O_2$ is obviously higher than that on $Ce_{0.95}Ni_{0.05}O_2$ or $Ce_{0.95}Ru_{0.05}O_2$. The difference in catalytic performance *between* the catalyst with both singly dispersed Ni_1 and Ru_1 *and* that with either only Ni_1 or only Ru_1 was confirmed with kinetic studies. These kinetics studies show the activation barrier for steam reforming on $Ce_{0.95}Ni_{0.025}Ru_{0.025}O_2$ is obviously lower than that on $Ce_{0.95}Ni_{0.05}O_2$ or $Ce_{0.95}Ru_{0.05}O_2$ (Figure 18.4). These catalytic evaluations suggest that the singly dispersed Ni_1 and Ru_1 sites play a synergetic role in the steam reforming of CH_4.[1]

Chemical states of Ni and Ru were tracked with AP-XPS when the catalyst was placed in the reaction cell of AP-XPS system. XPS spectra of Ce 3d, O 1s, Ni 2p, and Ru 3d were collected from the catalyst before catalysis and during catalysis at 550 °C (Figure 18.5). Based on the literature, one of the photoemission features of Ce^{3+}, v', is at 885.2 eV, which is distinct from the six peaks of Ce 3d of Ce^{4+} including v at 882.6 eV, v'' at 889.0 eV, v''' at 898.6 eV, u at 900.9 eV, u'' at 907.7 eV, and u''' at 916.9 eV. Thus, the intensity of the photoemission feature at 885.2 eV is an indicator of the concentration of Ce^{3+}. As shown in Figure 18.5a1, there is nearly a lack of the peak intensity of Ce^{3+} at 885.2 eV, suggesting the concentration of Ce^{3+} before catalysis is quite low. Through deconvolution of each Ce 3d

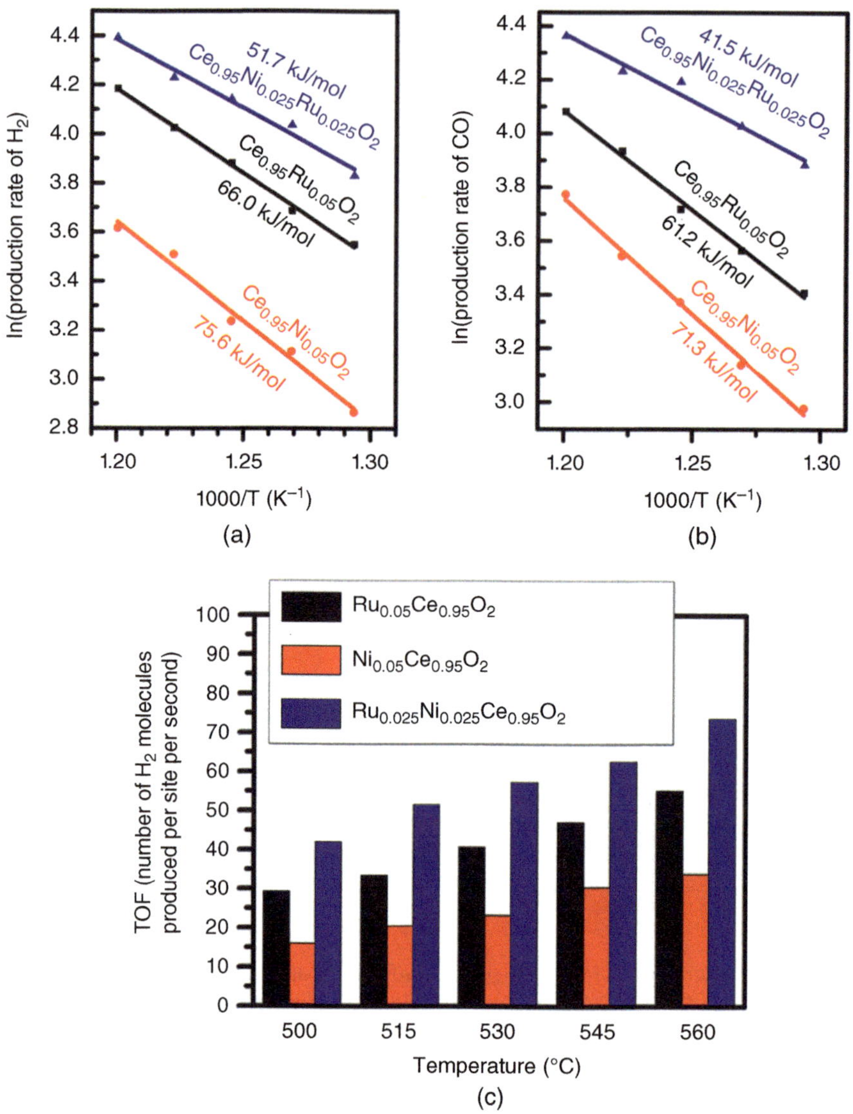

Figure 18.4 Arrhenius plots and measurement of turnover frequencies of the three catalysts through quantitative analysis of AP-XPS during catalysis. Arrhenius plot in terms of the rate of product of (a) H_2 and (b) CO catalyzed by the $Ce_{1-x-y}Ni_xRu_yO_2$ catalysts. Blue line: 5 mg of $Ce_{0.95}Ni_{0.025}Ru_{0.025}O_2$ diluted with 300 mg of purified quartz. Black line: 15 mg of $Ce_{0.95}Ru_{0.05}O_2$ diluted with 300 mg of purified quartz. Red line: 20 mg of $Ce_{0.95}Ni_{0.05}O_2$ diluted with 300 mg of purified quartz. Mixture of 40 ml/min of 1.0% CH_4 and 40 ml/min of 1.0% CO_2 was introduced during kinetic studies. Y-axis is the natural log of the production rate of H_2 (a) or CO (b) in a unit: the number of H_2 (or CO) molecules formed on a metal site per second. For $Ce_{0.95}Ni_{0.025}Ru_{0.025}O_2$, both Ni and Ru atoms of the topmost layer were counted as active sites in the calculation of production rate (the number of H_2 or CO molecules produced on a site per second). For $Ce_{0.95}Ru_{0.05}O_2$ and $Ce_{0.95}Ni_{0.05}O_2$, only Ni or Ru cations of the topmost layer were considered as active sites. The numbers of Ni_1 atoms of $Ce_{0.95}Ni_{0.05}O_2$, Ru_1 atoms of $Ce_{0.95}Ru_{0.05}O_2$, or both Ni_1 and Ru_1 of $Ce_{0.95}Ni_{0.025}Ru_{0.025}O_2$ were measured with AP-XPS during catalysis. (c) Turnover frequency (TOF) of reforming CH_4 with CO_2 in terms of hydrogen production per Ni_1 site of $Ce_{0.95}Ni_{0.05}O_2$, per Ru_1 site of $Ce_{0.95}Ru_{0.05}O_2$, and per Ni_1 or Ru_1 site for $Ce_{0.95}Ni_{0.025}Ru_{0.025}O_2$ at 500–560 °C. In the calculation of TOFs, the number of catalytic sites for $Ce_{0.95}Ni_{0.025}Ru_{0.025}O_2$ is the total number of Ni and Ru atoms of the topmost surface layer of $Ce_{0.95}Ni_{0.025}Ru_{0.025}O_2$. The number of Ni atoms on the topmost layer of $Ce_{0.95}Ni_{0.05}O_2$, the number of Ru_1 atoms on the topmost layer of $Ce_{0.95}Ru_{0.05}O_2$, and the total number of Ni_1 and Ru_1 atoms on the topmost layer of $Ce_{0.95}Ni_{0.025}Ru_{0.025}O_2$ were measured with an ambient-pressure X-ray photoelectron spectrometer during catalysis under kinetics control regime. *Source:* Reproduced with permission from Tang et al.[1]/American Chemical Society.

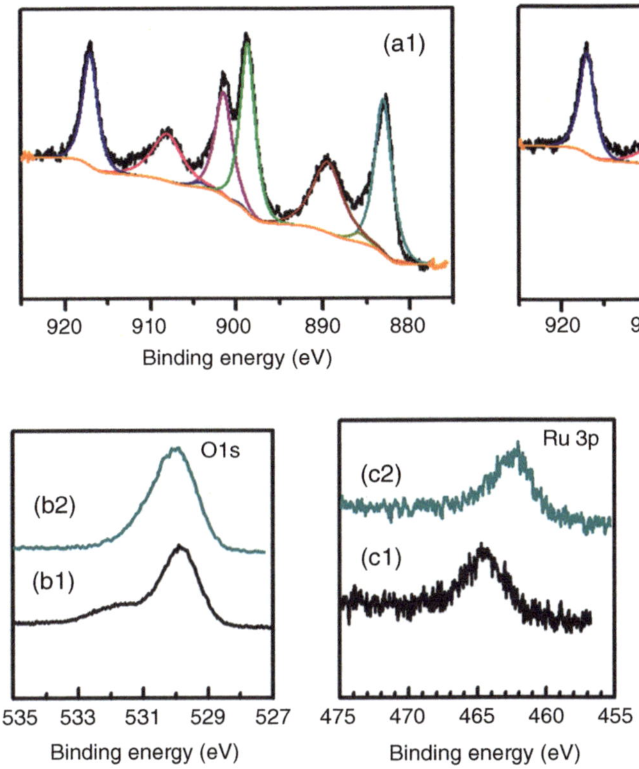

Figure 18.5 Ce 3d, O 1s, Ru 3p, and Ni 2p photoemission features from AP-XPS studies of $Ce_{0.95}Ni_{0.025}Ru_{0.025}O_2$ during reforming of CH_4 with CO_2. (a1, b1, c1, and d1) photoemission feature before catalysis. (a2, b2, c2, and d2) photoemission feature during catalysis at 550 °C. The molar fractions of CO_2 and CH_4 in the flowing mixture are 50 and 50%. *Source:* Reproduced with permission from Tang et al.[1]/American Chemical Society.

spectrum into ten peaks including v at 882.6 eV, v'' at 889.0 eV, v''' at 898.6 eV, u at 900.9 eV, u'' at 907.7 eV, and u''' at 916.9 eV for Ce^{4+} and v_o at 880.6 eV, v' at 885.2 eV, u_o at 899.2 eV, and u' at 904.0 eV for Ce^{3+}, the concentration of Ce^{3+} can be calculated with the following equation: $Ce^{3+}\% = \dfrac{A_{v_0} + A_{v'} + A_{\mu_0} + A_{\mu'}}{A_{v_0} + A_{v'} + A_{\mu_0} + A_{\mu'} + A_{v} + A_{v''} + A_{v'''} + A_{\mu} + A_{\mu''} + A_{\mu'''}}$. Figure 18.5a1 is the deconvoluted Ce 3d spectrum of the catalyst $Ce_{0.95}Ni_{0.025}Ru_{0.025}O_2$ before catalysis. The oxygen species of the catalyst in ultrahigh vacuum (UHV) before catalysis is surface lattice oxygen atoms. A close examination of the O 1s spectrum of the catalyst in UHV before catalysis (Figure 18.5b1) found a shoulder at 532.0 eV that is assigned to the oxygen atoms of OH groups on surface of CeO_2.[3,4] As Ru $3d_{5/2}$ peak overlaps C 1s, Rh 3p was collected for interpreting chemical state of Ru atoms of this catalyst (Figure 18.5c). Before catalysis, Ru atoms are obviously at an oxidizing state. Notably, compared to Ru 3p of the catalyst before catalysis, Ru 3p peak position downshifts when the catalyst was under the catalytic condition. A downshift of binding energy of Ru 3p subshell typically suggests an increase of electron density of Ru atoms. Compared to the catalyst before catalysis, the downshift of Ru 3p during catalysis suggests that Ru atoms were partially reduced or the coordination environment of Ru atoms was changed to an environment of Ru atoms having a higher electron density. It is critical to note that the Ru atoms during catalysis are still at cationic state although the Ru 3p binding energy downshifts. It suggests that Ru atoms did not form metallic Ru nanoparticles, consistent with in situ studies of EXAFS for $Ce_{0.95}Ni_{0.025}Ru_{0.025}O_2$ under a catalytic condition and ex situ studies using aberration-corrected STEM images for the used $Ce_{0.95}Ni_{0.025}Ru_{0.025}O_2$; both EXAFS and STEM-TEM suggest no metallic Ni nanoparticles were formed on $Ce_{0.95}Ni_{0.025}Ru_{0.025}O_2$.[1]

Figure 18.5d presents the XPS peak of Ni $2p_{3/2}$. There is no significant difference between the peak position of Ni $2p_{3/2}$ of the catalyst before catalysis and that during catalysis. As seen from Figure 18.5d2, the peak position of Ni $2p_{3/2}$ of $Ce_{0.95}Ni_{0.025}Ru_{0.025}O_2$ during catalysis is 854.8 eV. Definitely, the Ni atoms of $Ce_{0.95}Ni_{0.025}Ru_{0.025}O_2$ during catalysis are still at a cationic state if referred to the 853.8 eV of NiO. In terms of NiO nanoparticles, a very strong satellite peak of Ni $2p_{3/2}$ was clearly observed at 861.4 eV, which is attributed to the long-range scattering of ...O–Ni–O–Ni–O–Ni–O–Ni–O... in the surface lattice of NiO nanoparticle.[5] Thus, whether the satellite peak of Ni $2p_{3/2}$ at 861.4 eV could be observed or not, can be used to distinguish the NiO surface lattice (...O–Ni–O–Ni–O–Ni–O–Ni–O...) from the singly dispersed Ni atoms that do not have a continuously periodically packed —$(Ni–O)_n$—lattice.[5] The lack of the satellite peak of Ni $2p_{3/2}$ of NiO in the Ni $2p_{3/2}$ of $Ce_{0.95}Ni_{0.025}Ru_{0.025}O_2$ before catalysis (Figure 18.5d1) and during catalysis (Figure 18.5d2) suggests that Ni atoms are singly dispersed on the catalyst before and during catalysis, consistent with the EXAFS studies.[1]

The operando studies of the surface of $Ce_{0.95}Ni_{0.025}Ru_{0.025}O_2$ during catalysis using AP-XPS allows evaluating the turnover frequencies (TOFs) appropriately. The surface number of metal sites has been typically examined with CO chemisorption at room temperature. As a catalytic condition in terms of temperature or/and gas of reactant(s) could induce a significant change of the catalyst surface, the number of exposed metal atoms in CO chemisorption measurement at room temperature could not represent the number of metal atoms during catalysis at high temperature in the mixture of reactants. Thus, an evaluation of the number of metal atoms exposed to the surface during catalysis at high temperature would be more appropriate for calculating TOF. To check the difference in catalytic activity between $Ce_{0.95}Ni_{0.025}Ru_{0.025}O_2$ and $Ce_{0.95}Ni_{0.05}O_2$ (or $Ce_{0.95}Ru_{0.05}O_2$), TOFs were calculated by using the number of Ni_1 or/and Ru_1 atoms on the topmost layer of the catalyst *measured during catalysis* as the denominator. Specifically, in terms of $Ce_{0.95}Ni_{0.025}Ru_{0.025}O_2$, both the Ni_1 and Ru_1 atoms on the topmost layer of this catalyst during catalysis were counted as active sites. If only Ni_1 or Ru_1 atoms are taken as active sites of $Ce_{0.95}Ni_{0.025}Ru_{0.025}O_2$, the TOF must be larger. However, here we would give a conversative evaluation to the TOF of $Ce_{0.95}Ni_{0.025}Ru_{0.025}O_2$ so that the conclusion of synergistic effect is solid. Thus, both Ni_1 and Ru_1 were counted as active sites. The total number of the Ni_1 and Ru_1 sites on surface during catalysis was estimated by finding the atomic ratio of Ni/Ce and Ru/Ce of the topmost layer of a catalyst with the following equation:

$$n_{surface\ Ni\ and\ Ru} = n_{surface\ Ce} \times \frac{n_{Ni+Ru}}{n_{Ce}} = n_{surface\ Ce} \times \frac{\frac{A_{Ni\,2p}}{C_{Ni\,2p}} + \frac{A_{Ru\,3p}}{C_{Ru\,3p}}}{\frac{A_{Ce\,3d}}{C_{Ce\,3d}}}$$

Here n, A, and C stand for the number of metal atoms, the measured area of AP-XPS peak such as Ni 2p, and the cross section of a subshell level, respectively. Peak area, A was obtained from CasaXPS. The cross sections of Ni 2p and Ru 3p under Al Kα irradiation were taken from literature.[6] With the aid of the AP-XPS studies, the TOFs of the three catalysts $Ce_{0.95}Ni_{0.025}Ru_{0.025}O_2$, $Ce_{0.95}Ni_{0.05}O_2$, and $Ce_{0.95}Ru_{0.05}O_2$ in the temperature range of 500–560 °C were calculated with the yield of H_2 when the conversion of CH_4 was under kinetics-controlled regime. As shown in Figure 18.4c, $Ce_{0.95}Ni_{0.025}Ru_{0.025}O_2$ obviously exhibits higher TOFs than $Ce_{0.95}Ni_{0.05}O_2$ and $Ce_{0.95}Ru_{0.05}O_2$ in this temperature range, further confirming the synergistic effect of single-atom sites Ni_1 and Ru_1 for steam reforming of CH_4.

Obviously, quantitative analysis using AP-XPS for surface composition of a catalyst during catalysis can be crucial in the case where the surface composition varies under reaction conditions or during catalysis compared to that before or after catalysis. The use of surface composition in terms of the number of assumed catalytic sites measured through ex situ studies could give inappropriate information and even incorrectly concluded its catalytic mechanism. This work shows that AP-XPS studies allow for measuring surface composition of a catalyst during catalysis, by which the authentic catalytic activity can be evaluated to establish a reasonable base for elucidating catalytic mechanism.

18.2 Steam Reforming of CH_4 on CeO_2 Anchored with Ni_1 and Ru_1 Single-atom Sites

Oxidation of H_2 to protons on anode at low temperature generates electrical current, which is the base of low-temperature hydrogen fuel cell. Compared to H_2, CH_4 is much more inexpensive and more safe and available at household or even any places. Thus, oxidation of CH_4 to generate electrical current is an alternative way to utilize CH_4. Although a direct oxidization of CH_4 to CO_2 can provide eight electrons, it is an eight-step reaction with extremely unfavorable kinetics. Compared to the high kinetic barrier encountered by CH_4 oxidation, oxidation of H_2 requests a relatively low barrier. Thus, a compromised strategy is to first transform CH_4 to H_2 through thermal catalysis and then electrooxidize H_2 to form protons to generate electrical current. Thus, the integration of reforming CH_4 to form H_2 through thermal catalysis and a subsequent electrochemical oxidation of H_2 to generate electrical current can decrease the kinetic barrier of direct electrochemical oxidation of CH_4 to generate electrical current.

Electrochemical oxidation of H_2 to H^+ (or H_2O) on an anode can be performed at a low temperature even at room temperature. To utilize the low temperature electrochemical oxidation of H_2, CH_4 has to be first transformed to H_2. Thus, to make the conversion of chemical energy of CH_4 to electrical energy at an intermediate temperature feasible, a thermal catalyst is necessary for transforming CH_4 to H_2 at an intermediate temperature before a subsequent electrooxidation of H_2 to H^+ (or H_2O) at intermediate temperature or even low temperature to generate electrical current. With the aid of AP-XPS, an effective thermal catalyst $Ce_{0.90}Ni_{0.05}Ru_{0.05}O_{5+\delta}$ for transforming CH_4 to H_2 through steam reforming was developed.[7]

As shown in Figure 18.6, the thermal catalyst was the topmost layer of an anode to interact with the reactant gases CH_4 and H_2O.[7] The products of the steam reforming ($CH_4 + H_2O \rightarrow CO + H_2$), CO, and H_2 flow through the inter-particle space in the layer

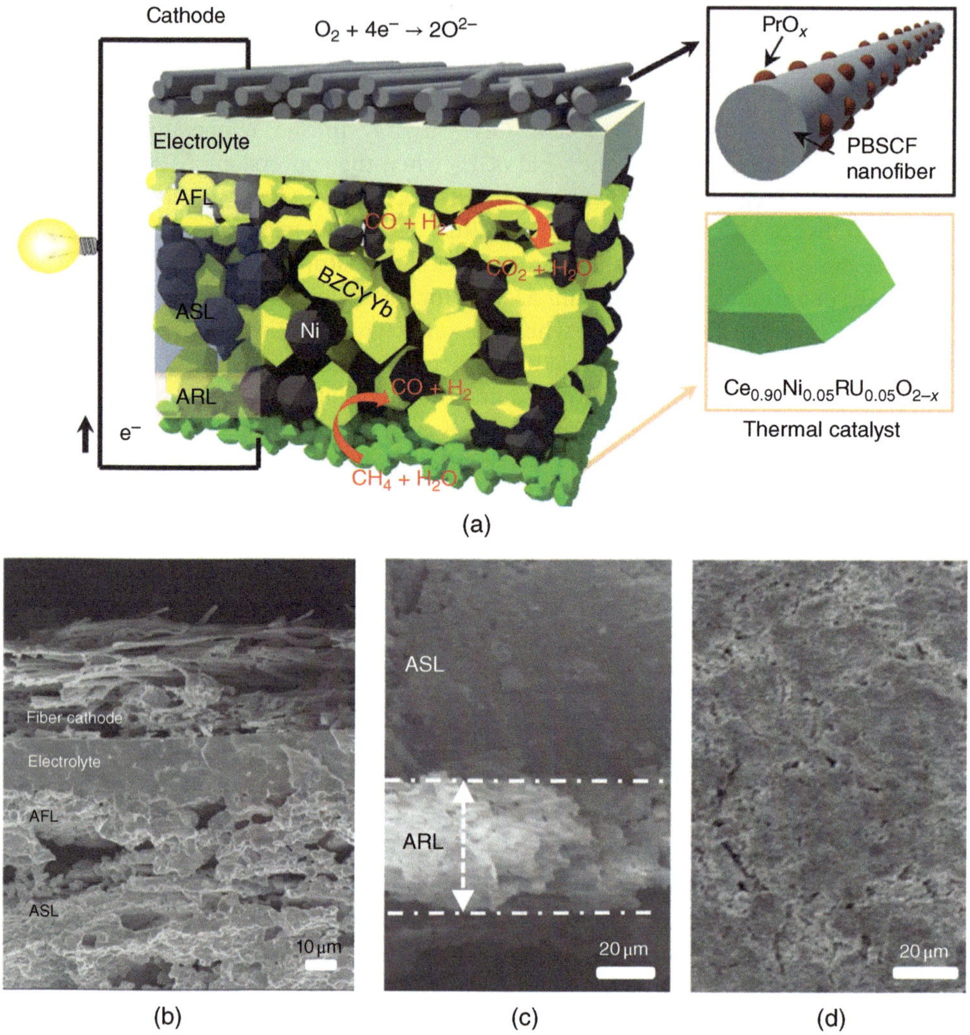

Figure 18.6 Structure of an intermediate-temperature fuel cell. (a) Schematics of a single cell (the yellow and gray grains represent the BZCYYb and the Ni phase, respectively). (b) A scanning electron micrograph of a single cell, including a fiber cathode, a dense electrolyte, an anode function layer (AFL), and an anode supporting layer (ASL). (c) A cross-sectional view of the ASL and the ARL. (d) A top-down view of the ARL. *Source:* Reproduced with permission from Chen et al.[7]/ Springer Nature.

of thermal catalyst and then are oxidized in the under layer of electrocatalysis to generate electrical current. Nanoparticle catalysts, $Ce_{0.90}Ni_{0.10}O_5$, $Ce_{0.90}Ru_{0.10}O_5$, and $Ce_{0.90}Ni_{0.05}Ru_{0.05}O_5$ were synthesized with the same methods.[7] Notably, $Ce_{0.90}Ni_{0.05}Ru_{0.05}O_5$ consists of the support CeO_2 and the anchored singly dispersed Ni^{n+} and Ru^{m+} cations. As shown in Figure 18.7b, $Ce_{0.90}Ni_{0.10}O_5$, $Ce_{0.90}Ru_{0.10}O_5$, and $Ce_{0.90}Ni_{0.05}Ru_{0.05}O_5$ are active for steam reforming of CH_4 in the temperature range of 400–500 °C. An obvious difference in catalytic activity among the three catalysts is found. Obviously, the TOF of $Ce_{0.90}Ni_{0.05}Ru_{0.05}O_5$ is higher than $Ce_{0.90}Ni_{0.10}O_5$ and $Ce_{0.90}Ru_{0.10}O_5$ (Figure 18.7b). The highest activity of

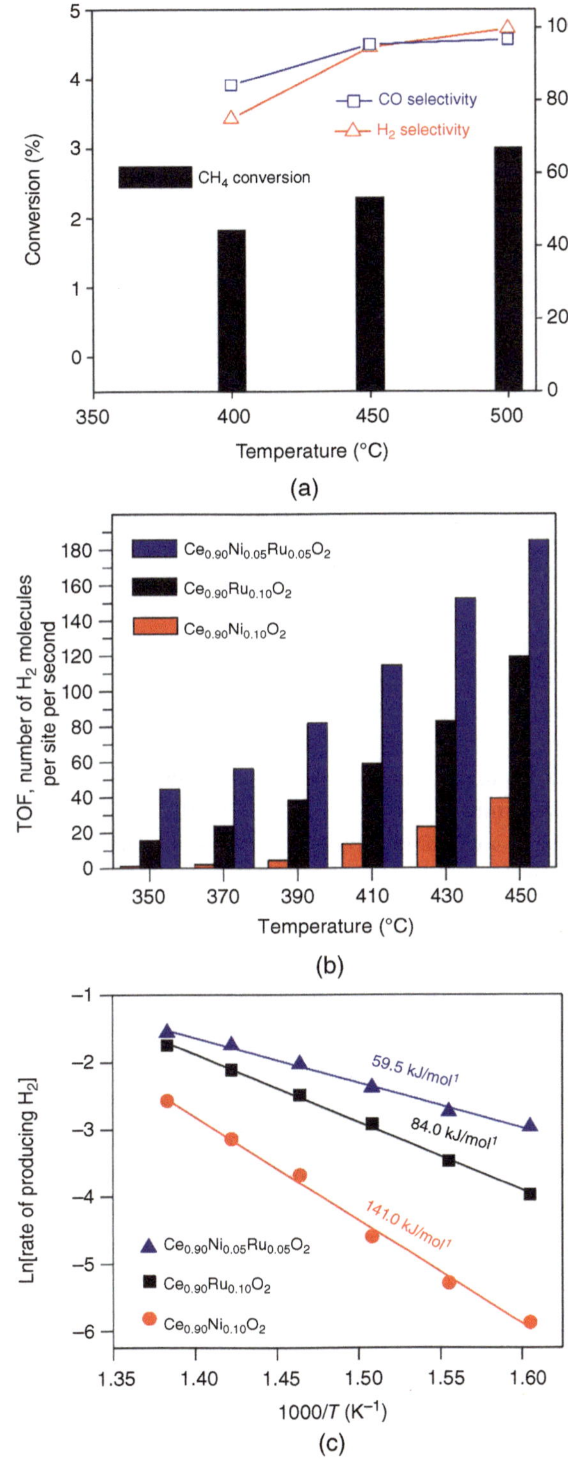

Figure 18.7 Catalytic performances of steam reforming of CH_4 on $Ce_{0.90}Ni_{0.05}Ru_{0.05}O_2$, $Ce_{0.90}Ni_{0.10}O_2$, and $Ce_{0.90}Ru_{0.10}O_2$ catalysts. (a) Conversion of methane and selectivity for hydrogen as a function of catalysis temperature for steam reforming on $Ce_{0.90}Ni_{0.05}Ru_{0.05}O_2$. (b) Number of produced H_2 molecules per second per Ni atom for $Ce_{0.90}Ni_{0.10}O_2$, per Ru atom for $Ce_{0.90}Ru_{0.10}O_2$, and per Ru and Ni atom for $Ce_{0.90}Ni_{0.05}Ru_{0.05}O_2$. (c) Arrhenius plots for producing H_2 on the three catalysts. Pure CH_4 (99%, Matheson) with a flow rate of 50 ml/min was introduced into a bubbling bottle containing deionized water kept at room temperature (~3.5% H_2O). These linear lines of these Arrhenius plots were generated through least-square fitting of data points for each catalyst. *Source:* Reproduced with permission from Chen et al.[7]/ Springer Nature.

$Ce_{0.90}Ni_{0.05}Ru_{0.05}O_5$ among the three catalysts was further confirmed by the kinetic studies (Figure 18.7c). Compared to 141.0 kJ/mol for $Ce_{0.90}Ni_{0.10}O_5$ and 84.0 kJ/mol for $Ce_{0.90}Ru_{0.10}O_5$, $Ce_{0.90}Ni_{0.05}Ru_{0.05}O_5$ exhibits a much lower barrier, 59.5 kJ/mol.

Quantitative and qualitative analyses of AP-XPS data played a significant role in calculating TOFs of these catalysts and tracking oxidation states of metal atoms during catalysis, respectively. The Ni 2p, Ru 3p, Ce 3d, and O 1s XPS spectral features of $Ce_{0.90}Ni_{0.05}Ru_{0.05}O_5$ in UHV (Figure 18.8a1, b1, and c1), during steam reforming of CH_4 at 450 °C (Figure 18.8a2, b2, and c2) and 500 °C (Figure 18.8a3, b3, and c3). There is no obvious change of binding energy of Ni 2p although Ru 3p obviously downshifts upon the characterization condition was switched from UHV to steam reforming of methane at 450 °C. Clearly, both Ni and Ru remain a cationic state instead of a metallic state during catalysis at 450 and 500 °C. The nature of cations of Ni and Ru atoms of the catalyst during catalysis was supported by the coordination environments of Ni and Ru atoms uncovered in in situ studies using EXAFS.[7] EXAFS studies confirmed that both Ni and Ru cations are singly dispersed. In other words, there is a lack of nanoparticles of Ni, NiO_x, Ru, or RuO_x.[7] Compared to Ce 3d of $Ce_{0.90}Ni_{0.05}Ru_{0.05}O_5$ in UHV (Figure 18.8c1), the obvious increase of intensity of the ν' peak of Ce^{3+} of $Ce_{0.90}Ni_{0.05}Ru_{0.05}O_5$ at 885.2 eV during catalysis (Figure 18.8c2 and c3) shows creation of molecular oxygen vacancies during catalysis (Figure 18.8c). Figure 18.8d is XPS peaks of O 1s. The high binding energy peak of O 1s at 535–536 eV is assigned to H_2O molecules in gas phase. The observation of gas phase peak of H_2O in the AP-XPS spectra (Figure 18.8d2 and d3) confirmed that the AP-XPS data of Figure 18.8a2 and d2 and a3 and d3 were collected during catalysis.

TOF of methane steam reforming on the three catalysts were calculated through dividing reaction rate of a catalyst (number of product molecules formed on the catalyst per second) by the number of sites on surface of this catalyst. The measurement of reaction rate is requested to perform under kinetics-controlled regime in terms of a low conversion of reactants.

$$TOF = \frac{r_{formed\ \mathrm{H_2}}}{n_{catalytic\ site}} = \frac{S_{selectivity\ to\ \mathrm{H_2}} \times X_{conversion\ of\ \mathrm{CH_4}} \times F_{molar\ flow\ rate\ of\ \mathrm{CH_4}} \times N_A \times 3}{m_{catalyst} \times A_{surface\ area\ per\ gram} \times d_{surface\ density\ of\ Ce} \times \dfrac{n_{Ni\ or\ Ru\ or\ both\ Ni\ and\ Ru}}{n_{\mathrm{Ce}}}}.$$

Here $r_{formed\ \mathrm{H_2}}$ is the rate of forming product H_2 on the catalyst. Strictly speaking, the number of catalytic sites ($n_{catalytic\ site}$) is the number of effective sites on the catalyst surface measured during catalysis. It could be the same, similar, or different from that of the catalyst before or after catalysis. Regardless of any potential difference between $n_{catalytic\ site}$ during catalysis or $n_{catalytic\ site}$ before catalysis or after catalysis, a measurement of the number of active sites during catalysis is necessary. Here AP-XPS exhibits its unique function compared to many other techniques as it has high surface sensitivity and it can collect data during catalysis. As the active sites for $Ce_{0.90}Ni_{0.10}O_5$, $Ce_{0.90}Ru_{0.10}O_5$, and $Ce_{0.90}Ni_{0.05}Ru_{0.05}O_5$ are Ni cations, Ru cations, and both Ni and Ru cations, respectively, the number of Ni cations, Ru cations, or both Ni and Ru cations were measured. In XPS and AP-XPS, it is not necessary to measure the absolute number of Ni, Ru, or both Ni and Ru cations. In fact, we only need to measure the ratio, $\frac{n_{Ni\ or\ Ru\ or\ both\ Ni\ and\ Ru}}{n_{Ce}}$, which is requested in the

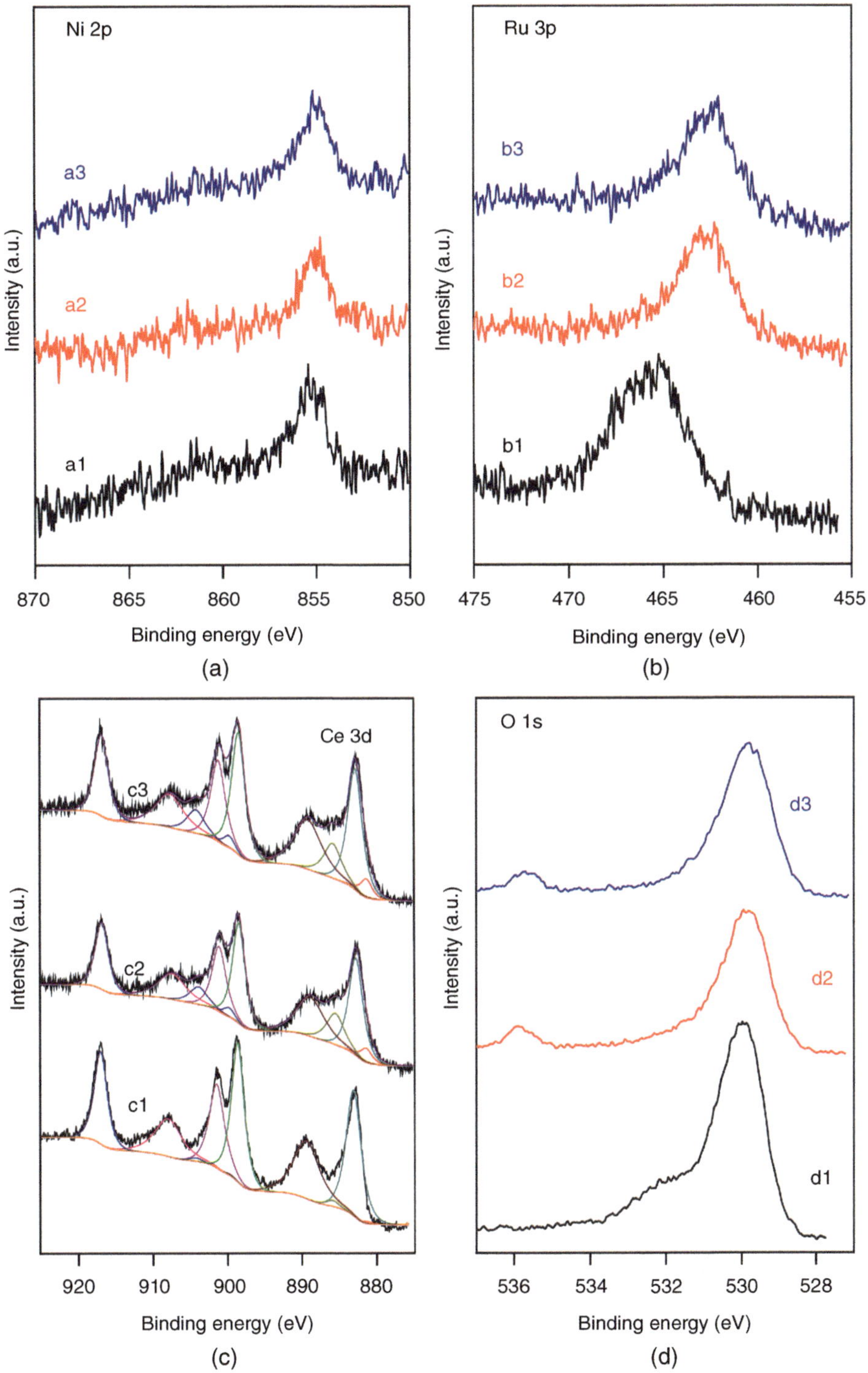

Figure 18.8 AP-XPS analysis of singly dispersed Ni atoms (Ni_1) and Ru atoms (Ru_1) of $Ce_{0.90}Ni_{0.05}Ru_{0.05}O_2$ during catalysis. (a) Ni 2p XPS peaks of Ni_1 atoms on CeO_2. (b) Ru 3p XPS peaks of Ru_1 atoms on CeO_2. (c) Ce 3d XPS peaks of Ce atoms of CeO_2 collected by AP-XPS. (d) O 1s XPS peaks of O atoms of CeO_2 collected by AP-XPS. Spectra a1, b1, c1, and d1 were collected from synthesized catalyst at 200 °C in ultrahigh vacuum to avoid charging. Spectra a2, b2, c2, and d2 were collected when $Ce_{0.90}Ni_{0.05}Ru_{0.05}O_2$ was in the mixture of CH_4 and H_2O during steam reforming at 450 °C. Spectra a3, b3, c3, and d3 were collected when $Ce_{0.90}Ni_{0.05}Ru_{0.05}O_2$ was in the mixture of CH_4 and H_2O during steam reforming at 500 °C. The deconvolution of experimental photoemission features in (c) is explained in the Methods (AP-XPS and in situ/operando EXAFS studies). *Source:* Reproduced with permission from Chen et al.[7]/Springer Nature.

formula of calculating TOF since the number of Ce atoms on surface of CeO_2 of the same amount of a catalyst, n_C, is almost the same for the three catalysts. Based on the principle of AP-XPS, $\frac{n_{Ni}}{n_{Ce}}$, $\frac{n_{Ru}}{n_{Ce}}$, or $\frac{n_{both\,Ni\,and\,Ru}}{n_{Ce}}$ can be calculated with $\frac{A_{Ni2p}/C_{Ni2p}}{A_{Ce3d}/Ce_{3d}}$, $\frac{A_{Ru3p}/C_{Ru3p}}{A_{Ce3d}/Ce_{3d}}$, or $\frac{A_{Ni2p}/C_{Ni2p}+A_{Ru3p}/C_{Ru3p}}{A_{Ce3d}/Ce_{3d}}$, respectively. The cross sections of Ni 2p, Ru 3p, and Ce 3d excited by Al Kα are available in reference.[6]

Through the quantitative analysis of these AP-XPS data (Ni 2p, Ru 3p, and Ce 3d) collected during catalysis, the TOFs of the three catalysts were calculated and thus authentically reflected the number of active sites. As shown in Figure 18.7b, the accurately measured TOFs based on the authentic number of active sites measured during catalysis with AP-XPS show that the catalyst consisting both Ni and Ru cations, $Ce_{0.90}Ni_{0.05}Ru_{0.05}O_{5+\delta}$, exhibits definitely higher activity than $Ce_{0.90}Ni_{0.10}O_{5+\delta}$ or $Ce_{0.90}Ru_{0.10}O_{5+\delta}$, by which a synergic effect between Ni and Ru cations was uncovered for the first time. The participation of both Ni and Ru cations of $Ce_{0.90}Ni_{0.05}Ru_{0.05}O_{5+\delta}$ and the synergic effect were supported by computational studies reported in literature.[7] This work demonstrated the crucial role of AP-XPS in giving an appropriate TOF since an accurate measurement of TOF is the base for comparing catalytic performances of different catalysts for fundamental understanding of catalytic mechanism at a molecular level.

References

1 Tang, Y., Wei, Y., Wang, Z. *et al.* 2019. "Synergy of single-atom Ni1 and Ru1 sites on CeO_2 for dry reforming of CH_4." *J. Am. Chem. Soc.* 141, 7283–7293.

2 Yang, J. C., Small, M. W., Grieshaber, R. V. *et al.* 2012. "Recent developments and applications of electron microscopy to heterogeneous catalysis." *Chem. Soc. Rev.* 41, 8179–8194.

3 Mullins, D., Overbury, S., and Huntley, D. 1998. "Electron spectroscopy of single crystal and polycrystalline cerium oxide surfaces." *Surf. Sci.* 409, 307–319.

4 Newberg, J. T., Starr, D. E., Yamamoto, S. *et al.* 2011. "Formation of hydroxyl and water layers on MgO films studied with ambient pressure XPS." *Surf. Sci.* 605, 89–94.

5 Shan, J., Huang, W., Nguyen, L. *et al.* 2014. "Conversion of methane to methanol with a bent mono(mu-oxo)dinickel anchored on the internal surfaces of micropores." *Langmuir* 30, 8558–8569.

6 Yeh, J. J. and Lindau, I. 1985. "Atomic subshell photoionization cross sections and asymmetry parameters: $1 \leqslant Z \leqslant 103$." *At. Data Nucl. Data Tables* 32, 1–155.

7 Chen, Y., deGlee, B., Tang, Y. 2018. "A robust fuel cell operated on nearly dry methane at 500 °C enabled by synergistic thermal catalysis and electrocatalysis." *Nat. Energy* 3, 1042–1050.

19

Reduction of NO with CO

NO is one of the most poisonous gasses. It can be transformed to a nonpoisonous gas, N_2, through reduction by CO to form N_2 and CO_2. Reduction of NO with CO is one of the steps in removal of wastes generated from gasoline engines in automobiles. Rh is the metal most active for this reaction due to its low activation barrier of dissociating NO. Unfortunately, it is one of the least abundant metals on the Earth (Figure 3.1). Search of replacement of Rh has been a task for decades. The abundance of Co is 10^5 of Rh on the Earth (Figure 3.1). Recently, Co_3O_4 and Rh_1Co_3 singly dispersed bimetallic sites supported on Co_3O_4 were reported as a potential replacement with high activity and selectivity.[1]

19.1 Reduction of NO with CO on Co_3O_4

Co_3O_4 nanorods were synthesized through a hydrothermal method.[1, 2] They are nanorods with a diameter of 6 nm or so (Figure 19.1b). It is active for the reduction of NO with CO. As shown in Figure 19.2, it exhibits high activity in conversion of NO through reduction with CO and high selectivity for production of N_2 under the given catalytic condition. AP-XPS uncovered the authentic active phase of Co_3O_4 for reducing NO with CO for producing N_2 and CO_2. In AP-XPS studies, Co_3O_4 nanorods were placed on a sample holder of AP-XPS instrument. The sample holder was subsequently transferred to the sample stage of a reaction cell where the mixture of NO and CO was flowing through the loaded catalyst powder; then, the catalyst was heated to a specific temperature in the flow of mixture of NO and CO. Both Co 2p and O 1s photoemission features were collected from the catalyst surface at certain temperature in the flow of mixture of NO and CO (Figure 19.3). Notably, at a temperature lower than 250 °C, there is no shoulder at 786.9 eV which is the characteristic peak of Co $2p_{3/2}$ at 786.9 eV for CoO (Figure 19.3a4–a12). The lack of the characteristic peak of Co $2p_{3/2}$ at 786.9 eV suggests that the surface phase of cobalt oxide at < 250 °C is lack of CoO. Notably, the Co^{2+} in Co_3O_4 coordinates with four oxygen atoms in a tetrahedral coordination sphere, which does not give a satellite peak at 786.9 eV. However, it does give a quite weak satellite peak of $2p_{3/2}$ at 791.5 eV contributed from Co^{3+} in an octahedral coordination. Thus, the observation of a weak satellite peak of Co $2p_{3/2}$ at 789.8 eV suggests the cobalt oxide is Co_3O_4.

At 250 °C, the shoulder at 786.9 eV can be obviously observed (Figure 19.3a3). This shoulder is assigned to Co^{2+} in an octahedral coordination with six oxygen atoms in CoO.

Application of Ambient Pressure X-ray Photoelectron Spectroscopy to Catalysis, First Edition. Franklin Tao.

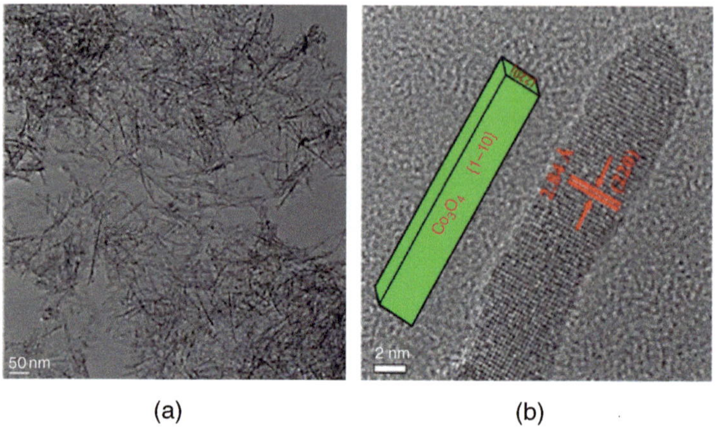

Figure 19.1 TEM images of as-synthesized Co_3O_4 nanorods. (a) Large-scale TEM image. (b) High-resolution TEM image of a representative Co_3O_4 nanorod. *Source:* Reproduced with permission from Zhang et al.[1]/American Chemical Society.

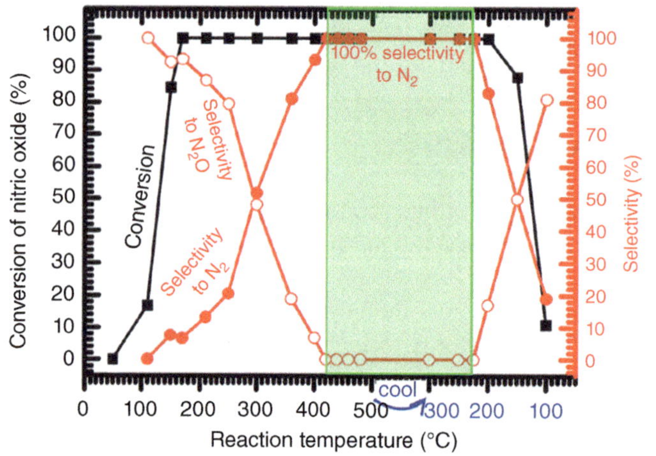

Figure 19.2 Catalytic activity (the left Y-axis, black line in figure) and selectivity (the right Y-axis, red line in figure) of cobalt oxides at different temperatures in the mixture gas of NO (5%) and CO (15%) followed by cooling down to 300 °C and then 100 °C. *Source:* Reproduced with permission from Zhang et al.[1]/American Chemical Society.

This appearance of the characteristic peak of Co^{2+} in the octahedral coordination suggests the transformation of surface phase from Co_3O_4 to CoO. Figure 19.4b schematically presents a CoO rocksalt structure in which each Co^{2+} coordinates with six oxygen atoms in an octahedral coordination sphere. This specific coordination gives a satellite peak for Co $2p_{3/2}$ at 786.9 eV. Thus, an indicator for the reduction of Co_3O_4 to CoO is the appearance of the specific and obvious satellite peak of Co^{2+} of CoO at 786.9 eV along the increase of temperature as exemplified in Figure 19.3. In addition, the main peak of Co $2p_{3/2}$ of Co_3O_4 in Figure 19.3b1–b4 upshifts by about 1.5 eV since Co_3O_4 is restructured to CoO (Figure 19.3b5–b12). The upshift of the main peak also supports the restructuring of Co_3O_4 surface to CoO.

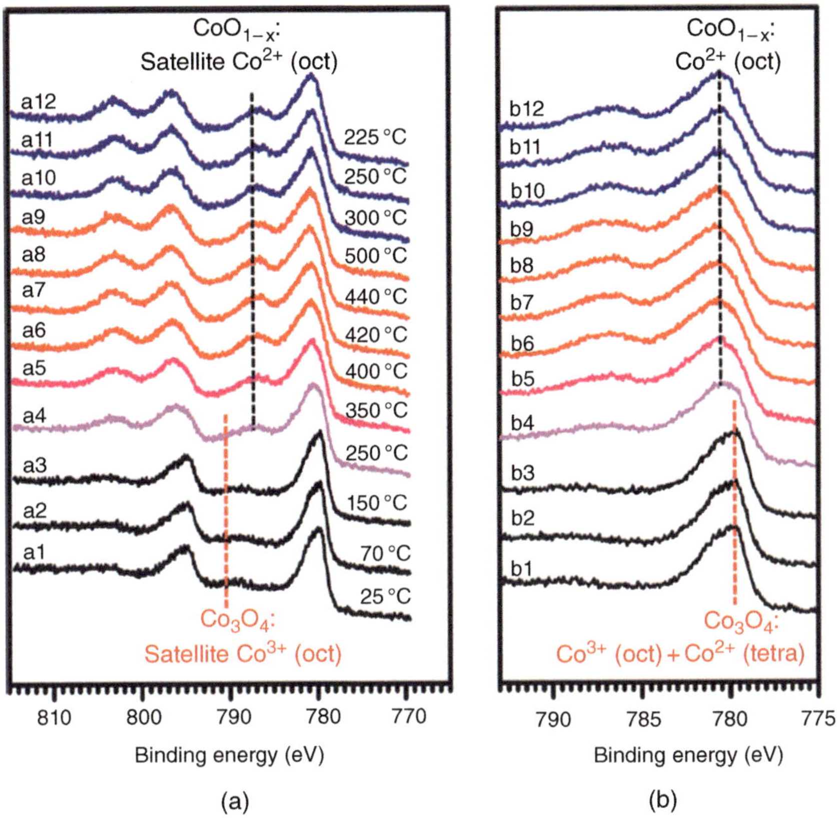

Figure 19.3 AP-XPS studies of Co 2p (a) and Co $2p_{3/2}$ (b) of Co_3O_4 in the mixtures of CO and NO at different catalysis temperatures. The evolution from low temperatures to high temperatures shows the restructuring of spinel Co_3O_4 to rock salt CoO_{1-x} under reaction conditions. The composition of reactant gases over the catalyst during data acquisition is 1 Torr NO and 3 Torr CO. *Source:* Reproduced with permission from Zhang et al.[1]/American Chemical Society.

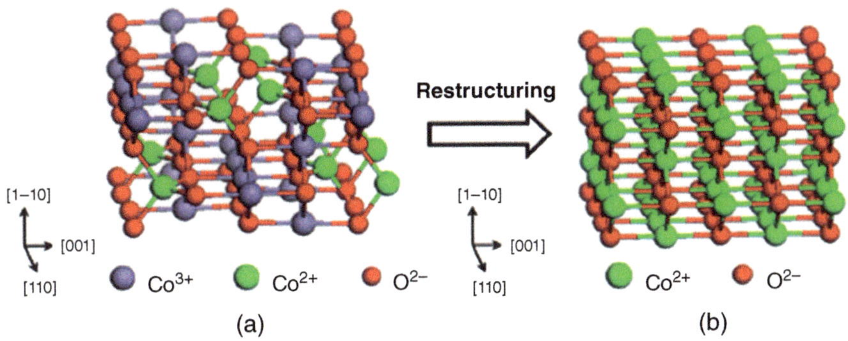

Figure 19.4 Schematic of Co_3O_4 nanorods and the CoO formed through restructuring from Co_3O_4 under reaction conditions. *Source:* Reproduced with permission from Zhang et al.[1]/American Chemical Society.

As shown in Figure 19.2, Co_3O_4 is active for reduction of NO with CO at a temperature as low as 100 °C. The selectivity for producing N_2 significantly increases in the temperature range of 200–400 °C. At 400 °C, the selectivity reaches 100%. By correlating with the increase of catalytic selectivity for producing N_2 with increase of catalytic temperature from 150 to 400 °C and the gained high selectivity of nearly 100% for producing N_2 at 400 °C, the transformation of surface Co_3O_4 to CoO phase in the temperature range of 250–350 °C suggests that the CoO is the active phase for the selective production of N_2 in reduction of NO with CO.

Quantitative analyses of AP-XPS to Co 2p and O 1s peaks collected during catalysis uncovered that active phase is in fact oxygen-deficient CoO. Atomic ratios of O to Co during catalysis at different temperatures were calculated by assuming the O to Co atomic ratio of Co_3O_4 at room temperature in the mixture of NO and CO is 1.33, which is the stoichiometric ratio of O to Co of Co_3O_4. With this assumption, the atomic ratios O/Co of cobalt oxide in the structural transformation from 25 to 500 °C were measured and plotted in Figure 19.5. As shown in Figure 19.5, O/Co atomic ratio progressively decreases along the increase of catalysis temperature in 25–350 °C, consistent with the transformation of Co_3O_4 with high O/Co ratio at 1.33 to CoO with low O/Co ratio. More accurately, the O/Co ratio in the 400–500 °C region during reduction of NO with CO is obviously lower than 1.0 as marked in Figure 19.5, showing the existence of oxygen vacancies of surface of CoO in this temperature range (marked with green shadow in Figure 19.5). In terms of catalytic performance, this catalyst exhibits the highest selectivity (100%) in the same temperature range (marked with green in Figure 19.2). Thus, we can conclude that the nonstoichiometric CoO_{1-x} is the active phase for reduction of NO with CO.

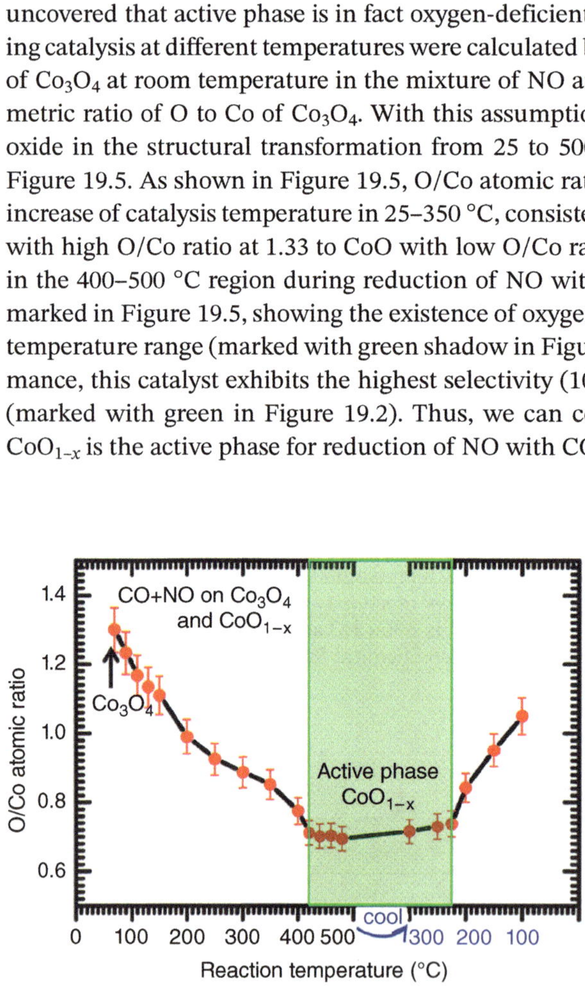

Figure 19.5 Atomic ratio of oxygen to cobalt of Co_3O_4 and CoO_{1-x} (formed through a restructuring of Co_3O_4 at a temperature above 250 °C) during catalysis of CO + NO (1 Torr NO and 3 Torr CO) in the same temperature regime as catalytic measurements in Figure 19.2. The atomic ratios of O to Co at different temperatures were calculated through a strict implementation of the same set of XPS analytical parameters of O 1s and Co 2p photoemission features to all spectra obtained at different temperatures for minimization of a potential error. *Source:* Reproduced with permission from Zhang et al.[1]/American Chemical Society.

Different from alumina and silica, most transition metal oxides are not insulator. Surface charging effect is negligible or at least manageable for these transition metal oxides. Thus, XPS is an appropriate tool to perform an accurate analysis for surfaces of these transition metal oxides. Oxygen vacancies of these oxides can activate a great number of molecules such as O_2, NO, and H_2O. In addition, their metal atoms could activate C—H bonds. The mostly studied transition metal oxides are CeO_2, TiO_2, Co_3O_4, CoO, NiO, ZrO_2, MoO_3, WO_3, V_2O_5, Fe_2O_3, CuO, Cu_2O, RuO_2, ZnO, PdO, Rh_2O_3, Ir_2O_3, etc. Thus, AP-XPS is a significant technique in studying catalysis on transition metal oxides. Analysis of these AP-XPS data allows us to correlate surface defects of these transition metal oxides such as the density of oxygen vacancies of CoO_{1-x} with their corresponding catalytic performances. Through such a correlation, the authentic catalytic sites can be identified.

19.2 Reduction of NO with CO on Rh_1Co_3 Clusters Supported on CoO

Another catalyst active for reducing NO with CO to form N_2 and CO_2 is Rh_1Co_3/CoO as reported by Tao group. It is a catalyst of singly dispersed bimetallic sites anchored on cobalt oxide. The feature of this catalyst is its high activity and high selectivity in transforming poisonous NO to nonpoisonous N_2 with CO *at a temperature as low as 110 °C* (Figure 19.6a). The feature of this type of catalyst is its unique structure of a catalytic site at atomic level (Figure 19.7d). This catalyst consists of singly dispersed bimetallic sites anchored on surface of an oxygen-deficient CoO. In brief, it was prepared by oxidizing the catalyst precursor at 150 °C in 5% O_2 and then reduction at 300 °C in 5% H_2.[3,4]

Preparation of this catalyst consists of wet and dry processes. The wet process includes deposition precipitation of Rh cations to Co_3O_4 nanorods and then drying the precipitate at 60 °C in open air for 24 hours. The dry process covers an annealing at certain temperature such as 150 °C in air and then annealing it at an appropriate temperature in H_2 (Figure 19.7). This preparation process was schematically shown in Figure 19.7. AP-XPS was used as a "video camera" to watch the evolution of surface of the catalyst precursor in the dry process and find the optimum reduction temperature for formation of the singly dispersed bimetallic sites (Figure 19.7d).[4] After Rh cations were loaded to surface of Co_3O_4, the precursor was dried at 60 °C in air for 24 hours. After the wet chemistry process, the catalyst precursor was placed in the reaction cell of an AP-XPS system to track how surface of the catalyst precursor evolves as a function of pretreatment gases and temperatures. By tracking surface chemistry of the catalyst precursor at different temperatures in different pretreatment gases with AP-XPS, an ideal preparation condition (temperature and gas) for forming a catalyst consisting of singly dispersed bimetallic sites Rh_1Co_3 on an oxide support was found. As shown in the black spectrum in Figure 19.8a, during oxidation at 150 °C with O_2, Rh atoms are at cationic state evidenced by the high binding energy of Rh $3d_{5/2}$ at 309.4 eV. Based on the corresponding EXAFS studies (black spectrum in Figure 19.9a),[3] Rh atoms are singly dispersed and bonds with five oxygen atoms on average, designated as Rh_1O_5. Upon oxidation at 150 °C, Rh atoms are anchored on surface of Co_3O_4. To make the singly dispersed Rh_1 atom bond with Co atoms directly, reduction with H_2 to remove certain amount of surface lattice oxygen atoms is necessary. AP-XPS and EXAFS studies of

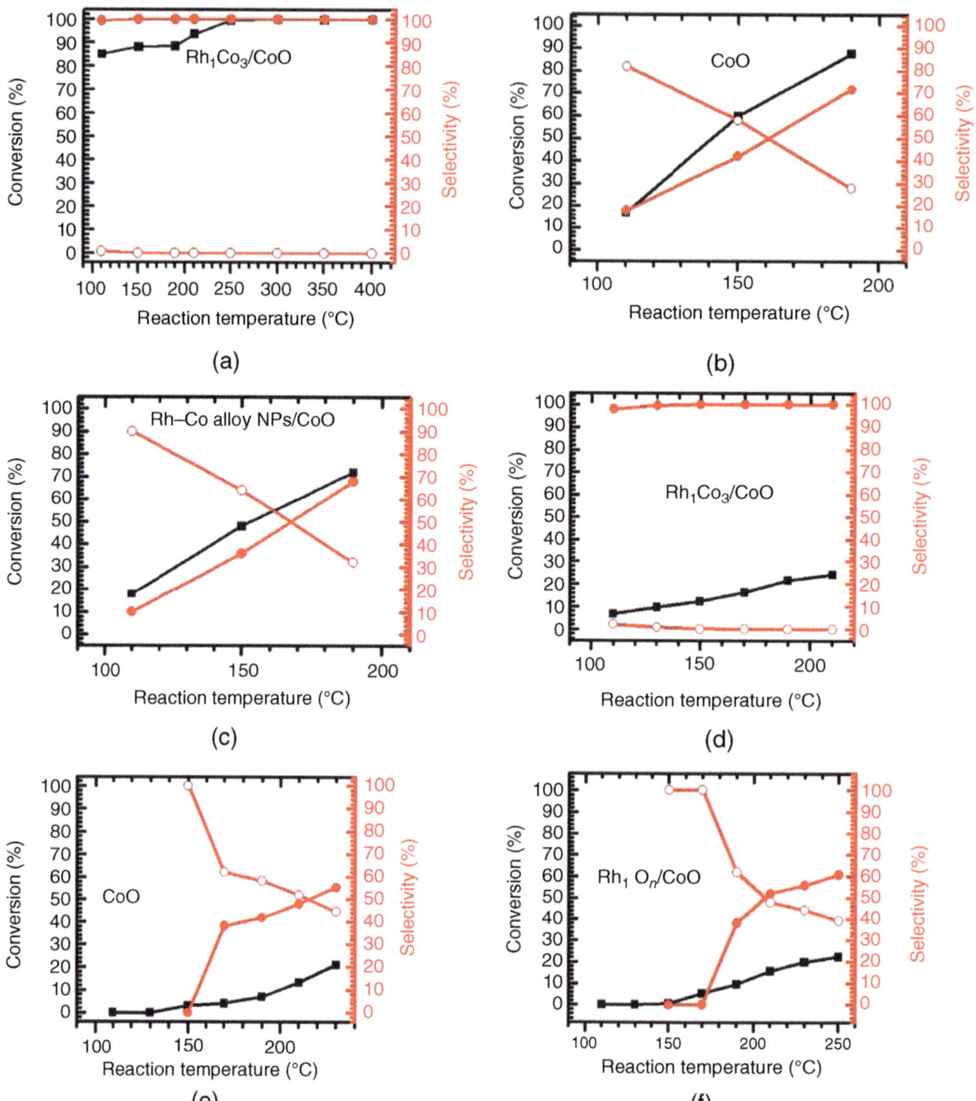

Figure 19.6 Studies of catalytic performances under two different catalytic conditions. Catalytic condition 1: catalytic activity and selectivity of these catalysts were measured with the exact same condition: mixture of 20 ml/min 15% NO balanced with Ar, 30 ml/min 10% CO and 10 ml/min 99.999% Ar; 10 mg of a catalyst was used in each catalytic measurement; the three catalysts are (a) Rh_1Co_3/CoO, (b) pure CoO, and (c) Rh–Co bimetallic nanoparticles supported on CoO. Catalytic condition 2: it is a condition for kinetics-controlled regime; the reactant is the mixture of 80 ml/min 15% NO (balanced with Ar) and 12 ml/min pure CO. (d) 5 mg of Rh_1Co_3/CoO, (e) 10 mg of CoO, and (f) 5 mg of Rh_1O_n/CoO. Each of these catalysts was mixed with purified quartz before evaluation of catalytic performance or kinetics studies. *Source:* Reproduced with permission from Zhang et al.[3]/Springer Nature.

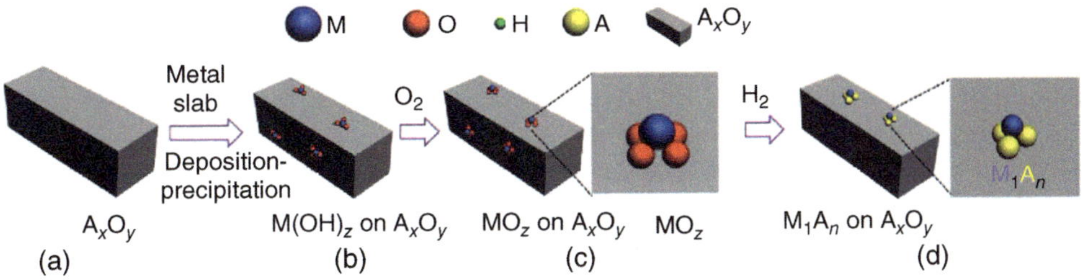

Figure 19.7 Schematic of synthesis of a catalyst of singly dispersed bimetallic sites M_1A_n. (a) Well-crystallized reducible oxide support, A_xO_y. (b) Deposition of hydroxide species of guest metal, $M(OH)_z$, to surface of the oxide support, A_xO_y; a low concentration of M cations (0.5 wt% or less) allows them to be singly dispersed on the surface of A_xO_y; the atomic ratio of M atoms to A atoms of the topmost layer of atoms A of A_xO_y is typically <5–10%; $M(OH)_z$ species are weakly bound to the surface of A_xO_y. (c) Formation of singly dispersed M_1O_n species upon oxidation in 5% O_2 at 150 °C; M cations are bonded to A_xO_y surface through M—O—A bonds. (d) A controlled reduction in 5% H_2; removal of oxygen atoms between M and A atoms by reduction in H_2 makes M directly bond with A atoms of oxide support, forming an M_1A_n site; as M atoms are singly dispersed on A_xO_y (due to the low concentration of guest atoms M), the formed M_1A_n bimetallic sites are isolated. They are called singly dispersed bimetallic sites. *Source:* Reproduced with permission from Zhang et al.[3]/Springer Nature.

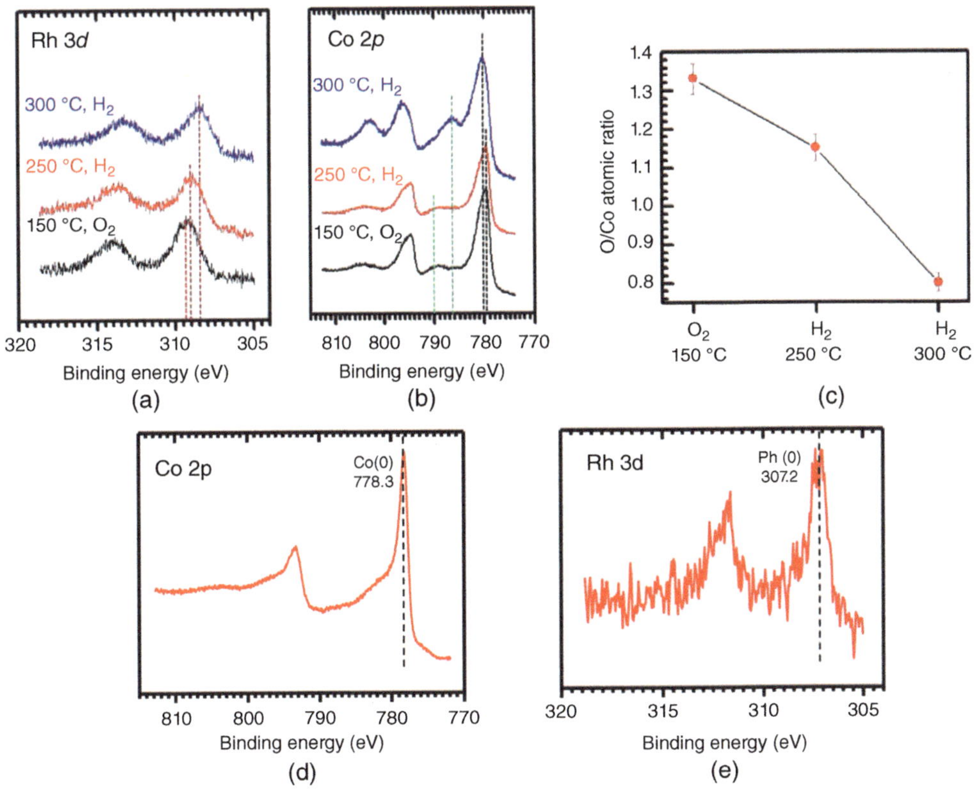

Figure 19.8 AP-XPS studies of surface chemistry of a catalyst prosecutor under different dry pretreatment steps. (a–c) In situ studies of surface chemistry of the catalyst precursor in the preparation of the catalyst of singly dispersed Rh_1Co_3 bimetallic sites with ambient pressure XPS; (a and b) Rh 3d and Co 2p photoemission features during oxidation or reduction step as marked; (c) evolution of O/Co atomic ratio under different reaction conditions; the error bar is ±5% of the measured data. (d, e) Co 2p and Rh 3d XPS peaks in H_2 at 400 °C. *Source:* Reproduced with permission from Zhang et al.[3]/Springer Nature.

Co_3O_4 in the pretreatment with H_2 suggest a reduction to the Rh_1O_5/CoO is necessary for formation of singly dispersed bimetallic sites.

Under the reduction at 250 °C, the Rh $3d_{5/2}$ downshifted to 309.0 eV (red spectrum in Figure 19.8a). Obviously, Rh atoms of the catalyst precursor upon reduction at 250 °C are still at cationic state. Along the increase of reduction temperature to 300 °C in 5% H_2, Rh $3d_{5/2}$ further downshifted by to 308.5 eV (blue spectrum in Figure 19.8a). The downshift of Rh $3d_{5/2}$ suggests that the chemical environment of Rh atoms has been changed to some extent. Notably, the Rh atoms upon the reduction of catalyst precursor at 300 °C are still at a cationic state. EXAFS studies confirmed that Rh atoms of the catalyst precursor at 300 °C in H_2 are singly dispersed and Rh atoms bond with Co atoms and there are no Rh nanoparticles formed (red spectrum in Figure 19.9a).[3] Upon the reduction at 300 °C, the formed surface of the catalyst is highly active in catalyzing the reduction of NO with CO to form N_2 at a temperature as low as 110 °C (Figure 19.6a). The selectivity for producing N_2 on this catalyst at 110 °C is 100%.

Upon the reduction of the catalyst precursor at a temperature as high as 400 °C, the chemical environment of Rh_1 atoms experienced significant change as shown in Figure 19.8d and e. At 400 °C in 5% H_2, both Rh and Co atoms in surface region of the catalyst precursor were reduced to metallic states, forming a Rh–Co ultrathin alloy film. Obviously, the reduction temperature of 400 °C in 5% H_2 is too high to form the hypothesized, singly dispersed bimetallic site Rh_1Co_n. Other than tracking the oxidation sates of Rh atoms, AP-XPS studies found that the temperature of phase transformation of the support Co_3O_4 to CoO is between 250 and 350 °C (Figure 19.8b) and quantitively measured the

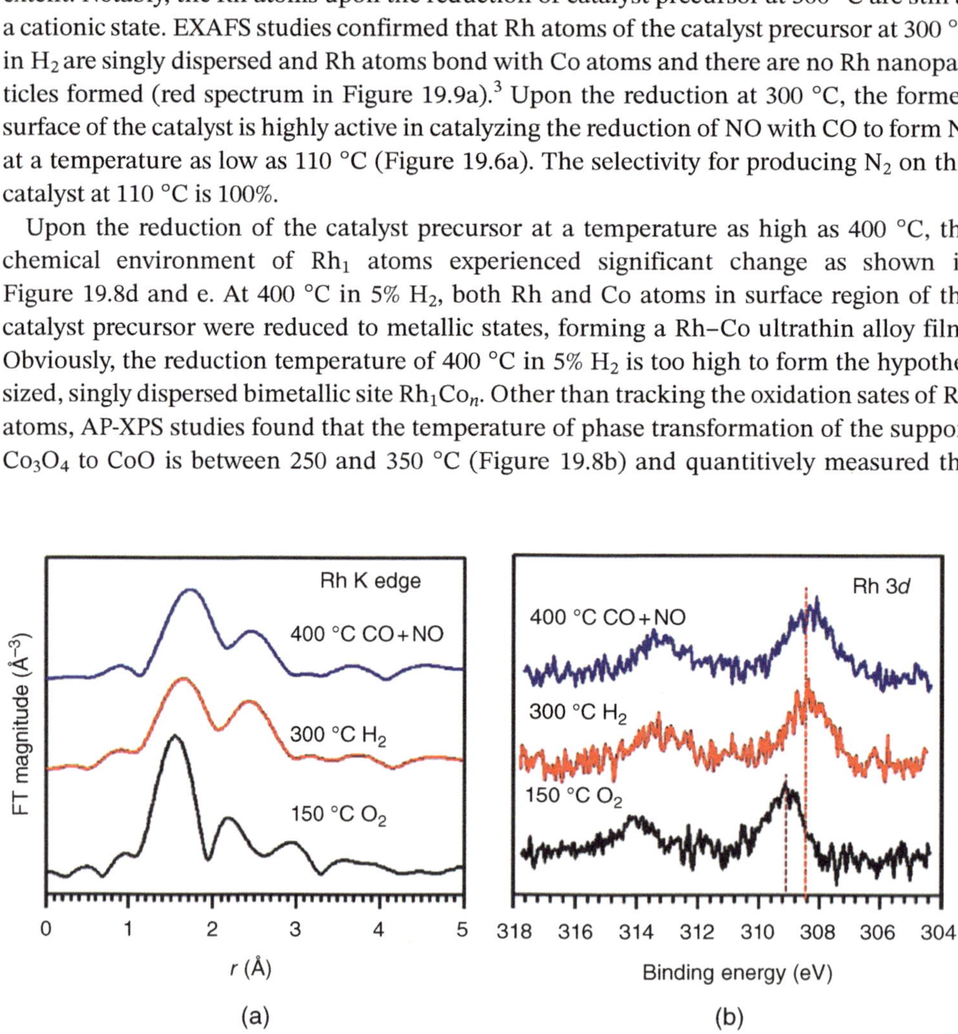

Figure 19.9 Characterizations in the preparation of isolated bimetallic sites. Characterizations in the preparation of isolated bimetallic sites Rh_1Co_3/CoO and during catalysis by in situ EXAFS and in situ AP-XPS. (a) r-Space spectra of Rh K edge of catalyst precursor at 150 °C in O_2 (black line), a following reduction at 300 °C in H_2 (red line) and during catalysis at 400 °C from in situ EXAFS studies (blue line). (b) Photoemission feature of Rh 3d of catalyst precursor at 150 °C in O_2 (black line), a following reduction at 300 °C in H_2 (red line) and during catalysis at 400 °C (blue line) from in situ AP-XPS studies. *Source:* Reproduced with permission from Zhang et al.[3]/Springer Nature.

density of oxygen vacancies of the restructured support (CoO_{1-x}) of the catalyst (Figure 19.8c) at each step of dry process of pretreatment of catalyst precursor. The blue spectra in Figure 19.9a and b show that Rh atoms are singly dispersed and bond with Co atoms and O atoms *during catalysis* at 400 °C, suggesting the formed singly dispersed bimetallic site anchored on CoO_{1-x} is stable up to 400 °C.

Clearly, by using AP-XPS as a video camera-like analytical tool we can track surface of the catalyst precursor at different temperatures, in different gases, or gases with different compositions. By changing pretreatment temperature, type of gas, or gas composition, an appropriate pretreatment condition forming an ideal catalyst surface structure can be found through observing the whole pretreatment process with AP-XPS. This work demonstrated that AP-XPS can significantly aid in finding an optimum pretreatment condition including annealing temperature, type of gas, or gas composition for preparing a catalyst surface with hypothesized catalytic sites such as singly dispersed bimetallic sites.

References

1 Zhang, S., Shan, J., Zhu, Y. *et al.* 2013. "Restructuring transition metal oxide nanorods for 100% selectivity in reduction of nitric oxide with carbon monoxide." *Nano Lett.* 13, 3310–3314.

2 Xie, X., Li, Y., Liu, Z.-Q. *et al.* 2009. "Low-temperature oxidation of CO catalysed by Co_3O_4 nanorods." *Nature* 458, 746–749.

3 Zhang, S., Nguyen, L., Liang, J. *et al.* 2015. "Catalysis on singly dispersed bimetallic sites." *Nat. Commun.* 6, 7938.

4 Tao, F., Li, Y., 2023. "A new type of catalysts: catalysts of singly dispersed bimetallic sites." *Trend. Chem.* 5, 486–499.

20

Tuning Catalyst Surfaces for Developing Catalysts

Bimetallic nanoparticle catalyst is one of the important types of catalysts. Many bimetallic catalysts exhibit high activity and selectivity for a great number of chemical transformations since a second metal, also called guest metal (Z) can modulate the electron state of the first metal, also termed the host metal through a ligand effect or/and a geometric effect.[1–3] A ligand effect and a geometric effect can be realized through variation of surface composition of a bimetallic catalyst. Variation of surface composition gives multiple consequences to surface structures including significant changes of the coordination environment of the host metal (A) and the availability of specific catalytic sites such as dimer (A_2), trimer (A_3), pair of guest and host metal atoms (A–Z). Literature recently reported that a post-synthesis reaction or pretreatment in a specific reactant gas or a series of sequential reactions can restructure the surface composition of an as-synthesized bimetallic catalyst and thus vary catalytic performance largely. AP-XPS has demonstrated its significant role in this development.

20.1 Capability of Compositional Restructuring Checkable with AP-XPS

AP-XPS has been powerful in tracking surface and finding an optimum condition to form an expected surface. It can check surface composition in a real-time mode since atomic fractions of constituting metals (A and Z) could vary by a perturbation. Here the perturbation can be the annealing temperature of a bimetallic catalyst, type of gas (reducing gas versus oxidizing gas), strength of reduction or oxidation (strong reducing agent versus weak reducing agent, or strong oxidizing agent versus weak oxidizing agent), or composition of a gas mixture.

One example is the Rh-Pd bimetallics. Rh and Pd can form an alloy at macroscopic scale due to the similarity of the two elements in many aspects. As demonstrated by Tao et al. in Figure 20.1a, the surface composition of Rh and Pd of a $Rh_{0.2}Pd_{0.8}$ nanoparticle catalyst changes obviously when the gas composition around the catalyst nanoparticles changes from a mixture of NO and CO to pure NO or changes from pure NO to pure H_2.[4,5] Similar to $Rh_{0.2}Pd_{0.8}$, both $Rh_{0.5}Pd_{0.5}$ and $Rh_{0.8}Pd_{0.2}$ can be restructured through changing gas composition or type of gas surrounding the catalyst nanoparticles as shown in Figure 20.1b and c.[4]

Application of Ambient Pressure X-ray Photoelectron Spectroscopy to Catalysis, First Edition. Franklin Tao.

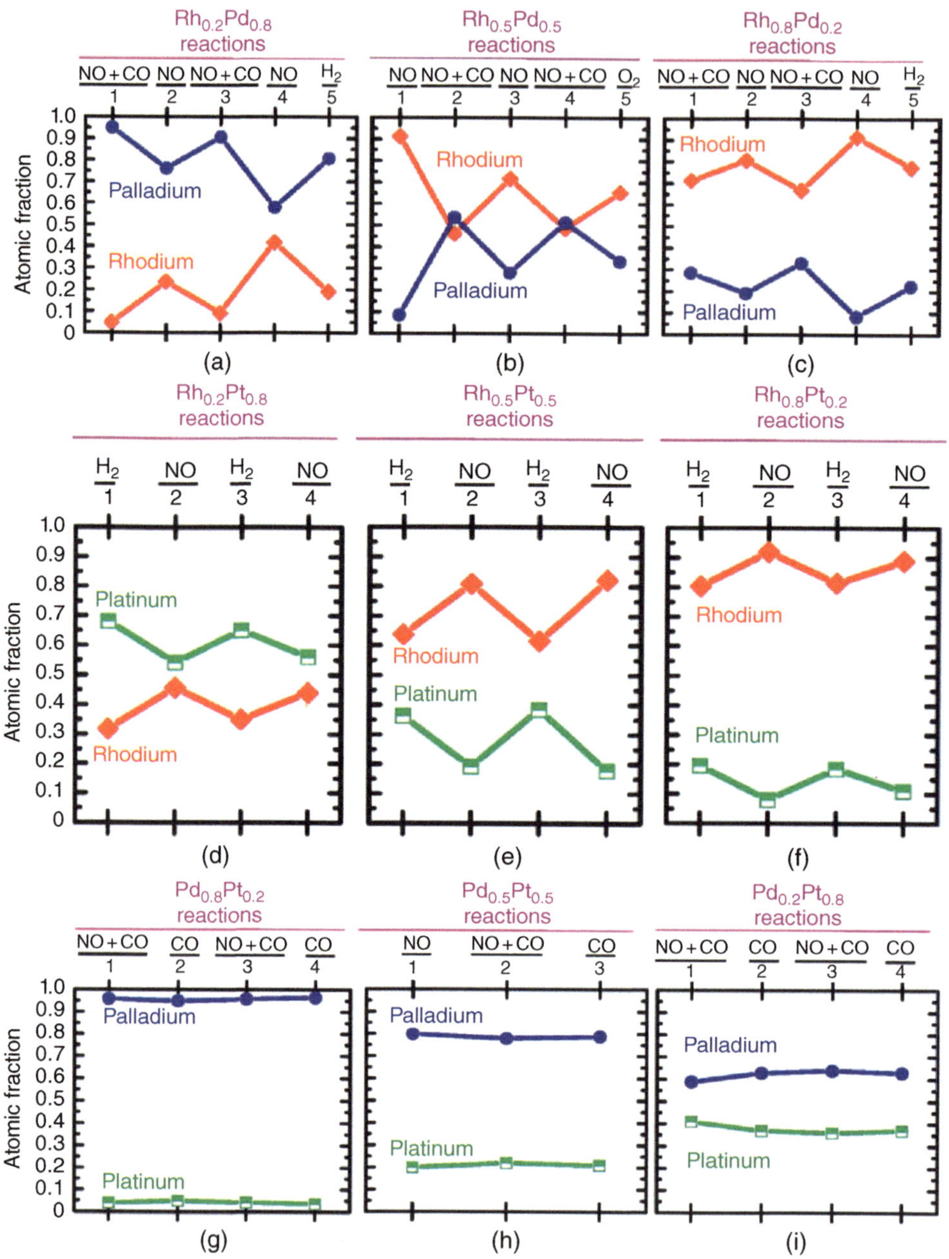

Figure 20.1 Evolutions of surface compositions of Rh–Pd, Rh–Pt, and Pd–Pt bimetallic nanoparticles. (a–c) Evolution of rhodium and palladium atomic fractions in Rh_xPd_{1-x} nanoparticles (~15 nm) at 300 °C under oxidizing conditions (100 mTorr NO or O_2), reducing conditions (100 mTorr H_2), and catalytic conditions (100 mTorr NO and 100 mTorr CO), as denoted on the horizontal axis. All of the atomic fractions were obtained using an X-ray energy of 645 eV for Rh 3d and Pd 3d, which generated photoelectrons with a mean free path of ~0.7 nm. (d–f) Evolution of rhodium and palladium atomic fractions in Rh_xPd_{1-x} nanoparticles (~15 nm) at 300 °C under oxidizing conditions (100 mTorr NO or O_2), reducing conditions (100 mTorr H_2), and catalytic conditions (100 mTorr NO and 100 mTorr CO), as denoted on the horizontal axis. All of the atomic fractions were obtained using an X-ray energy of 645 eV for Rh 3d and Pd 3d, which generated photoelectrons with a mean free path of ~0.7 nm. (g–i) Evolution of palladium and platinum atomic fractions for Pd_xPt_{1-x} nanoparticles (~16 nm) at 300 °C under oxidizing (100 mTorr NO), reducing (100 mTorr CO), and catalytic (100 mTorr NO and 100 mTorr CO) conditions. The horizontal axis represents different reaction conditions. All of the atomic fractions were obtained using X-ray energies of 645 eV for Pd 3d and 380 eV for Pt 4f, which generated photoelectrons with a mean free path of ~0.7 nm. *Source:* Reproduced with permission from Tao et al.[4]/American Chemical Society.

Pd–Pt is another bimetallics. $Pd_{0.8}Pt_{0.2}$, $Pd_{0.5}Pt_{0.5}$, and $Pd_{0.2}Pt_{0.8}$ are three Pd–Pt alloy. Unfortunately, the above perturbations driving the restructuring of Rh-Pd (Figures 21a–c) and Rh-Pt alloy nanoparticles (Figures 21a–c) could not drive Pd-Pt alloy nanoparticles restructured (Figure 20.1g–i).[4] The difference in responding to the perturbation between Rh–Pd (or Rh–Pt) and Pd–Pt can be rationalized by the surface free energy of a metal at a metallic state and the formation energy of a metal oxide. If a metal A of a bimetallics A–Z (such as Pd of Pd–Pt alloy), has lower surface free energy than Z (such as Pt) at metallic states and the oxide of metal A (such as PdO) has higher formation energy than oxide of metal Z (such as Pt_xO_y), the element A (such as Pd of bimetallic Pd-Pt) will preferentially remain in surface region regardless of the oxidizing gaseous environment (such as NO or O_2) or reducing environment (such as CO or H_2). For this type of bimetallic nanoparticles (such as Pd_xPt_{1-x}), switching from a reducing gas (such as CO or H_2) to an oxidizing gas (such as NO or O_2) could not change the locations of the constituting metals (such as Pd or Pt of Pd_xPt_{1-x}) in the bimetallic nanoparticles.[4] This prediction is consistent with the finding of AP-XPS studies of these Pd_xPt_{1-x} catalysts in different gases (Figure 20.1g–i), where the atomic fractions of Pt and Pd in surface region of Pd–Pt bimetallic nanoparticles do not change along switching the reactant gas from reducing to oxidizing or from oxidizing to reducing.

The above factors of surface energy of constituting metals and formation energy of metal oxide can well explain the restructuring of Rh–Pd bimetallic nanoparticles uncovered by AP-XPS by Tao et al.[5] For Rh–Pd, in reducing gas environment the metallic Pd is preferentially segregated to surface region because metallic Pd has low surface energy than Rh; if Pd is in subsurface or even core, it can continuously segregate to surface region to form a Pd-rich shell and Rh-rich core. In the oxidizing environment, however, Rh_2O_3 has a high formation energy than PdO. Thus, Rh is preferentially oxidized on surface and thus stays on surface in the form of RhO_y. If a surface is Pd-rich, its exposure to NO or O_2 will preferentially oxidize Rh and thus pull Rh atoms from subsurface or deeper layers to surface region and then be oxidized, forming a Rh-rich RhO_x shell. Obviously, by switching gas from reducing gas to oxidizing gas, the surface of a Rh-Pd bimetallic nanoparticle is restructured from a nanoparticle with a Pd-rich metallic shell to a nanoparticle consisting of RhO_x shell. Similar to Rh_xPd_{1-x}, Rh_xPt_{1-x} exhibits the same capability of restructuring when its surrounding gas is switched between a reducing gas and an oxidizing gas.

The significant takeaway of these works[4,5] is that AP-XPS is a unique technique in tracking both surface composition of a bimetallic catalyst and oxidation states of the constituting metals of a bimetallic catalyst in different gases or at different temperatures to uncover any potential restructuring of a bimetallic catalyst under either a reaction condition in terms of having a single reactant or a catalytic condition in terms of a mixture of all reactants of a reaction.

20.2 Tracking Restructuring of Bimetallic Surface under Reaction and Catalytic Conditions for Tuning Catalytic Performance of a Bimetallic Catalyst

In terms of metal catalysts, the electronic state of metal atom A on surface is determined by both the number of metal atoms Z directly bonding to the atom of element A and the geometric arrangement of atoms of element Z around the atom A. This atomic-scale coordination environment of the atom A determines adsorption of a reactant molecule or an intermediate of a catalytic reaction on the atom A. Numerous alloys can be formed through adding one or two guest metal element Z or X to a host metal A through various physical and chemical methods. Different from a crystal of ionic compound such as NaCl, the tolerance of a host metal A to size and electron configuration of guest atom Z in forming an alloy A_xZ_{1-x} is quite high since the free electron sea consisting of valence electrons of all constituting metal atoms A and Z acts as a flowing "glue" in the lattice of a bimetallic nanoparticle to hold atoms of elements Z and A even though atom Z could have physical size and electron configuration obviously different from atoms of element A. This durability leads to high flexibility in forming bimetallic catalysts. The high flexibility provides numerous possibilities of tuning catalytic performance of a host metal A through varying molar fraction of the guest metal Z, type of a guest metal (Z or X), or location of a guest metal Z in terms of formation of a core-shell structure, subsurface alloy, intermetallic, or random alloy.

Section 20.1 demonstrated that bimetallic nanoparticle catalysts could be restructured by several types of perturbations. Gas treatment of a bimetallic catalyst has been demonstrated as an effective method of tuning surface composition and even oxidation state of constituting metal elements on surface of the bimetallic catalyst.[4,5] A recent work has demonstrated that the gas-induced restructuring of bimetallic nanoparticle catalysts is a significant approach to designing new catalysts through modulating surface composition of catalysts.[6] The base of this work is that the catalytic performance of a bimetallic catalyst is determined by surface composition and atomic packing of constituting metals. AP-XPS tracked the surface restructuring triggered by a perturbation such as a change of reactive gas surrounding the bimetallic catalyst nanoparticles. Through correlating with the catalytic kinetics of these restructured catalysts, the capability of tuning catalytic performance through gas driving-surface restructuring was demonstrated.[6] By analyzing electronic states of the restructured catalyst surface, modulation of surface electronic state through surface restructuring was confirmed.

In this demonstration, two different Pt–Cu bimetallic nanocubes were used as starting materials by Tao group.[6] Figure 20.2 schematically shows the as-synthesized Pt–Cu regular nanocube (RNC) [Pt–Cu RNC-as synthesized] and Pt–Cu concave nanocube (CNC) [Pt–Cu CNC-as

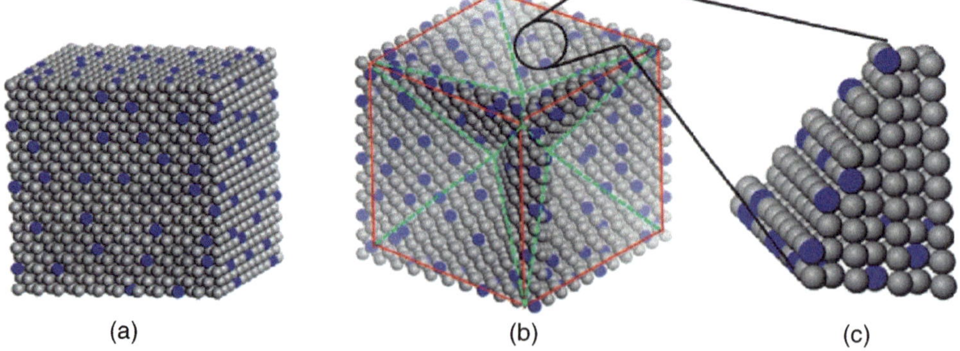

Figure 20.2 Structural models of (a) the as-synthesized Pt–Cu regular nanocube (Pt–Cu RNC), (b) the as-synthesized Pt–Cu concave nanocube (Pt–Cu CNC), (c) the stepped surface of Pt–Cu CNC. *Note:* the locations of blue and gray spots of these structural models only schematically show positions of Cu and Pt atoms on the surface of these nanocube catalysts. *Source:* Reproduced with permission from Shan et al.[6]/American Chemical Society.

synthesized]. Notably, the surface of RNC is $Pt_{1-x}Cu_x$ (100). Different from Pt–Cu RNC (Figure 20.2a), each facet of a Pt–Cu CNC consists of four triangle-like concave surfaces (Figure 20.2b); the four triangles have a common apex that is the lowest point on the concave surface. Figure 20.2c schematically shows a triangular alloy surface. Each of these triangular surfaces is a stepped surface. Obviously, the coordination number of metal atoms Cu to a targeted metal atom Pt on the triangular surface of CNC in Figure 20.2b is lower than that to a Pt atom of the (100) surface of an RNC (Figure 20.2a). Both Pt–Cu CNC-as synthesized (Figure 20.2b) and Pt–Cu RNC-as synthesized (Figure 20.2a) were used as probes by Tao group.[6]

Figure 20.3 briefs the experimental methods for restructuring Pt–Cu RNC and Pt–Cu CNC catalysts toward tuning catalytic performance through annealing in different gas or mixtures with different gas composition. CO oxidation was chosen to test whether a restructured surface could exhibit a different catalytic performance. The as-synthesized Pt–Cu RNC was annealed at 200 °C in H_2 in Figure 20.3a. Notably, a catalyst (Pt–Cu RNC) treated in H_2 is termed Pt–Cu RNC–H_2; after this H_2 treatment, kinetics study of CO oxidation on Pt–Cu RNC–H_2 was performed as noted in Figure 20.3a. Compared to the H_2 pretreatment at 200 °C in Figure 20.3a, an additional step, annealing in CO gas was introduced for another potential restructuring (Figure 20.3c). Thus, this catalyst experienced the treatments of first H_2 and then CO is termed Pt–Cu RNC–H_2–CO in Figure 20.3c. Kinetics studies of CO oxidation on the three catalysts Pt–Cu RNC, Pt–Cu RNC–H_2, and Pt–Cu RNC–H_2–CO were performed. Table 20.1 lists the apparent activation barriers for CO oxidation performed on these restructured catalysts. Based on the kinetic studies, the apparent activation energy of CO oxidation on Pt–Cu RNC-as synthesized is 128 ± 7 kJ/mol. Upon it was reduced in 5% H_2, Pt–Cu RNC–H_2 was formed. Kinetics study of CO oxidation on the formed Pt–Cu RNC–H_2 was done. The activation barrier on the Pt–Cu RNC–H_2 is as low as 39 ± 4 kJ/mol, significantly lower than the barrier (128 kJ/mol) of the catalyst without the restructuring (Pt–Cu RNC-as synthesized). Then, the Pt–Cu RNC–H_2 was

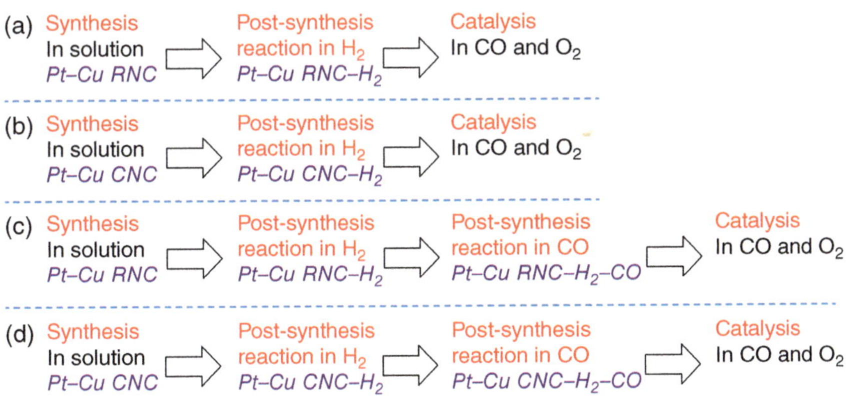

Figure 20.3 Procedures to form a new catalyst surface through a single or sequential reaction(s) in gas phases. (a) The protocol for Pt-Cu RNC consisting of synthesis, post-synthesis reaction in H_2, and catalysis. (b) The protocol for Pt-Cu CNC consisting of synthesis, post-synthesis reaction in H_2, and catalysis. (c) The protocol for Pt-Cu RNC consisting of synthesis, post-synthesis reaction in H_2, post-synthesis reaction in CO, and catalysis. (d) The protocol for Pt-Cu CNC consisting of synthesis, post-synthesis reaction in H_2, post-synthesis reaction in CO, and catalysis. *Source:* Reproduced with permission from Shan et al.[6]/American Chemical Society.

Table 20.1 List of measured activation barrier of restructured bimetallic catalysts.

Entry	Catalysts	Post-synthesis reaction(s)	Activation energy (kJ/mol) in CO oxidation after post-synthesis reaction(s)
1	Pure Pt RNC-H_2	H_2, 200 °C	92 ± 5
2	Pt − Cu RNC-as synthesized	N/A	128 ± 7
3	Pt − Cu RNC − H_2	H_2, 200 °C	39 ± 4
4	Pt − Cu RNC − H_2 − CO	H_2 at 200 °C, CO at 350 °C	173 ± 11
5	Pt − Cu CNC-as synthesized	N/A	109 ± 10
6	Pt − Cu CNC − H_2	H_2, 200 °C	72 ± 6
7	Pt − Cu CNC − H_2 − CO	H_2 at 200 °C, CO at 350 °C	80 ± 8

Source: Reproduced with permission from Shan et al.[6]/American Chemical Society.

restructured into another structure by annealing in pure CO at 350 °C, forming the Pt–Cu RNC–H_2–CO. A significant change is that the apparent activation barrier for CO oxidation on Pt–Cu RNC–H_2–CO was dramatically increased to 173 ± 11 kJ/mol. Thus, for Pt–Cu RNC, a treatment at 200 °C in H_2 offers a surface with a quite low activation barrier (39 ± 4 kJ/mol). The significantly improved catalytic performance confirmed the feasibility of developing bimetallic catalysts with better catalytic performances through restructuring in gaseous phase. More importantly, this restructuring is trackable with AP-XPS. In other words, the process of developing a new bimetallic catalyst can be observed with AP-XPS.

In terms of Pt–Cu CNC, the apparent activation barrier for CO oxidation is 109 ± 10 kJ/mol. The barrier is decreased to 72 ± 6 kJ/mol on Pt–Cu CNC–H_2 after the Pt–Cu CNC was reduced in 5% H_2 (entry 5 versus 6 in Table 20.1). Similar to Pt–Cu RNC, the significant decrease of apparent activation barrier of Pt–Cu CNC-H_2 demonstrates that H_2 treatment at 200 °C to Pt–Cu–CNC-as synthesized can dramatically increase catalytic activity of the Pt–Cu–CNC-as synthesized. A further treatment to Pt–Cu CNC–H_2 in CO at 350 °C formed Pt–Cu CNC–H_2–CO (Figure 20.4f). Kinetic studies show that the treatment in CO does not change the catalytic performance of Pt–Cu CNC–H_2 obviously (entry 6 versus 7 in Table 20.1). Likely, this is related to the AP-XPS observation (Figure 20.4f) that CO treatment to Pt–Cu RNC–H_2 at 350 °C could not largely change the surface composition.

As summarized in Table 20.1, the above restructurings generated different catalysts. Some of the restructured catalysts including Pt–Cu RNC–H_2 and Pt–Cu CNC–H_2 exhibit lower activation barriers for CO oxidation than Pt–Cu RNC-as synthesized and Pt–Cu CNC-as synthesized, respectively. However, some restructuring give higher activation barriers. How could we understand these differences fundamentally? We need to uncover an intrinsic correlation between catalytic activity and its corresponding surface structure during catalysis. How could we establish the correlation among (1) these gas treatments, (2) the surfaces created by them, and (3) the corresponding catalytic performances? The key is the identification of the authentic catalyst surface formed under treatment conditions and during catalytic conditions with AP-XPS.

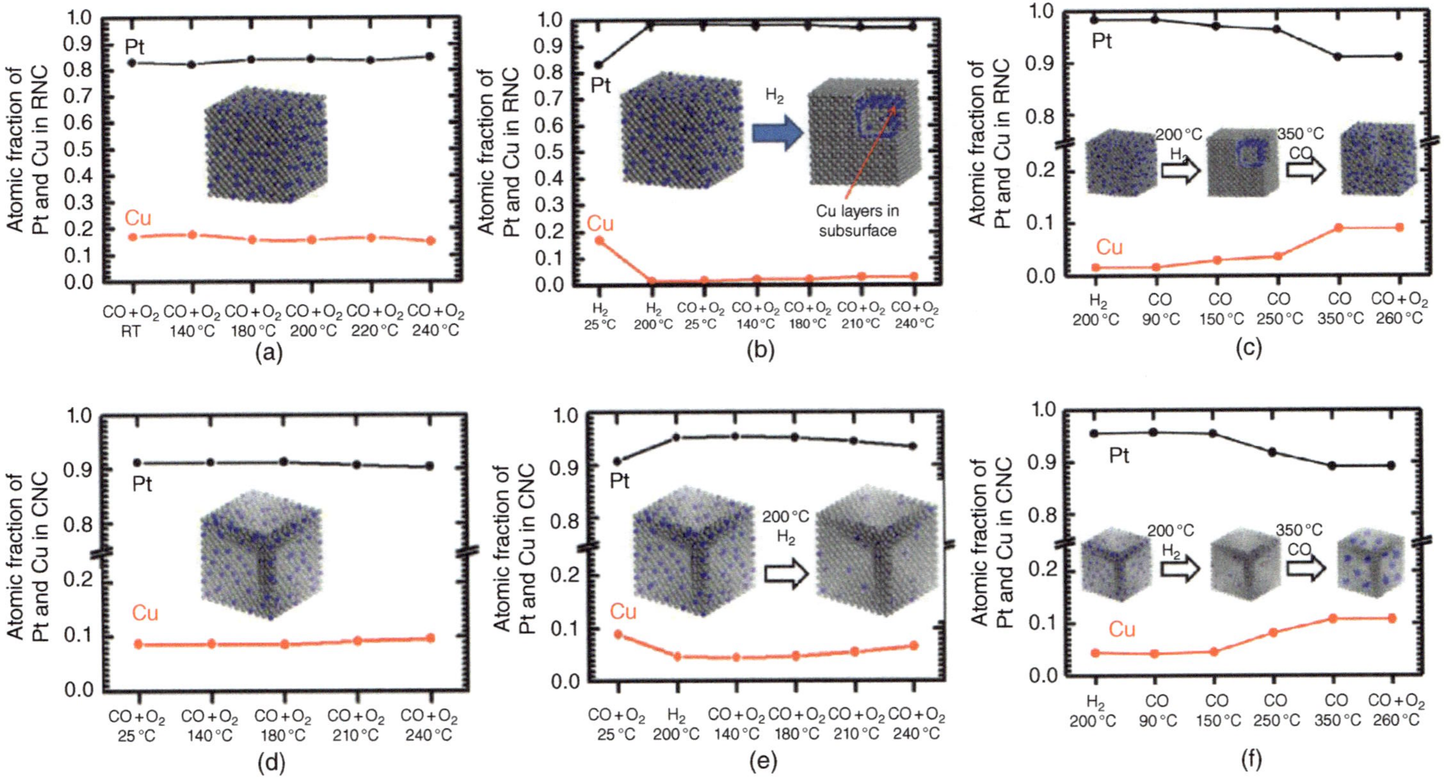

Figure 20.4 Evolution of surface composition (Pt and Cu atomic fractions) of Pt–Cu RNC-as synthesized or Pt–Cu CNC-as synthesized and those after post-synthesis reaction(s) in gas phase(s). (a) Pt–Cu RNC-as synthesized (without a reaction in H_2) in CO and O_2. (b) Pt–Cu RNC in H_2 at 200 °C and then CO oxidation. (c) Pt–Cu RNC in H_2 at 200 °C, CO up to 350 °C, and then in the mixture of CO and O_2. (d) Pt–Cu CNC-as synthesized (without a reaction in H_2) in CO and O_2. (e) Pt–Cu CNC in H_2 at 200 °C and then in CO and O_2. (f) Pt–Cu CNC in H_2 at 200 °C, in CO up to 350 °C, and then in the mixture of CO and O_2. *Note*: the locations of blue and gray spots of these structural schematics inserted to these figures do not represent the positions of Cu and Pt atoms on surface of these nanocube catalysts. The error bars of these measurements of atomic fractions are ±5% of the given atomic fractions. *Source:* Reproduced with permission from Shan et al.[6]/American Chemical Society.

AP-XPS studies under each condition can rationalize these observed tuning of catalytic performance. Figure 20.4 presents the quantitative analyses of Pt 4f and Cu 2p collected by AP-XPS under the different post-synthesis treatment conditions listed in the X-axis including temperature and gaseous environment. Based on the AP-XPS measurements at 25 °C in the mixture of CO and O_2, the atomic fractions of Pt and Cu of the Pt–Cu RNC-as synthesized are 82 and 18%, respectively (Figure 20.4a). They are 92 and 8% for Pt–Cu CNC (Figure 20.4d). As shown in Figure 20.4a, the fractions of Pt and Cu in Pt–Cu RNC-as synthesized do not change in the mixture of CO and O_2 in the temperature range of 25–240 °C.

The reduction in H_2 at 200 °C to Pt–Cu RNC-as synthesis led to a significant change of Cu and Pt fractions (Figure 20.4b). The AP-XPS studies suggest that Cu atoms in the surface region completely migrated into deep layers during H_2 pretreatment. Thus, the surface region of Pt–Cu RNC–H_2 is nearly a pure Pt shell in terms of a Pt skin layer. With the aid of in situ studies of EXAFS,[6] a surface alloy consisting of skin layers of nearly pure Pt atoms and underneath Cu atomic layers was deduced from AP-XPS studies. Based on reference[7], the "floating" Pt layer on Cu subsurface in a surface alloy typically exhibits quite low adsorption energy to CO and thus CO oxidation on such a surface alloy of Pt–Cu RNC–H_2 can be performed with a quite low activation barrier (39 ± 4 kJ/mol) (entry 3 in Table 20.1). The significant decrease of activation barrier for CO oxidation is definitely a demonstration that a new catalyst with higher activity can be developed through gas-driving restructuring of an as-synthesized bimetallic catalyst. Obviously, AP-XPS played a significant role in understanding how a bimetallic catalyst can be restructured and in elucidating what drives a better catalytic performance for a bimetallic catalyst.

It is also found that the activity of CO oxidation on Pt–Cu RNC–H_2–CO is largely decreased upon an additional treatment in CO to Pt–Cu RNC–H_2. By annealing the Pt–Cu RNC–H_2 in CO at 350 °C, another restructured catalyst was formed. It is designated as Pt–Cu RNC–H_2–CO. The activation barrier for CO oxidation on Pt–Cu RNC–H_2–CO is as high as 173 ± 11 kJ/mol. Again, AP-XPS played an important role in identification of the surface structure of Pt–Cu RNC–H_2–CO for understanding why the activity of CO oxidation on Pt–Cu RNC–H_2–CO decreases largely in contrast to the high activity on Pt–Cu RNC–H_2. AP-XPS uncovered that the difference in surface composition and structure between Pt–Cu RNC–H_2 and Pt–Cu RNC–H_2–CO results from the segregation of Cu of the subsurface region to form Pt–Cu alloy on the surface of Pt–Cu RNC–H_2–CO. Again, the sharp decrease of catalytic activity of CO oxidation upon Pt–Cu RNC–H_2 was treated in CO is due to the formation of a PtCu alloy. Computational studies rationalized why the Pt–Cu alloy surface has an extremely sloppy kinetics for CO oxidation.[6] Based on density functional theory (DFT) calculation, the adsorption energy of CO on Pt of Pt–Cu alloy is obviously higher than that on Pt atoms of pure Pt(111) due to the charge transfer from Cu to Pt.[7] A high adsorption energy of CO indicates a high activation barrier of CO oxidation. Thus, the finding of AP-XPS in Figure 20.4c rationalized the change of catalytic activity of Pt–Cu RNC–H_2–CO.

Definitely, AP-XPS is a unique technique in observing how the surface is evolved in gas or even liquid phase. More importantly, it played crucial roles in establishing a base to understand the observed change of catalytic performance of a bimetallic nanoparticle catalyst during or upon a perturbation such as a different pretreatment.

References

1 Chen, M., Kumar, D., Yi, C. W. *et al.* 2009. "The promotional effect of gold in catalysis by palladium-gold." *Science* 310, 291–293.

2 Tao, F., Zhang, S., Nguyen, L. *et al.* 2012. "Action of bimetallic nanocatalysts under reaction conditions and during catalysis: evolution of chemistry from high vacuum conditions to reaction conditions." *Chem. Soc. Rev.* 41, 7980–7993.

3 Yu, W., Porosoff, M. D., and Chen, J. G. 2012. "Review of Pt-based bimetallic catalysis: from model surfaces to supported catalysts." *Chem. Rev.* 112, 5780–5817.

4 Tao, F., Grass, M. E., Zhang, Y. *et al.* 2010. "Evolution of structure and chemistry of bimetallic nanoparticle catalysts under reaction conditions." *J. Am. Chem. Soc.* 132, 8697–8703.

5 Tao, F., Grass, M. E., Zhang, Y. *et al.* 2008. "Reaction-driven restructuring of Rh-Pd and Pt-Pd core-shell nanoparticles." *Science* 322, 932–934.

6 Shan, J., Zhang, S., Choksi, T. *et al.* 2017. "Tuning catalytic performance through a single or sequential post-synthesis reaction(s) in a gas phase." *ACS Catal.* 7, 191–204.

7 Andersson, K. J., Calle-Vallejo, F., Rossmeisl, J. *et al.* 2009. "Adsorption-driven surface segregation of the less reactive alloy component." *J. Am. Chem. Soc.* 131, 2404–2407.

21

Photocatalysis

A unique photocatalysis system, $(Rh_{2-y}Cr_yO_3)/(Ga_{1-x}Zn_x)(N_{1-x}O_x)$, was reported recently.[1–4] It consists of photocatalyst $(Ga_{1-x}Zn_x)(N_{1-x}O_x)$ and cocatalyst $Rh_{2-y}Cr_yO_3$.[1–4] It is highly active for splitting water to generate H_2 and O_2 under the excitation of visible light. Photocatalyst $(Ga_{1-x}Zn_x)(N_{1-x}O_x)$ was prepared with a nitridation method.[1,5] In this preparation, Ga_2O_3 and ZnO were mixed in a 1 : 2 molar ratio and then the mixture was calcinated at 850 °C in the flow of NH_3 (250 ml/min) for 15 hours. Upon $(Ga_{1-x}Zn_x)(N_{1-x}O_x)$ being prepared, Cr and Rh were loaded on surface of prepared $(Ga_{1-x}Zn_x)(N_{1-x}O_x)$ through impregnation. $(Ga_{1-x}Zn_x)(N_{1-x}O_x)$ was dispersed in the aqueous solution of $Cr(NO_3)_3 \cdot 9H_2O$ and $Na_3RhCl_6 \cdot 2H_2O$ for obtaining 1.0 wt% Rh and 1.5 wt% Cr loading. Upon all water had been evaporated, the powder was collected. A calcination of the powder at 350 °C in air for one hour formed the unique photocatalysis system, $(Rh_{2-y}Cr_yO_3)/(Ga_{1-x}Zn_x)(N_{1-x}O_x)$.[1–4] The introduced Rh–Cr–O behaves as a cocatalyst. The role of the co-catalyst during photocatalysis was studied with AP-XPS.[6]

In general, the size of the photocatalyst particles is typically >100 nm and appears as the high contrast bump in Figure 21.1a. The supported co-catalyst $Rh_{2-y}Cr_yO_3$ nanoparticles are about 10 nm.[6] These cocatalyst nanoparticles appear as small high-contrast dots in transmission electron microscopy (TEM) images as marked with red arrows in Figure 21.1a. Energy dispersive X-ray spectrum mapping confirmed that Cr and Rh are distributed on the semiconductor particle $(Ga_{1-x}Zn_x)(N_{1-x}O_x)$ (Figure 21.1c–f). As shown in Figure 21.1f the Rh and Cr mappings overlap very well. It indicates that each area in terms of each pixel of the mapping image has both Rh and Cr. The exact overlap between Rh mapping and Cr mapping suggests that a mixed oxide $Rh_{2-y}Cr_yO_3$ was formed on $(Ga_{1-x}Zn_x)(N_{1-x}O_x)$.[6] Ultrahigh vacuum (UHV) X-ray photoelectron spectroscopy (XPS) studies of the $(Rh_{2-y}Cr_yO_3)/(Ga_{1-x}Zn_x)(N_{1-x}O_x)$ show that N and O are the main nonmetallic elements in surface region of the catalyst. In surface region of the catalyst particles, the concentrations of Ga and Cr are higher than other metal elements. Thus, surface composition of the photocatalyst $(Rh_{2-y}Cr_yO_3)/(Ga_{1-x}Zn_x)(N_{1-x}O_x)$ is very different from its bulk composition.

Compositions of the topmost atomic layers of the surface region of catalyst $(Rh_{2-y}Cr_yO_3)/(Ga_{1-x}Zn_x)(N_{1-x}O_x)$ were measured by Wachs group with high sensitivity-low energy ion scattering (HS-LEIS) technique in UHV as it has the highest surface sensitivity. Combining with Ne^+ gas ion sputtering, HS-LEIS established the depth profiling of Ga, Zn, Rh, and Cr.

Application of Ambient Pressure X-ray Photoelectron Spectroscopy to Catalysis, First Edition. Franklin Tao.

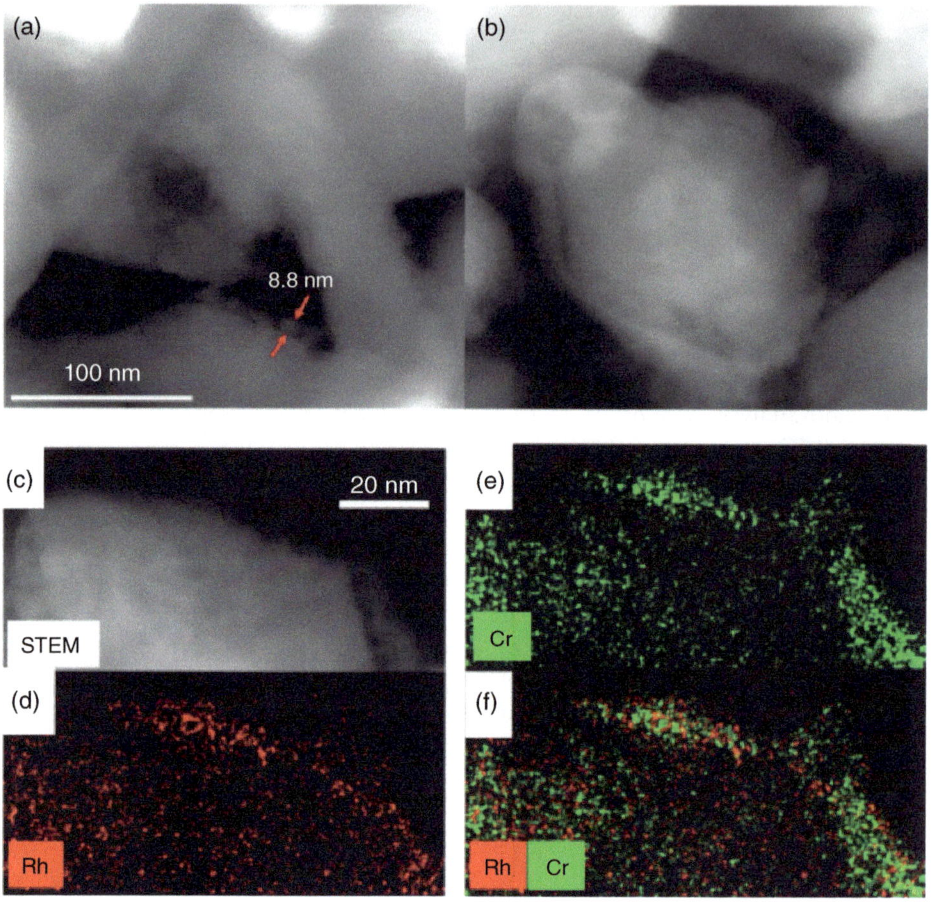

Figure 21.1 STEM images of $(Rh_{2-y}Cr_yO_3)/(Ga_{1-x}Zn_x)(N_{1-x}O_x)$ before water splitting (a) and $(Rh_{2-y}Cr_yO_3)/(Ga_{1-x}Zn_x)(N_{1-x}O_x)$ after water splitting (b), small area of $(Rh_{2-y}Cr_yO_3)/(Ga_{1-x}Zn_x)$ $(N_{1-x}O_x)$ before water splitting (c) and the corresponding EDX mappings of Rh (d), Cr (e), and overlapping of Rh and Cr (f) of the particle. *Source:* Phivilay et al.[6]/American Chemical Society.

As shown in Figure 21.2a, along the increase of depth of the analyzed region, the signals of Zn and Ga significantly increase. The outmost layer of the catalyst is slightly rich in both Rh and Cr since $Rh_{2-y}Cr_yO_3$ nanoparticles were loaded on the photocatalyst $(Ga_{1-x}Zn_x)$ $(N_{1-x}O_x)$ particle. In addition, the concentration of Rh in the surface region slightly decreases along the increase of depth (Figure 21.2a). It suggests that the cocatalyst $Rh_{2-y}Cr_yO_3$ is supported on the semiconductor particles $(Ga_{1-x}Zn_x)(N_{1-x}O_x)$. After $(Rh_{2-y}Cr_yO_3)/(Ga_{1-x}Zn_x)(N_{1-x}O_x)$ was used for photocatalysis in water, Cl^- was not found on the used catalyst by HS-LEIS (Figure 21.2b) although a large amount of Cl^- was detected from the as-prepared catalyst before photocatalysis (Figure 21.2a). Clearly, Cl^- anions were washed out by water during photocatalysis. There are comparable amounts of Rh and Cr on the topmost surface of the used photocatalyst as shown in Figure 21.2b, showing that both Rh and Cr remained on surface of the photocatalyst during photocatalysis in water. These studies suggest that Rh–Cr–O cocatalyst remained on surface of the photocatalyst.

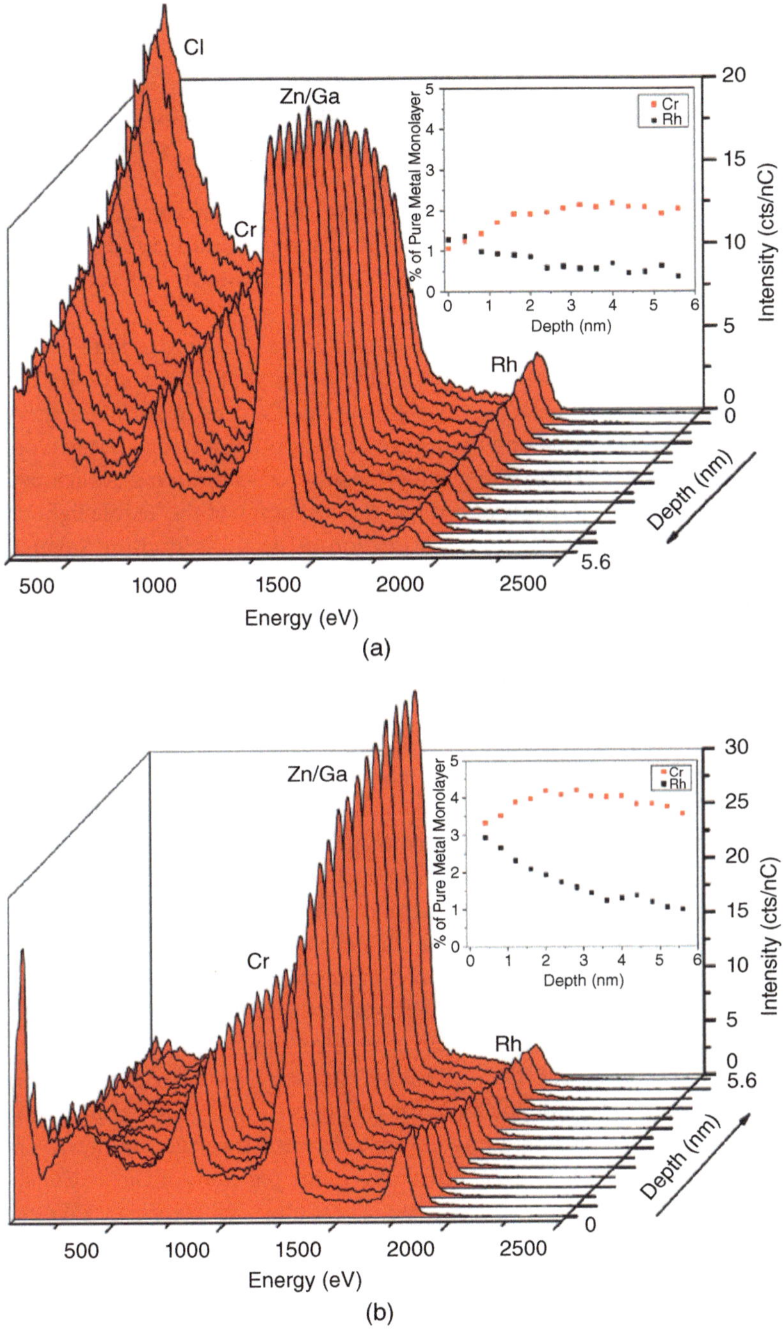

Figure 21.2 HS-LEIS depth profile for photocatalysts using (a) He^+ ion gas and (b) Ne^+ ion gas. Inset shows Rh and Cr depth profiles. (Arrow indicates increasing depth from the surface.) (a) The $(Rh_{2-y}Cr_yO_3)/(Ga_{1-x}Zn_x)(N_{1-x}O_x)$ before photocatalysis. (b) The $(Rh_{2-y}Cr_yO_3)/(Ga_{1-x}Zn_x)(N_{1-x}O_x)$ after photocatalysis. *Source:* Reproduced with permission from Phivilay et al.[6]/American Chemical Society.

Identification of the active sites accumulating electrons and holes on surface of the photocatalyst $(Ga_{1-x}Zn_x)(N_{1-x}O_x)$ is one of the most important tasks in photocatalysis. In terms of AP-XPS studies of photocatalysis on $(Rh_{2-y}Cr_yO_3)/(Ga_{1-x}Zn_x)(N_{1-x}O_x)$, the following describes the steps. The photocatalyst was first loaded on the sample holder and then placed in a reaction cell of AP-XPS. Water vapor flew through the catalyst.[6] Photoirradiation was realized by turning on the visible light source whenever the photon in terms of visible light needs to be introduced. Regarding to the XPS process, photoelectrons were excited from subshells (Ga 2p, Ga 3d, Rh 3d, Cr 2p, and O 1s) of the catalyst surface by Al Kα X-ray beam. During AP-XPS studies of this catalyst, the gas around the photocatalyst is the continuously flowing water vapor.

The mechanism of generation of photoelectrons by X-ray-based photoionization for XPS measurement is different from the process of photocatalysis. In photoionization of XPS, X-ray with energy of 1486.3 eV excites electrons on subshells of atoms in surface region of the catalyst. The excited electrons leaving atoms are called photoelectrons. They typically have kinetic energy of 100–1400 eV. The photoionization cross-section significantly decreases along with the increase of energy of photons. As the energy of the visible light of photocatalysis is much lower than Al Kα, the cross-section of Al Kα in exciting valence electrons is much lower than that of visible light. Thus, the contribution of the continuously irradiating Al Kα to the photocatalyst is negligible. In terms of irradiation to the catalyst using a visible light, only electrons in valence band of the semiconductor particles were excited along with the generation of holes. These excited electrons traveled through the photocatalyst particles and then reached surface of the semiconductor. Majority of the photoexcited electrons recombine with holes or are trapped by defects in the traveling process and thus cannot arrive at the surface of the semiconductor. These electrons and holes arriving at the surface can reduce H_2O to H_2 or oxidize H_2O to O_2, respectively. To minimize the recombination of electrons and holes on surface of the photocatalyst, cocatalysts of electrons or/and holes were loaded on the surface of the photocatalyst particles to collect electrons or holes for promoting the oxidation and reduction of H_2O to generate H_2 and O_2, respectively. In addition, these electrons excited from valence bands of the semiconductor have quite low kinetic energy. They are not a part of the interested photoelectrons being collected and analyzed by AP-XPS. Thus, the AP-XPS process and water-splitting process do not interfere with each other at all although they occur simultaneously.

Compared to the peak positions of Ga 2p, Ga 3d, Rh 3d, Cr 2p, and O 1s collected under the condition of having water vapor but no photoirradiation ("light off" marked in Figure 21.3), these peak positions downshifted to lower binding-energy side when the visible light was on ("light on" marked in Figure 21.3). Downshifts of these XPS peak positions of Ga 2p, Ga 3d, Rh 3d, Cr 2p, and O 1s suggest that Ga, Rh, Cr, and O were partially negatively charged during photocatalysis. It further suggests the supported $Rh_{2-y}Cr_yO_3$ is a collector of electrons in terms of a cocatalyst collecting electrons to promote the reduction of H_2O to H_2. However, N 1s did not shift (Figure 21.3e). This observation achieved through AP-XPS is supported by density functional theory (DFT) calculations.[6] The DFT calculations show that the top of valence band of the semiconductor $(Ga_{1-x}Zn_x)(N_{1-x}O_x)$ consists of N 2p and Zn 3d and the bottom of its conduction band is composed of Ga 4s and 4p.[7–9] When photons are introduced to the semiconductor, electrons of valence band mainly consisting of N 2p and Zn 3d are excited to the conduction band mainly consisting of Ga 4s and 4p. Thus, this visible

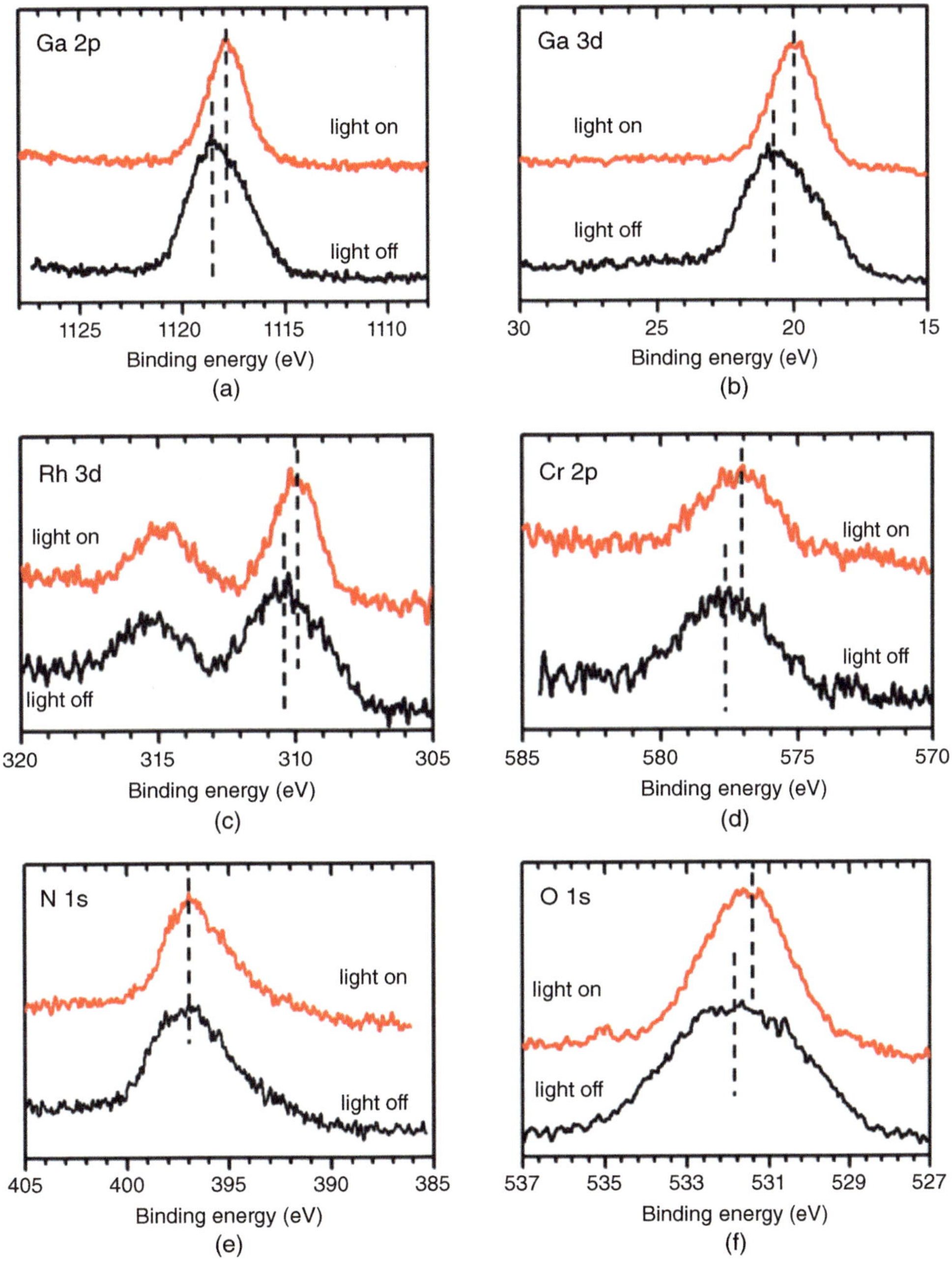

Figure 21.3 AP-XPS studies of surface of $(Rh_{2-y}Cr_yO_3)/(Ga_{1-x}Zn_x)(N_{1-x}O_x)$ in water vapor (0.1 Torr) at room temperature when light from 1000 W xenon arc lamp (Newport) was irradiating (light on) or not. (a) Ga 2p. (b) Ga 3d. (c) Rh 3d. (d) Cr 2p. (e) N 1s. (f) O 1s. *Source:* Reproduced with permission from Phivilay et al.[6]/American Chemical Society.

light excitation-induced electron transition results in the increased electron density on Ga atoms and decreased electron densities on N and Zn atoms. The lack of upshift of N 1s peak observed in Figure 21.3e suggests that these positive charges in terms of holes on N atoms formed in photoexcitation of visible light are consumed through oxidizing H_2O to O_2. As the concentration of Zn is low in surface region of the catalyst and photoelectrons of Zn 2p (KE < 500 eV) have a short mean free path, intensity of XPS peak Zn 2p is quite weak. The low intensity of Zn 2p XPS peak cannot exclude a potential role of Zn atom in oxidizing H_2O to O_2. Notably, compared to peak positions of Rh 3d and Cr 2p of the catalyst in water vapor without light irradiation, the peak positions of Rh 3d and Cr 2p in water vapor while the catalyst was irradiated, downshift. The downshifts of Rh 3d of Rh atoms and Cr 2p of Cr atoms of $Rh_{2-y}Cr_yO_3$ suggest that (i) $Rh_{2-y}Cr_yO_3$ is the cocatalyst of the semiconductor $(Ga_{1-x}Zn_x)(N_{1-x}O_x)$ and (ii) $Rh_{2-y}Cr_yO_3$ has a function of trapping the electrons generated through photoexcitation of electrons of valance band of the semiconductor $(Ga_{1-x}Zn_x)(N_{1-x}O_x)$ by visible light.

In this work, the AP-XPS was used for identifying the cocatalyst where the excited electrons from valence band of the semiconductor particles are transferred to and then trapped at a cocatalyst since the AP-XPS system allows for introducing photons of visible light and water vapor to perform photocatalysis of water while surface of the photocatalysis system is being observed with AP-XPS. The trapped electrons or holes on cocatalysts are expected to downshift or upshift binding energies of the subshells of atoms of the cocatalyst, respectively. Compared to the cocatalyst nanoparticles on a semiconductor particle, the surface region of a semiconductor particle without a cocatalyst nanoparticle is expected to be neutral since the electrons and holes can readily recombine on the surface region even if electrons and holes generated in the bulk of a semiconductor particle could arrive at its surface. Thus, through checking whether the position of an XPS peak of an element of an expected cocatalyst shifts during photocatalysis or not, the role of the potential cocatalyst nanoparticles is expected to be uncovered through AP-XPS as demonstrated here.

References

1 Maeda, K., Teramura, K., Lu, D. *et al.* 2006. "Photocatalyst releasing hydrogen from water." *Nature* 440, 295.

2 Maeda, K., Teramura, K., Lu, D. *et al.* 2007. "Roles of Rh/Cr_2O_3 (core/shell) nanoparticles photodeposited on visible-light-responsive $(Ga_{1-x}Zn_x)(N1-_xO_x)$ solid solutions in photocatalytic overall water splitting." *J. Phys. Chem. C* 111, 7554–7560.

3 Lee, Y., Terashima, H., Shimodaira, Y. *et al.* 2007. "Zinc germanium oxynitride as a photocatalyst for overall water splitting under visible light." *J. Phys. Chem. C* 111, 1042–1048.

4 Maeda, K., Teramura, K., Saito, N. *et al.* 2006. "Overall water splitting using (oxy) nitridephotocatalysts." *Pure Appl. Chem.* 78, 2267–2276.

5 Maeda, K., Teramura, K., Takata, T. *et al.* 2005. "Overall water splitting on $(Ga_{1-x}Zn_x)(N_{1-x}O_x)$ solid solution photocatalyst: relationship between physical properties and photocatalytic activity." *J. Phys. Chem. B* 109, 20504–20510.

6 Phivilay, S. P., Roberts, C. A., Gamalski, A. D. *et al.* 2018. "Anatomy of a visible light activated photocatalyst for water splitting." *ACS Catal.* 8, 6650–6658.

7 Hirai, T., Maeda, K., Yoshida, M. *et al.* 2007. "Origin of visible light absorption in GaN-rich $(Ga_{1-x}Zn_x)(N_{1-x}O_x)$ photocatalysts." *J. Phys. Chem. C* 111, 18853–18855.

8 Maeda, K., Teramura, K., Masuda, H. *et al.* 2006. "Efficient overall water splitting under visible-light irradiation on $(Ga_{1-x}Zn_x)(N_{1-x}O_x)$ dispersed with Rh–Cr mixed-oxide nanoparticles: effect of reaction conditions on photocatalytic activity." *J. Phys. Chem. B* 110, 13107–13112.

9 Wei, W., Dai, Y., Yang, K. *et al.* 2008. "Origin of the visible light absorption of GaN-rich $Ga_{1-x}Zn_xN_{1-x}O_x$ ($x = 0.125$) solid solution." *J. Phys. Chem.* 112, 15915–15919.

Index

Application of Ambient Pressure X-ray Photoelectron Spectroscopy to Catalysis, First Edition. Franklin Tao.

Printed in the USA/Agawam, MA
September 16, 2026

910661.185